SPECTRAL CLASSIFICATION AND MULTICOLOUR PHOTOMETRY

INTERNATIONAL ASTRONOMICAL UNION
UNION ASTRONOMIQUE INTERNATIONALE

SYMPOSIUM No. 50

HELD IN VILLA CARLOS PAZ, ARGENTINA, OCTOBER 18–24, 1971

SPECTRAL CLASSIFICATION AND MULTICOLOUR PHOTOMETRY

EDITED BY

CH. FEHRENBACH

Observatoire de Haute-Provence, Saint-Michel l'Observatoire, France

AND

B. E. WESTERLUND

European Southern Observatory, Santiago de Chile

D. REIDEL PUBLISHING COMPANY

DORDRECHT-HOLLAND / BOSTON-U.S.A.

1973

Published on behalf of
the International Astronomical Union
by
D. Reidel Publishing Company, P.O. Box 17, Dordrecht, Holland

Sold and distributed in the U.S.A., Canada, and Mexico
by D. Reidel Publishing Company, Inc.
306 Dartmouth Street, Boston,
Mass. 02116, U.S.A.

Library of Congress Catalog Card Number 72–87471

ISBN 90 277 0280 2

TABLE OF CONTENTS

PART I / CLASSIFICATION OF SLIT SPECTRA

PART II / CLASSIFICATION OF OBJECTIVE-PRISM SPECTRA

PART III / PHOTOMETRIC CLASSIFICATION

PART IV / CATALOGUES AND DOCUMENTATION

INTERNATIONAL ASTRONOMICAL UNION
ORGANIZING COMMITTEE

B. E. Westerlund (Chairman), D. L. Crawford, Ch. Fehrenbach,
C. Jaschek, E. K. Kharadze, and J. Landi Dessy

LOCAL ORGANIZING COMMITTEE

Dr J. Landi Dessy (Chairman), M. Jaschek, R. Sistero,
M. C. Sistero, J. R. Colazo, and A. Puch

PREFACE

Dr J. Landi Dessy, Director of the Astronomical Observatory, Córdoba, Argentina, invited the International Astronomical Union to hold a Symposium in Córdoba in connection with the celebration of the Centennial of the Córdoba Observatory; the date of foundation is October 24, 1871. He proposed that the Symposium should deal with Spectral Classification and Multicolour Photometry as seven years had elapsed since the Symposium No. 24 in Saltsjöbaden, and much development had occurred in the field.

The invitation and the proposal were accepted by the IAU, and the Symposium was held in Villa Carlos Paz, near Córdoba, between October 18 and October 24, 1971. It was attended by about 50 scientists representing Argentina, Canada, Chile, Denmark, France, Germany, Italy, Mexico, Sweden, Switzerland, U.K., U.S.A., Vatican City State and Venezuela.

The Symposium was divided into four sessions:
1. Classification of slit spectra,
2. Classification of objective-prism spectra,
3. Photometric classification,
4. Catalogues and documentation.

It was decided not to include *problems of calibration of absolute magnitudes and temperatures of stars* but to refer this to a separate symposium.

The contents of the present volume follow closely the programmes of the individual sessions of the Symposium.

All the participants at the Symposium appreciated the excellent hospitality of their Argentinian hosts and the efficient help given by the staff of the Córdoba Observatory.

Financial assistance was provided by the Executive Committee of the International Astronomical Union for travel grants to certain participants. We are indepted to the Universidad Nacional de Córdoba, Consejo Nacional de Investigaciones Científicas y Técnicas, Comisión Nacional de Estudios Geo-Heliofísicos and the Observatorio Astronómico de la Universidad Nacional de Córdoba for defraying all the costs of the Symposium in Villa Carlos Paz.

BENGT E. WESTERLUND

LIST OF PARTICIPANTS

A. A. Ardeberg, ESO, Santiago, Chile
R. A. Bell, Md., U.S.A.
A. Cowley, Ann Arbor, Mich., U.S.A.
L. Divan, Paris, France
Ch. Fehrenbach, Haute-Provence, France
A. Feinstein, La Plata, Argentina
R. F. Garrison, Ontario, Canada
W. Gliese, Heidelberg, F. R. Germany
M. Golay, Sauverny, Switzerland
J. Graham, AURA, La Serena, Chile
L. Gratton, Rome, Italy
K. Haramundanis, Cambridge, Mass., U.S.A.
B. Hauck, Sauverny, Switzerland
R. Herman, Mcudon, France
N. Houk, Ann Arbor, Mich., U.S.A.
C. Jaschek, La Plata, Argentina
M. Jaschek, La Plata, Argentina
P. C. Keenan, Del., U.S.A.
J. Landi Dessy, Córdoba, Argentina
E. Maurice, ESO, Santiago, Chile
M. F. McCarthy, Castel Gandolfo, Vatican City State
R. D. McClure, New Haven, Conn., U.S.A.
E. Mendoza, Mexico City, Mexico
L. Milone, Córdoba, Argentina
A. G. Moreno, Santiago, Chile
H. Moreno, Santiago, Chile
S. C. Morris, Victoria, Canada
K. Nandy, Edinburgh, U.K.
B. Nordström, Stockholm, Sweden
W. Osborn, Merida, Venezuela
M.-N. Perrin, Saint Genis-Laval, France
Ch. L. Perry, Baton Rouge, La., U.S.A.
A. G. D. Philip, Albany, N.Y., U.S.A.
H. Richer, Vancouver, Canada
N. Roman, Washington, D.C., U.S.A.
P. Rybski, Evanston, Ill., U.S.A.
W. Seitter, Bonn, F. R. Germany

M. E. Sisteró, Córdoba, Argentina
R. F. Sisteró, Córdoba, Argentina
F. Smriglio, Rome, Italy
U. W. Steinlin, Binningen, Switzerland
C. B. Stephenson, East Cleveland, Ohio, U.S.A.
J. Stock, Merida, Venezuela
B. Strömgren, Copenhagen, Denmark
R. West, ESO, Hamburg, F. R. Germany
B. Westerlund, ESO, Santiago, Chile
R. F. Wing, Columbus, Ohio, U.S.A.
K. M. Yoss, Evanston, Ill., U.S.A.

PART I

CLASSIFICATION OF SLIT SPECTRA

THE ROLE OF CLASSIFICATION OF SLIT SPECTROGRAMS

PH. C. KEENAN

Perkins Observatory, The Ohio State and Ohio Wesleyan Universities, Ohio, U.S.A.

I take it as evident that the aims of stellar classification are the same whether we work by direct photoelectric photometry through filters, by spectrophotometry, or by visual classification of spectrograms. In all cases we are trying to make direct estimates of as many physical properties of the stars as possible, and as nearly as I can judge, no one method has such exclusive advantages that it seems likely to replace all others in the immediate future. The problems are so difficult – so many subtle differences between stars that we once thought of as identical are turning up now – that we must use every effective means of attack. The present challenge is to combine these methods most effectively.

Where spectroscopy through slits can contribute most is in the establishment of accurate standards which the photometric people can use to relate their precise readings made through various band passes to differences in particular molecular or atomic features in the spectrum. This morning I can give you only a few examples of recent refinements in classification – and must omit important contributions, such as the completion of the Córdoba Atlas by the groups headed by Dr Landi and the Jascheks.

I. Since Dr Morgan was not able to be here himself, I want to begin with the revision of the classification of A- and F-type stars that Morgan and Abt (1971) have just completed. They have kindly allowed me to summarize their results in advance of publication. The spectrograms are a very uniform set taken by Abt at 125 Å mm^{-1} with the small prismatic spectrograph on the 36-in. telescope at Kitt Peak. For this spectral range they consider the relatively small scale actually an advantage, for on such spectrograms the appearance of the G-band can be judged readily. I should mention that for earlier spectral classes (B-stars), Morgan and his students have been using higher dispersions (20 Å mm^{-1}) to bring out fainter lines than could be used before.

For types A7–G0 and luminosity classes II to IV, Morgan and Abt established the zero points from original MK standard stars, but within that framework a real gain in precision has been attained by using as many lines as possible between $\lambda 3850$ and Hγ, plus the apparent structure of the G band. They employed criteria usual for that scale and in addition took into account the general appearance of the strong features, blended and unblended; particularly the relative strength of the stronger lines of ions (Fe II, Ti II, Sr II), and of neutral atoms (Fe, Mn). This total classification is important, and I think that most experienced spectroscopists do it almost unconsciously, though it is difficult to describe the process. It is much easier to tell how one sensitive line ratio is related to, say, luminosity class. The importance of the overall appearance of the spectrum is at least twofold. First, any slight but general weakening or strengthening of the absorption lines, as compared to a normal star of the type, is likely to be detected. Second, a peculiarity may be shown by some feature that is not normally used as a criterion of classification; whereas, if one just mechanically records the intensity ratios of the standard criteria, that peculiarity can easily be overlooked.

In their work, Morgan and Abt ended up with 32 standard F-stars above the main sequence. As a particular application, they then made use of these accurate standards to examine a group of 14 δ Scuti stars. These are the white variable stars, presumably pulsating, with very small amplitudes, about 0.1 mag. There has been some question about their position in the HR diagram with respect to the main sequence. The authors find that all of these δ Scuti stars are either subgiants or giants – they lie above the main sequence. These luminosities, at least for the 10 variables that show no marked peculiarities in their spectra, should be reliable. Four of the stars do have their H and K lines weakened by conspicuous amounts, and the range in both H and K strengths and in luminosities leads the authors to conclude that the δ Scuti stars do not form a homogeneous group.

Morgan and Abt find also that Δm_1, the differences between the Strömgren metallic line indices for the δ Scuti stars and for standard stars of the same luminosity and temperature, show a rather smooth, increasingly negative, run with increasing type. But since Strömgren some years ago found Δm_1 to be less sensitive to metal type for F stars earlier than F4 than those of later type, the authors conclude: "…The interpretation in terms of a progressively higher metal content with advancing MKA spectral type, for the δ Scuti stars of Table II, seems doubtful".

II. For the second example of a classification problem I shall not discuss one particular investigation, but instead consider an area of the H-R diagram that has been of increasing interest. This is the region of the giant branch near types G8–K0, which is important as a locus where evolutionary tracks for stars of different ages, populations, etc., either cross or come close together. To illustrate stellar spectra in this region Figure 1 shows a set of grating spectrograms for which the original scale was in the

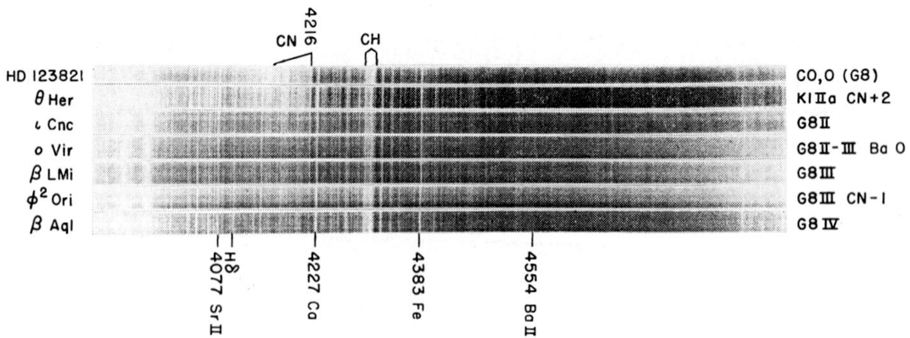

Fig. 1.   Spectral peculiarities in giants, G8–K1. HD 123821 taken at Lowell Observatory with Perkins 72-inch telescope. Remaining stars taken with 32-inch reflector at Perkins Observatory.

range from 75 to 85 Å mm^{-1}. This is just enough greater than the scale of the original Yerkes Atlas (120 Å mm^{-1} at Hγ) to give better resolution of some of the metallic lines (e.g. Cr 4254) which are emerging through the subdivisions of type G.

Except for the top spectrum, the slide shows nothing new or very striking. I have deliberately avoided the more spectacularly peculiar objects in order to bring out instead the more subtle differences that we must be able to recognize if we hope to improve our analyses of the differences between populations.

In the illustration, the spectra of β Aql, β LMi, and ι Cnc are G8III standards showing the normal luminosity effects in Sr II, CN, and the ratio 4376/4383. The second star from the bottom, ϕ^2 Ori, is a well-known weak-CN star – a moderate population-II star. Everyone is used to looking for this peculiarity. In the opposite sense the second spectrum from the top – that of θ Her – shows the opposite departure from normality. Here CN is unusually strong, even for a K1IIa star, and, indeed, θ Her has an intensity break at 4216 Å which is the largest of any in the catalogue of Griffin and Redman (1960). Its CN absorption is about that of a normal K1 supergiant of luminosity class Iab, but if the star really were that luminous the metallic lines at $\lambda\lambda$ 4376 and 4386 would appear about equal.

These stars with a positive CN anomaly do appear to have ratios of metals to hydrogen that are higher than normal, as given by the means of model-atmosphere analyses of abundance. The disagreements in the data for individual stars are so great relative to their means, however, that it seems an exaggeration to call them 'super metal-rich' stars (Spinrad and Taylor, 1969). It would be more realistic to refer to them merely as metal-rich, though observers who prefer to describe what they actually see in the spectra will probably continue to refer to these objects as 'strong-CN stars'.

Clearly, the luminosity classification of the strong CN-stars must be done very carefully to avoid estimating their luminosities as too bright, just as there is the well-known opposite tendency to underestimate the luminosities of the weak-CN stars. An interesting recent determination of statistical parallaxes by John Schmitt (1971) gave lower absolute magnitudes for G9–K4 class III stars with strong CN than for standard giants of the same types as given in Blaauw's calibration. Since, however, his spectrograms had rather low dispersion and the methods of calibration of the two groups were not identical, I mention the comparison merely to emphasize how likely we are to make systematic errors in estimating spectroscopic absolute magnitudes for stars which have even slight peculiarities in their spectra. We can minimize these errors by reviewing the general appearance of a spectrum after it has been classified by the usual criteria. Then, if CN or other features look abnormal, the classification is revised by means of those criteria which are least affected by the evident abnormality. This sounds obvious enough, but it is a precaution that we have all probably neglected at times.

For an instance of another and more subtle peculiarity, look at the middle spectrum (υ Vir). At first glance this spectrum might be classified as G8II from the ratios Sr II 4077/Fe 4063, 4071 and 4216/Ca 4226. Evidently, however, the CN depression is abnormally weak for this type and the Balmer lines do not show the slight enhancement usually observed at luminosity class II as compared to a normal class III giant. The discrepancy can be largely resolved by noting the slight enhancement of Ba II 4554, which is normally only marginally visible in a G8 spectrum at this dispersion. The line is weaker than in most of the well-known barium stars, and *o* Vir is accordingly classified as G8 II-III Ba0. Alternatively, one could term such objects 'semi-barium' stars. Incidentally, *o* Vir has been recognized as a barium star by Williams (1971) also, being included in a list of 5 stars which he observed as having weak but definite enhancement of Ba II 4554. This classification accounts for the enhancement of Sr II in *o* Vir, though CN does not show as much enhancement as is usual in barium stars. The recognition of such stars, in which the lines of the heavy elements are only slightly strengthened, offers a major challenge to either low-dispersion spectroscopy or photometry.

III. The top spectrum in this plate forms the bridge to the last group of stars that I want to discuss – for it is a newly-recognized carbon star. HD 123821 appeared in two recent lists of G-type stars with strong CN, one by McClure (1970) and one by Schmitt (1971), but the latter remarked that it was possibly a very early carbon star, with weak lines. In the illustration of an 80 Å mm^{-1} Flagstaff plate the 1.0 Swan band of C_2 at $\lambda 4737$ is barely visible, but it is clearly present on our 40 Å mm^{-1} spectrogram, leaving no doubt of the C0 classification. The atomic line ratios indicate a temperature equal to that of the ordinary G8 giants.

It is among the carbon stars of lower temperature that the serious problems of classification occur. When the classification using C-types was introduced (Keenan and Morgan, 1941), it was admittedly just a first approximation to a temperature sequence. Since then many attempts to estimate temperatures of the carbon stars have led to somewhat inconclusive results. Most observers have concluded that the C-sequence represents temperatures fairly well from C0 through C3, but is not a very consistent temperature sequence in the later subdivisions. For example, WZ Cas seems definitely hotter than its original C9.1 type implied, though this star has never fitted well into any suggested sequence.

Numerous suggestions for new sequences or groupings have been made. First Bouigue (1954) and later Mrs Gordon (1971) proposed a separate class J for the stars in which a prominent band at $\lambda 6168$ shows that the C^{13} isotope is unusually abundant. Richer (1971) devised a classification based on the photographic infrared in which the same symbols as in the old C-system were used, but with the numbers having entirely different significance. Gordon (1968) and others have suggested also that most of the middle C-stars form a sequence connecting directly to the Ba stars rather than to the early carbon stars. Dolidze (1969) and Kholopov (1971) at Abastumani have proposed still other subgroups based upon color-spectrum relations.

Most of the suggestions involve some good ideas that should not be forgotten, but the general picture is one of confusion. I wish that I could give you a simple solution that would make all the carbon stars fall neatly into place, but all that I can do at this time is to call your attention to some of the difficulties that stand in our way.

Consider first that there are seven composition parameters that appear to vary within the group of carbon stars:

(1) C/O. Carbon stars are sharply defined by the condition that $C/O > 1$, and the calculations of dissociation equilibrium by Tsuji (1964), Schadee (1968), and Greene (1971) show that only a very slight carbon excess is needed to give observable C_2 bands.

(2) Metals/H. By metals we mean all the elements heavier than helium. Anomalously large values of this ratio were recognized as early as 1940 by Wurm in HD 182040 and found in other early carbon stars by Bidelman (1953).

(3) Heavy Metals/Light Metals. Enhancement of the S-process elements Zr, Y, Sr, etc., is well known to be much greater in some carbon stars than in others.

(4) C^{13}/C^{12}. Differences in the relative strengths of the C_2 bands involving the two isotopes of carbon were studied most extensively by McKellar (1949).

(5) Li/Other Metals. That the great differences in the intensity of Li I 6707 among the carbon stars represent differences in the relative abundance of lithium was confirmed by Torres-Peimbert and Wallerstein (1966) who measured [Li/Ca].

(6) Tc/Other Metals. Although technetium is difficult to observe in carbon stars because the strong ultimate lines occur in the faint blue region of their spectra, its definite presence in several stars (Merrill, 1956), and absence in others (Peery, 1971) suggests marked differences in abundance.

Quite possibly we should add the abundance of nitrogen as a variable, but the difficulty of observing the NH bands in the $\lambda 3360$ region has kept us from having certain information on this point. If some of these abundances were tightly coupled we could simplify the problem by using only one or two abundance parameters in classification, but the available evidence suggests that they are only partially correlated; e.g., (1) with (2), and (4) with (5).

Some of the practical problems are illustrated in Figure 2, which shows a fairly typical set of red spectra of southern carbon stars. The plates were exposed to show the $\lambda 6200$ region well, since the scale of about 106 Å mm^{-1} was too small to permit reliable estimates of the Li 6707 intensity except when the line is unusually strong. The top spectrum is somewhat earlier than the next three, but all have strong C_2 bands. Three heads of the $\Delta v = -2$ sequence of $C^{12}C^{12}$ are marked at the top, and there appears to be some increase in their intensity gradient as we go from top to bottom in the set. Of the network of $C^{12}N^{14}$ features which cover this whole region, only the strong heads near 5749 and 6206 Å are marked.

The bottom spectrum in Figure 2 is noteworthy in showing $\lambda 6168$ and the other $C^{12}C^{13}$ bands very strongly, together with the depression due to $C^{13}N^{14}$ near $\lambda 6260$. TZ Car is thus a strong $\lambda 6168$ star, and its features involving C^{13} are as strong as in the spectrum of any carbon star known to me. A comparable star in the northern hemisphere is T Lyr.

In spite of the obvious overlapping of bands, a rough estimate of the C^{12}/C^{13} ratio can be made with the aid of the synthetic CN absorption spectra computed by Marenin. Figure 3 is a revision of a computer plot presented by Marenin and Greene (1971) at the August, 1971, meeting of the AAS. The absorption curves were computed for a simple Milne-Eddington atmosphere, using the known f-values for the $C^{12}N^{14}$ molecule, and smoothed by giving each molecular band line a Doppler half-value width of 2.5 Å to match approximately the resolution of the spectrograms of Figure 2. The lower curve is that of a pure $C^{12}N^{14}$ molecule, while the upper represents a mixture of $C^{12}N^{14}$ for $C^{12}/C^{13}=4$. It can be seen that the model mixture predicts that the $\lambda 6206$ head of $C^{12}N^{14}$ (4,0) will be slightly stronger than the corresponding $C^{13}N^{14}$ feature near 6260 Å. In Figure 2, the $C^{13}N^{14}$ feature is more conspicuous, and further computations by Miss Marenin indicate that $C^{12}/C^{13}=2$ gives a reasonably good fit for TZ Car. This ratio is lower than has generally been suggested on theoretical grounds, but an iso-intensity argument should have some weight. The near equality of the C_2 heads, $C^{12}C^{12}$ at 6191 Å and $C^{12}C^{13}$ at 6168 Å tends to support our conclusion.

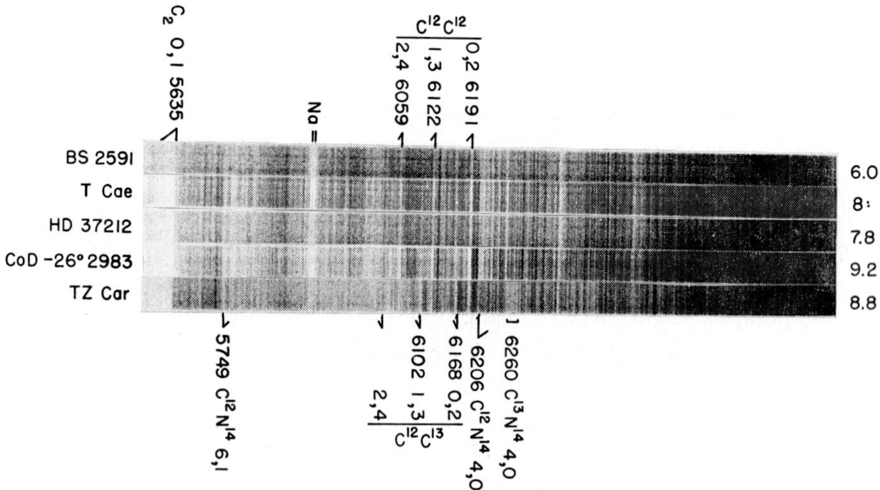

Fig. 2.   Typical carbon stars photographed at 106 Å mm^{-1} with the 60-in. reflector at Cerro Toloro Inter-American Observatory on Eastman 098-02 spectrographic plates.

Another rather obvious conclusion from Figure 3 is that when CN is fairly abundant there are no windows in the red region where a real band-free continuum can be observed. As long as only $C^{12}N^{14}$ is present the main heads at 5749 Å could be used as fairly good measures of total CN abundance, but as soon as the amount of $C^{13}N^{14}$ becomes appreciable the heads are so much smoothed that any photometric measurements, including those of spectrophotometric scanners or narrow-band photometry, would very much underestimate the abundance of CN. The only way to avoid the effect is to use only $\Delta v=0$ heads, for which vibrational isotopic shifts are not appreciable, but such bands are not available in the visual and near infrared regions.

The effect of general observation of features by both isotopic cyanogen molecules is roughly just as serious in the photographic infrared as in the red region shown in Figure 2. For the $\Delta v=+2$ sequence the isotopic shift remains between 40 and 45 Å,

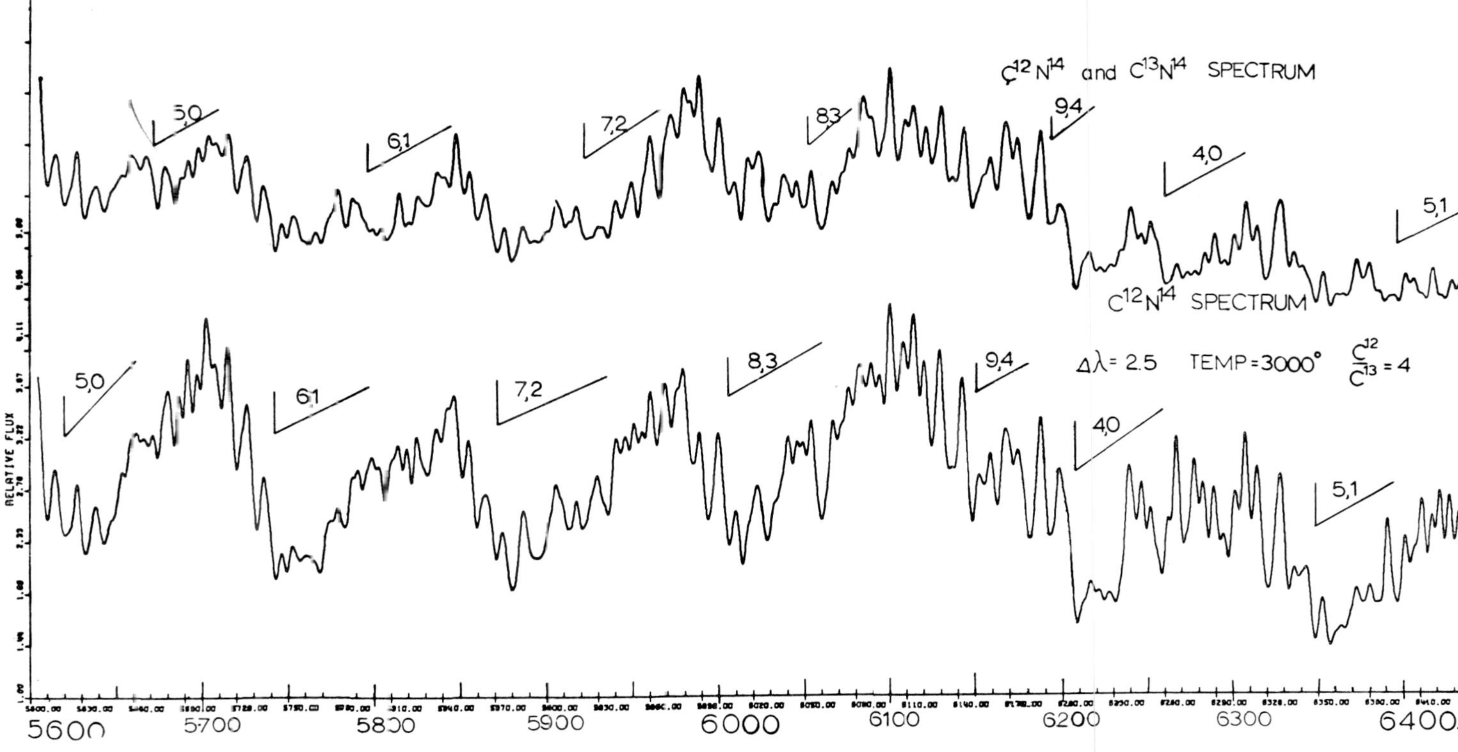

Fig. 3. Computer plots of expected absorption of $C^{12}N^{14}$ (upper) and a mixture of $C^{12}N^{14}$ with $C^{13}N^{14}$ in the ratio $C^{12}/C^{13} = 4$ (lower). The plots were prepared by Marenin from her computations.

almost as great as for the red bands. Only the first of the $\lambda 7800$ group (at 7850 Å) of $C^{12}N^{14}$ bands will be reasonably free from contamination by $C^{13}N^{14}$. For hundreds of angstroms redward from there nearly all other features, atomic and molecular, will be subject to heavy blanketing, which increases with both the C^{13}/C^{12} ratio and the total CN concentration.

The group of stars intermediate between the C-stars and the S-stars are particularly difficult to classify on any system that will be consistent with the types assigned to more heavily banded spectra. Figure 4 shows two of the more striking southern stars in this group. These are UY Cen, studied by Feast (1968) and Catchpole (1971) at Radcliffe, and the red star which was found to be strongly variable by Welch in New Zealand. It was actually noticed by Henize some years ago, and classified as a member of the intermediate group by Feast before I photographed it last winter, without knowing of their work. It is one of the most spectacularly red stars that I have seen through a telescope, and is almost certainly a Mira variable, although the strong $H\alpha$ emission on this plate is burned out in the enlargement.

That stars like these form a subgroup in which $C \approx 0$ has come to be generally accepted through the work of Bidelman, Feast, and particularly Stephenson (1967), who formally defined them as a group called CS stars. Recently, Feast and I have thought it might be useful to split them into two groups, calling them CS stars if C_2 bands were visible, implying $O/C < 1$, and SC stars if ZrO bands are present, indicating $O/C > 1$. Most of these stars vary appreciably in light (and temperature), and even if no bands (except the ubiquitous CN) are visible at maximum light, they are likely to exhibit the defining bands weakly when near minimum light. Actually, the same suggestion was made at least two years ago by Dolidze (1969). Admittedly, the question of notation would then prove embarrassing if we find stars with such exactly equal amounts of oxygen and carbon (within perhaps $\frac{1}{2}\%$), that neither set of bands ever appears. Such stars must be extremely rare, however, and can always be classified as peculiar! One possible star of this kind is CY Cyg, which is being analyzed by Culver (1971). Unfortunately, it is nearly constant in light, and its excitation temperature of 2600–2800 K is a little too high for a definite test.

Along with these comparatively rare CS and SC stars there are a considerable number of carbon stars that have somewhat stronger C_2 bands but show essentially the same characteristics, particularly the very strong atomic lines. The D-lines of sodium in Figure 4 are a striking example, and in such stars as WZ Cas they have total absorptions as great as 8A. In the original C-system we used the D-lines as one of the indicators of temperature, but it can scarcely be doubted now that they are enormously influenced by overlying molecular opacity. There may also be appreciable differences in the abundance of sodium itself. I am afraid that the net result is that we can put very little confidence in the D-lines as temperature parameters. The potassium resonance lines in the near infrared ($\lambda\lambda 7665$ and 7699) are more reliable, but they also show some anomalies connected with opacity, abundance, and luminosity.

I do not mean to be unduly pessimistic about the prospects for better classification of the carbon stars. All spectral classification can be thought of as essentially component analysis carried out either visually or by measurement and automatic reduction. For the carbon stars we must just face the fact that there are more variables to take into account. This probably means that in establishing standards we must combine the information from several spectral regions. If we had data from the far red and infrared regions comparable to the valuable set of intensities of blue and yellow lines and bands compiled by Yamashita (1967), considerable progress would be possible.

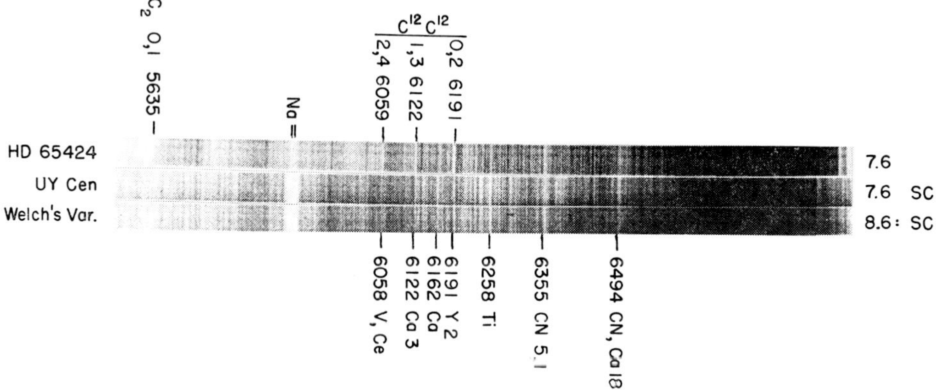

Fig. 4. Spectrograms of two southern SC stars compared with the typical middle carbon star HD 65424. Same scale and source as Figure 2.

Simultaneous photometric data on the blue-violet depression to augment the valuable visual estimates made originally by Shane (1928) would also be valuable. This depression is apparently due at least in part to the C_3 molecule, and dissociation curves (cf. Tsuji, 1964) suggest that the abundance of this molecule increases so rapidly with falling temperatures (wherever carbon is abundant), that the appearance or disappearance of these bands may serve to define a rather critical temperature.

References

Bidelman, W. P.: 1953, *Astrophys. J.* **117**, 25.
Bouigue, R.: 1954, *Ann. Astrophys.* **17**, 104.
Catchpole, R. M.: 1971, Not yet published.
Culver, R.: 1971, Ph.D. Thesis, Ohio State Univ.
Dolidze, M. R.: 1969, *Astron. Tsirkular*, No. 507, 6; No. 514, 4.
Feast, M. W.: 1968, in L. Perek (ed.), *Highlights of Astronomy*, D. Reidel Publ. Co., Dordrecht, Holland. p. 237.
Gordon, R. C. P.: 1968, *Astrophys. J.* **153**, 915.
Gordon, R. C. P.: 1971, *Publ. Astron. Soc. Pacific* **83**, 667.
Greene, A. 1971, Ph.D. Thesis, Ohio State Univ.
Griffin, R. F. and Redman, R. O.: 1960, *Monthly Notices Roy. Astron. Soc.* **120**, 287.
Janes, K. A. and McClure, R. D.: 1971, *Astrophys.* **165**, 561.
Keenan, P. C. and Morgan, W. W.: 1941, *Astrophys. J.* **94**, 501.
Kholopov, A. Z.: 1971, *Astron. Tsirkular*, No. 601, 6.
McClure, R. D.: 1970, *Astron. J.* **75**, 41.
McKellar, A.: 1949, *Publ. Dominion Astrophys. Obs., Victoria* **7**, No. 26, 395.

Marenin, I. and Greene, A.: 1971, *Bull. Am. Astron. Soc.* **3**, 379.
Merrill, P. W. 1956, *Publ. Astron. Soc. Pacific* **68**, 70.
Morgan, W. W. and Abt, H.: 1971, *Astron. J.* **77**, 35.
Peery, B. F., Jr.: 1971, *Astrophys. J. Letters* **163**, L1.
Richer, H. B.: 1971, *Astrophys. J.* **167**, 521.
Schadee, A.: 1968, *Astrophys. J.* **151**, 239.
Schmitt, J. L.: 1971, *Astrophys. J.* **163**, 75.
Shane, C. D.: 1928, *Lick Obs. Bull.* **13**, No. 396, 123.
Spinrad, H. and Taylor, B. J.: 1969, *Astrophys. J.* **157**, 1279.
Stephenson, C. B.: 1967, *Astrophys. J.* **150**, 543.
Torres-Peimbert, S. and Wallerstein, G.: 1966, *Astrophys. J.* **146**, 724.
Tsuji, T.: 1964, *Tokyo Obs. Annals* **9**, No. 1.
Williams, P. M.: 1971, *Observatory* **91**, 37.
Wurm, K.: 1940, *Himmelswelt* **50**, 20.
Yamashita, Y.: 1967, *Publ. Dominion Astrophys. Obs., Victoria* **13**, No. 5, 67.

SPECTROSCOPIC CRITERIA FOR THE CLASSIFICATION OF PECULIAR B-TYPE STARS WITH COLOUR-SPECTRUM DISCREPANCIES

R. F. GARRISON

David Dunlap Observatory, University of Toronto, Richmond Hill, Ontario, Canada

Abstract. The so-called 'weak helium' stars are discussed from a spectral classification point of view. Several authors have included different kinds of objects in this class. Often, the only similarity is the discrepancy between the spectra and the colours. It is desirable to decouple the discovery of new members of the class from this dependence on colours so that the colour can be used as an independent parameter.

I would like to use this opportunity to make some comments on the technique and philosophy of spectral classification, then give an example and a model. As the example, I have chosen the peculiar B stars with colour-spectrum discrepancies.

The first point concerns the relationship between photometry and spectral classification. It is assumed by many astronomers that there is a one-to-one relationship between the two and that photometry can replace classification or vice-versa. In my experience this is not so, and the best use of the two techniques is in taking advantage of their complementarity.

The second point concerns the relationship between high dispersion and low dispersion. It is usually assumed that if something can be done well with low dispersion spectra, it can be done that much better with higher resolution. In my experience, this is not always the case, because of the greater visibility of wings and blends with low dispersion as well as the ease of looking at the spectrum as a whole. Thus, high dispersion and low dispersion studies are also complementary rather than competitive.

The example of the so-called 'weak helium' peculiar B stars is a good illustration of the complementarity of these techniques. This is a class of stars which were picked out as spectroscopically peculiar, without reference to their colours, during a study of the members of the association in upper Scorpius (Garrison, 1967, see notes to Table III).

A superficial classification would assign them to the region of the B7–9 giants, but a closer examination would reveal several peculiarities which can be detected at MK dispersion.

(1) Broad, washed out H wings and He lines (by 'washed out', I mean apparently filled in by continuum emission).

(2) Cores of H are normal for either B8III or B3V. This with (1), gives the H lines a characteristic and easily recognized shape at low dispersion which is not apparent in high dispersion spectra.

(3) The presence, usually but not always, of C II and Si III lines, which are characteristic of B3 stars.

(4) The presence, occasionally, of faint lines of Fe II and Ti II, which are characteristic of A and late B-type stars.

When the MK type, say B8IIIp, (the p indicating the above characteristics) is compared with the colours, it is invariably found that the colours are those of a B3–5 star. This discrepancy is clearly shown by plotting $(U-B)_0$ vs MK type (Garrison, 1967, Figure 7). Such colour-spectrum discrepancies were noted earlier by Sharpless (1952), Jugaku and Sargent (1961), and McNamara and Larsson (1962). Later, the discrepancy was used to pick out such stars; e.g. Bernacca (1968), Jaschek et al. (1969). (Note that Table IV in the latter paper contains a large number of errors, which seem to have eluded both authors and editors.) However, it is necessary to use spectroscopic criteria to distinguish the 'weak helium' stars from stars of a rather different peculiar nature, e.g. well known Ap types, such as Si, Hg, Mn, etc., and the uniquely peculiar object 36 Lyn, which does not resemble the rest of the group.

There are several reasons for decoupling the colour from the spectrum, in discovery surveys at least. It may well be that these stars will turn out to be hot analogues of the Ap stars but it is desirable to know their colour characteristics as distinct from their spectral characteristics; therefore the relationship between colour and spectrum should not be assumed *a priori*. Secondly, it may well be that some 'normal' stars exhibit some of the same characteristics without the colour discrepancy. If the colour and spectrum were coupled *a priori*, then interesting examples would be missed. Thirdly, there is some indication that the presence of Fe II may be correlated with colour (Jaschek, 1971), but a pure sample, unrelated to colour, would be necessary to establish this.

This leads to a consideration of a model for Bp stars. The obvious model is that of a B3V star with helium underabundant by a factor of 100, or more. There are several objections to this interpretation, none of which alone is sufficient, but the weight of all of them makes the interpretation unsatisfactory.

(1) Many of these stars are physical members of young clusters and associations (Orion, α Persei, Sco-Cen, etc.) and coexist with stars of 'normal' helium abundance or even 'helium rich' stars (in the case of the σ Ori subgroup).

(2) The peculiar spectrum variable a Cen resembles a 'helium poor' star at one phase and a 'helium rich' star at another. It varies back and forth with a period of slightly more than one week. During this period there are differential velocity variations and spectral changes which indicate extreme density variations (Klinglesmith, Bernacca, and Frey, 1971). Figure 1 is an illustration of the variations in this remarkable star.

(3) The spectra at low dispersion have many of the characteristics of very mild shell stars and the variations in a Cen, described in (2), indicate non-LTE effects.

(4) When plotted according to their colours in well-determined HR diagrams, such as α Persei (Morgan, Hiltner, and Garrison, 1971), the Orion Belt, and Upper Scorpius (Garrison, 1967), they fall 0.5–0.75 mag. below the main sequence.

(5) The presence of faint lines of Ti II and Fe II in some spectra suggests that if B3

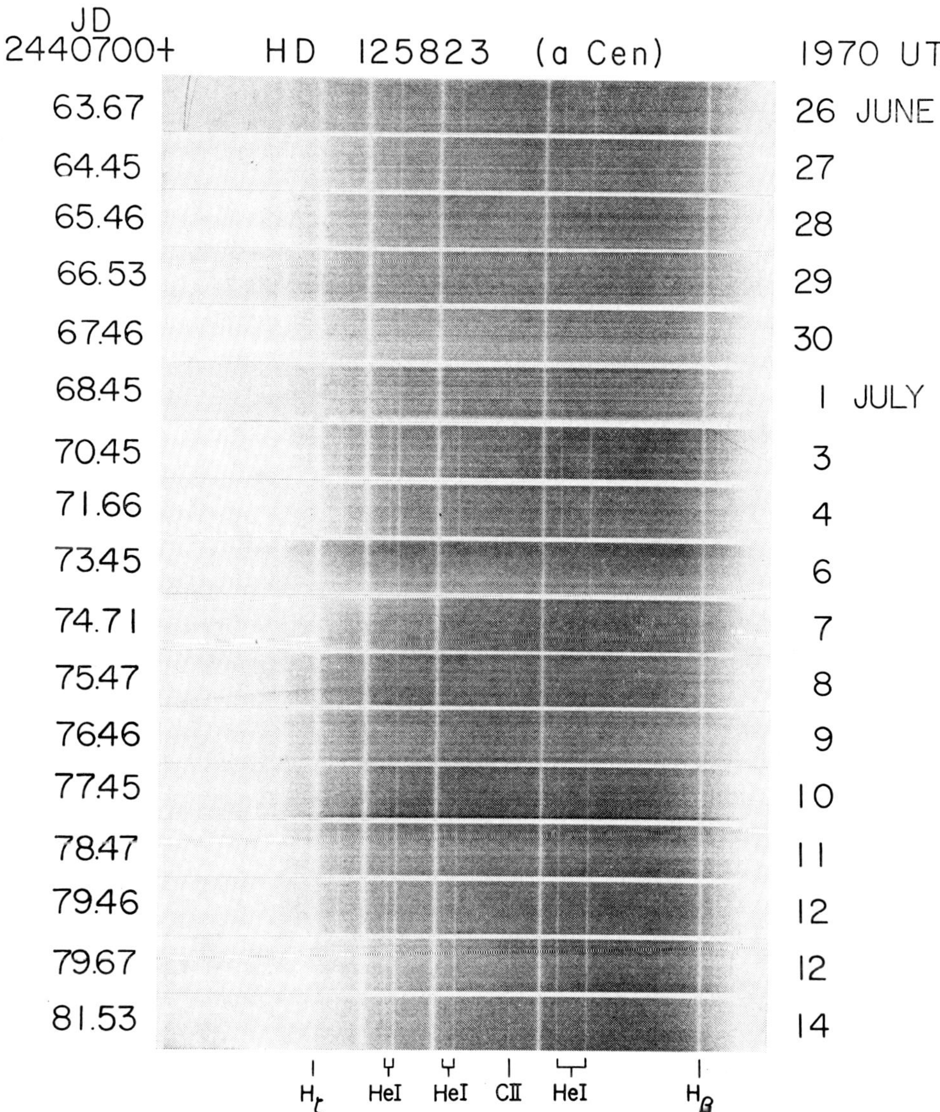

Fig. 1. Cerro Tololo spectrograms of a Cen (HD 125823). Original dispersion 91 Å mm⁻¹ at Hγ. Note the extreme changes in He I strength. Variations are observed in C II and possibly in H. The colour is that of a B2 star and changes very little during the cycle. At maximum He strength, the spectrum resembles the 'helium rich' stars and at minimum, the 'weak helium' group.

is assumed, then not only is He underabundant, but Fe and Ti are overabundant. This involves too many abundance anomalies.

The model which I would like to propose here, to stimulate discussion, is one in which the underlying star is B3–5 but with a semi-opaque electron-scattering shell located within or near the photosphere; in other words, large departures from the normal density-radius relation. This provides for the spectroscopic characteristics and explains the spectrum-colour discrepancy. Details have not been worked out quantitatively.

More observations are obviously needed for these stars. In particular, statistical studies and high dispersion studies of line identifications, line profiles, and velocities would be useful.

High dispersion studies are being undertaken by several groups. Norris, now at Yale, is studying HD 142301 in Upper Scorpius. One of my students, Bruckner, is studying HD 21699 in the α Persei cluster and another, Campbell, is studying a group of 5 stars in the Orion Belt region. Bernacca, Klinglesmith, Molnar, Underhill and others at NASA are studying a Cen and a number of other members of the group.

The statistical studies are very difficult and time consuming if done properly. What percentage of the stars with B3–5 colours have B7–9 spectral characteristics? How many are in clusters and how many are in the field? In order to answer these and other questions properly, it is necessary to take all the HD stars from B3 to A0 to some limiting magnitude at MK dispersion for classification. This region of the HR diagram is often neglected, because the OB-type stars are more interesting for galactic structure and the A-type stars are more interesting for their spectacular peculiarities. Such a survey, without reference to colours, is being undertaken at the University of Toronto and will cover both northern and southern hemispheres, beginning with the bright stars.

From the preliminary results, the indications are that up to 20% of the stars with B3–5 colours have these peculiar characteristics. The observation that their luminosities are systematically lower than 'normal' stars with similar colours indicates that photometry needs to be supplemented by MK spectral classification if one is to obtain accurate distance moduli.

Acknowledgements

This work is being supported by the National Research Council of Canada.

I am also indebted to the Director and staff of the Cerro Tololo Inter-American Observatory for telescope time and to Chile for the excellent weather (even in winter) which enabled me to obtain the series on HD 125823 illustrated in Figure 1.

References

Bernacca, P. L.: 1968, *Contr. Oss. Astrofiz. Univ. Padova*, No. 202.
Garrison, R. F.: 1967, *Astrophys. J.* **147**, 1003.
Jaschek, M., Jaschek, C., and Arnal, M.: 1969, *Publ. Astron Soc. Pacific* **81**, 650.
Jaschek, M.: 1971, Private communication.
Jugaku, J. and Sargent, W. L. W.: 1961, *Publ. Astron. Soc. Pacific* **73**, 249.
Klinglesmith, D. A., Bernacca, P. L., and Frey, H.: 1971, Bamberg Conference on Variable Stars, *Veroeffentl. Remeis-Sternwarte Bamberg* **9**, Nr. 100, p. 205.
McNamara, D. H. and Larsson, H. J.: 1962, *Astrophys. J.* **135**, 748.
Morgan, W. W. Hiltner, W. A., and Garrison, R. F.: 1971, *Astrophys. J.* **76**, 242.
Sharpless, S.: 1952, *Astrophys. J.* **116**, 251

CLASSIFICATION DES ÉTOILES B À PARTIR DES RAIES DE L'HYDROGÈNE – COMPARAISON AVEC D'AUTRES CLASSIFICATIONS

R. HERMAN

Observatoire de Paris, Section d'Astrophysique, 92190 Meudon, France

Résumé. Comparaison entre notre classification, basée sur les raies de l'hydrogène ($R_c \overline{\gamma, \delta, \varepsilon}$ $W \overline{\gamma, \delta, \varepsilon}$), et les classifications Chalonge, Barbier-Morguleff et J. Rountree-Lesh. Il semble que notre classification soit voisine de ces trois autres mais qu'elle est plus sûre dans le cas des étoiles Be.

1. Introduction

Il est important de classer les étoiles Be pour comprendre leur évolution. Ce problème n'est pas simple en raison de leurs variations rapides pouvant passer en quelques années (ou même en quelques mois) d'une émission à une absorption supplémentaire plus ou moins importante dans les raies de Balmer ('shell'). De plus, des variations ont lieu également dans le continu stellaire (Feinstein, 1968; Peton, communication privée). Généralement, la discontinuité de Balmer n'est pas affectée par l'émission car, la plupart du temps, l'émission ne concerne que les premières raies de Balmer, par contre, les grandes enveloppes modifient considérablement la discontinuité de Balmer. De plus, le continu et la discontinuité de Balmer risquent d'être perturbés par la qualité de l'atmosphère terrestre, c'est pourquoi nous avons été amenés (Rojas-Herman, 1955) à chercher une autre méthode qui utilise les mêmes quantités (T ou g) que la position et la grandeur de la discontinuité de Balmer.

2. Méthode Rojas-Herman

Celle-ci est basée uniquement sur les moyennes de mesures des largeurs équivalentes W et des profondeurs centrales R_c des raies de la série Balmer de l'hydrogène; soit $W \overline{\gamma, \delta, \varepsilon}$ et $R_c \overline{\gamma, \delta, \varepsilon}$. L'avantage est que l'on peut travailler avec un ciel moyen puisque les mesures sont monochromatiques. De plus, la dispersion est suffisante pour observer aisément une émission ou une 'shell' (dispersion à $H\gamma = 77$ Å mm^{-1}, $H\delta = 59$ Å mm^{-1}, $H\varepsilon = 50$ Å mm^{-1}).

3. Comparaison avec les classifications Chalonge-Divan, Morgan-Keenan-Lesh, Barbier-Morguleff

Nous donnerons tout d'abord 4 tableaux de comparaison. Le premier concerne la comparaison de notre classification avec les 3 autres classifications: Chalonge-Divan (communication privée), Lesh (1968), Barbier-Morguleff (communication privée).

Le deuxième compare notre classification avec celles de Chalonge-Divan et de Morgan-Keenan-Lesh. Le troisième compare notre classification avec celles de Chalonge-Divan et de Barbier-Morguleff et le quatrième compare notre classification avec celles de Morgan-Keenan-Lesh et de Barbier-Morguleff. On constate que, dans l'ensemble, les résultats sont comparables à $\pm$ une classe spectrale ou à $\pm$ une classe de luminosité, sauf pour HD 4180 (Tableau I), HD 171 406 (Tableau III), HD 164 284 (Rakotoarijimy et Herman, 1958) et HD 187 811 (Tableau IV). Les cas signalés ici sont caractéristiques. Ils correspondent à des variations importantes d'émission et d'absorption d'enveloppe, avec alternance entre le maximum d'émission et le maximum d'absorption dans les raies Balmer de l'hydrogène.

Dans les Tableaux I, II, III et IV toutes les étoiles indiquées (var.) se comportent probablement de la même façon. Le Tableau V donne d'autres étoiles variant également de la même façon.

Parfois, l'enveloppe est très peu développée comme c'est le cas pour HD 23 408 (Figure 1) et on peut penser, si on utilise la discontinuité de Balmer, que l'étoile est moins chaude qu'elle ne l'est en réalité. Huit étoiles (signalées par un astérisque) ont

TABLEAU I

Etoile	Auteur	Chalonge-Divan	Morgan-Lesh	Morguleff
HD 4180	B4IV–B9III (env. var.)	B5–6III	B5III	B3–4III
14818	B2Ia	B2Ia	B2Ia	B2I
22192	B5IV?	B3–4IV–III	B5V	B2V
42087	B3Ia	B2–3Ia	B2.5Ib	B2
91316	B1Iab	B1Iab	B1Iab	B1I
204172	B1Iab	B0Iab	B0Ib	B0

TABLEAU II

Etoile	Auteur	Chalonge-Divan	Morgan-Lesh	Morguleff
HD 10516	O9.5V?	O9Ia	B2Vp	–
23302	B8.5IV–III	B7–8III	B6III	–
23338	B9.5IV–III	B6IV	B6IV	–
23408	B7III (env.)	B7III	B8III	–
23480	B6.5IV–V	B6IV–III	B6IV	–
23630	B8IV–III	B8III	B7III	–
35439	B2III–IV	B1IV	B1Vn	–
37742	B1.5Ia	O9–8Ib–II	O9.5Ib	–

TABLEAU III

Etoile	Auteur	Chalonge-Divan	Morgan-Lesh	Morguleff
HD 24131	B4IV (env.?)	B1V		B1V
171406	B7V–B9IV (var.)	B4V		B5V

TABLEAU IV

Etoile	Auteur	Chalonge–Divan	Morgan–Lesh	Morguleff
HD 5394	B0II?		B0.5IV	B0
22780	B9V?(env.?)		B7Vn	B8V
32343	B3V?		B2.5V	B2II–III
34078	O8III		O9.5V	B1
37202	B2III		B4III	B2III
43285	B7V?		B6V	B5V
45995	B2V?		B2.5V	B1III
138749	B5–6V–IV		B6Vnn	B3
164284	B2V–B6V (var.)		B2V	B0
183362	B3V (env. var.?)		B3V	B2
187811	B4–5V–B8–9IV–III (var.)		B2.5V	B3V
194335	B3IV?		B2Vn	B2
197419	B0V?? var?		B2IV–V	B3V
198478	B3Ia		B3Ia	B3I
200120	O9V?		B1.5V	B0I
212076	B2IV		B2IV–V	B2
214168	B1–2V		B1V	B2
217891	B8.5 var.		B6V	B7
224544	B7IV?		B6IV	B3
224559	B4V?		B4V	B3

TABLEAU V

HD 24131	B2V–B4.5III	HD 177648	B3.5V–B7IV ou B8III
45542	B6V–B9III	179343	B6.5V–IV–A1III
164447	B7V–B9IV	191610	B3V–B6.5IV–III
168957	B6V–B9IV	193911	B5V–B8.5IV
171780	B6V–B9IV	217675	B4.5V–B9.5Ib
174105	B7V–B9IV		
174237	B3V–B8III		
175869	B7.5V–B9IV		

été étudiées par M. Lacoarret (1965) qui a discuté seulement les variations de la moyenne $W\gamma, \delta, \varepsilon$ sans indiquer les classes spectrales et de luminosité.

Un autre cas intéressant est celui de HD 162732 que nous considérons comme B6V. Cette étoile semble très stable. L'émission est peu importante alors que la 'shell' est forte. Des mesures récentes sur cette étoile, faites par N. Morguleff, donnent (B5) et A1 ou 2 pour les bandes D et K. L'importance des raies D et K n'est pas due aux raies interstellaires, mais bien à l'enveloppe.

4. Conclusion

On voit que notre classification, faite sur les raies de l'hydrogène, semble dans le cas des étoiles Be, apporter plus de renseignements que les autres, tout en étant cohérente avec elles. Il semble que cette classification, faite à partir d'une théorie grossière et améliorée par l'expérience, ne correspond pas aux résultats de Mihalas, qui trouve des

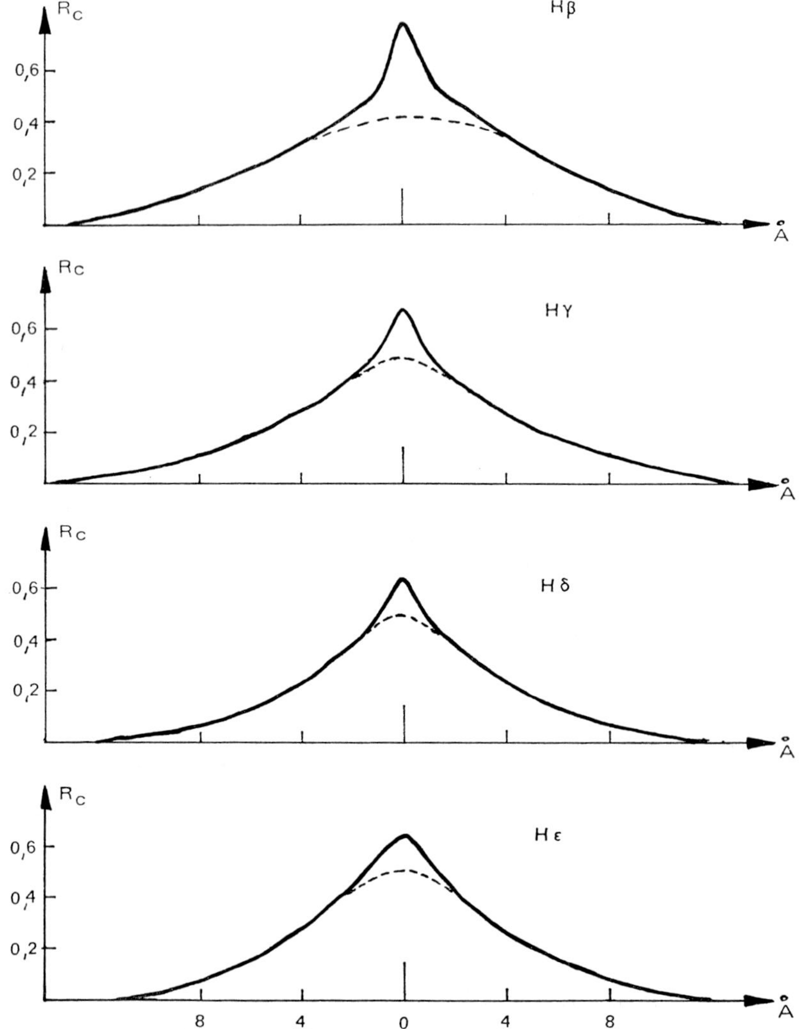

Fig. 1.   Profils des raies d'hydrogène de l'étoile HD 23 408 (W 3014).

ailes beaucoup plus grandes que ne les donnent les observations. De plus, les différences de profondeurs centrales de Hα et Hβ sont plus petites que celles données par Mihalas.

Bibliographie

Feinstein, A.: 1968, *Z. Astrophys.* **68**, 29.
Lacoarret, M.: 1966, *Ann. Astrophys.* **28**, 231.
Lesh, J.: 1968, *Astrophys. J. Suppl.* **151**, 17.
Rakotoarijimy, D. et Herman, R.: 1958, in 'Etoiles à raies d'émission', *Extr. Mém. in 8° Roy. Sci Liège* **XX**, *4ème sér.* p. 204.
Rojas, H. et Herman, R.: 1955, *Compt. Rend. Acad. Sci. Paris* **240**, 727.

ÉTUDE DES ÉTOILES SUPERGÉANTES DU GRAND NUAGE DE MAGELLAN, LEURS MAGNITUDES ABSOLUES ET COULEURS

CH. FEHRENBACH

Observatoire de Haute-Provence, St. Michel l'Observatoire, 04300 Forcalquier, France

Abstract. Nous indiquons une liste de 29 étoiles membres du Grand Nuage de Magellan que nous proposons comme standards de type spectral des Supergéantes. La situation de ces étoiles dans les diagrammes RH et couleur-couleur est indiquée. Nous proposons des valeurs de $(U-B)_0$ et $(B-V)_0$ pour ces étoiles.

La mesure des vitesses radiales au Prisme Objectif nous a permis de publier en 1970 une liste de 500 étoiles brillantes membres du Grand Nuage et un peu plus de 2000 étoiles de même magnitude apparente de la Galaxie et situées dans la région du Grand Nuage. Depuis cette publication, une centaine d'autres étoiles ont été ajoutées (Fehrenbach et Duflot, 1973).

Une étude systématique de ces étoiles au spectrographe à fente nous a paru très utile, d'autant plus que de nombreuses étoiles du Nuage paraissent beaucoup plus lumineuses que les Supergéantes de la Galaxie. E. Maurice, L. Prévot et moi-même avons pris de nombreux spectres à 73 Å mm^{-1} et J. P. Brunet et A. Ardeberg ont fait de nombreuses mesures photoélectriques *UBV*. Le catalogue ainsi obtenu sera décrit par Maurice (1972).

Je ne donne ici que la description des classifications spectrales des étoiles et leurs mesures photométriques dans le système *UBV*.

(a) Le diagramme HR ne montre pas dans la représentation V, $B-V$ des classes très marquées d'étoiles. La classification de Morgan et Keenan en classes Ia, Iab, Ib paraît donc à priori difficile.

(b) Nous avons essayé de classer ces étoiles en nous servant soit de standards MK, soit des étoiles classées par les Astronomes de Radcliffe Observatory (Feast *et al.*, 1960). Ceci nous a permis d'établir une classification provisoire en nous servant des critères classiques.

(c) Mais la difficulté provient du fait qu'il n'y a dans le Grand Nuage aucune étoile de type spectral déterminé entre A3 et F8, nous avons donc dû interpoler les types spectraux entre A3 et F8. Ceci nous a amené à proposer une liste (Tableau I) de 29 étoiles standards dont les types sont compris entre O6 et G8. La Figure 1 montre la place des 29 étoiles dans le diagramme HR. La position de nos étoiles dans ce diagramme provient d'un effet de sélection.

Ardeberg *et al.* (1972) publieront les spectres de ces 29 étoiles. Les types spectraux attribués sont en excellent accord avec ceux de Thackeray pour les 11 étoiles communes. La séquence proposée est donc en accord avec la séquence MK, mais il y a un certain nombre de difficultés.

TABLEAU I

Liste des étoiles standards proposées

1			V	$B-V$	$U-B$	Radcliffe	
1	G 355	O6f	12.27	−0.22	−1.01	R115	O5f
2	G 84	O9f	11.95	−0.21	−0.97	−	
3	G 414	O9.5 I	11.36	0.00	−0.86	R129	B0Ia:
4	G 420	O9.5:I	11.23	−0.02	−0.81	−	
5	G 22	B0Ia	11.34	−0.12	−0.95	R53	B0Ia:
6	G 481	B1Ia	11.59	−0.06	−0.88	R155	B1.5Ia
7	G 434	B2Ia	11.21	+0.04	−0.74	R152	B2.5Ia
8	G 384	B3Ia	11.04	−0.01	−0.73	−	
9	G 283	B5Ia	11.59	−0.01	−0.69	R100	B3I
10	G 486	B5Ia	11.73	−0.04	−0.67	−	
11	G 305	B6Ia	10.88	+0.00	−0.67	−	
12	C 10	B8Ia	(10.95)	(−0.01)	−	R68	B8I
13	G 64	B9Ia	12.52	+0.10	−0.41	−	
14	G 390	A0IaO:	10.81	+0.12	−0.37	R119	A0IaO
15	G 268	A0Ia	10.52	+0.13	−0.45	R98	A0IaO:
16	G 321	A1IaO	10.74	+0.13	−0.22	−	
17	G 339	A2IaO:	10.73	+0.19	−0.18	−	
18	G 437	A2I	11.70	+0.10	−0.09	−	
19	G 200	A3IaO	9.12	+0.19	−0.24	R76	A3:IaO(e)
20	G 377	A5Ia:	11.71	+0.16	+0.15	−	
21	G 303	A8:IaO	10.65	+0.38	0.00	−	
22	G 322	A9Ia	11.44	+0.27	−0.02	−	
23	G 247	F2Ia	11.20	+0.27	+0.10	−	
24	G 352	F6Ia	10.28	+0.41	+0.30	−	
25	G 296	F6Ia	9.90	+0.42	−0.10	−	
26	G 460	F8Ia	10.20	+0.68	+0.50	−	
27	G 454	G0Ia	10.38	+0.95	+0.65	−	
28	G 367	G2Ia	9.91	+1.15	+0.60	R117	G0Ia
29	C 3	G8Ia	10.09	+1.55	+1.29	−	

(1) La classification visuelle n'est pas suffisante pour attribuer une classe de luminosité et l'accord avec Thackeray n'est pas parfait.

OM

		IaO	Ia	I
	IaO	2	1	
Rad.	Ia		4	2
	I		1	

Et, fait plus grave, il n'y a pas une corrélation nette entre la magnitude absolue et la classe de luminosité. Par classification visuelle, nous obtenons ainsi pour deux étoiles:

$$A0Ia \qquad V = 10.52$$
$$A0IaO \qquad V = 10.74.$$

(2) Le diagramme HR montre qu'il y a continuité entre les classes de luminosité. Il vaut mieux remplacer les classes IaO, Iab, Ib par des indications directes de la

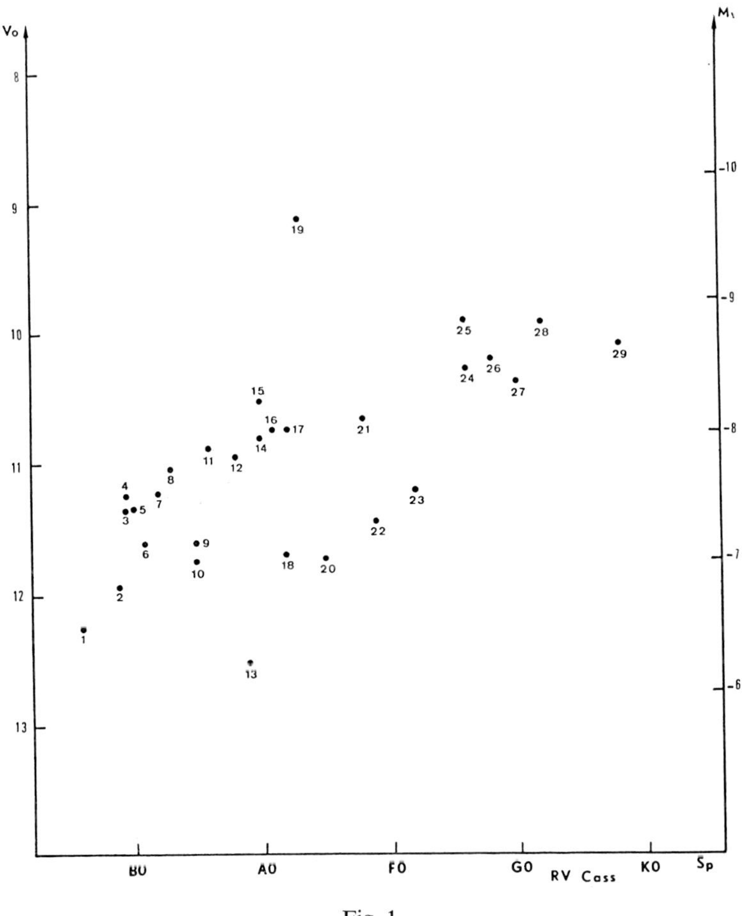

Fig. 1.

magnitude absolue. Il semble que pour les étoiles chaudes la mesure des valeurs de $W\lambda$ pour $H\gamma$ et $H\delta$ serait un excellent critère à l'intérieur de chaque classe spectrale. Ces mesures sont en cours (Maurice). Un critère analogue doit être utilisé pour les étoiles avancées.

(3) Il n'est pas possible de donner la magnitude absolue V en fonction du type spectral. Seule une étude quantitative le permettra. Par contre, les couleurs intrinsèques $(B-V)_0$ de ces étoiles sont bien déterminées (Figure 2). Les corrections de rougissement suivantes ont été appliquées.

Correction $(B-V)$	absorption galactique	-0.06
	absorption moyenne dans le GNM	-0.03
	Total $\Delta(B-V)$	$= -0.09$
Correction $(U-B)$	$0.7\Delta(B-V)$	$= 0.06$

En fait, on constate une différence systématique entre les valeurs de $(B-V)$ pour les étoiles de même type spectral mais de luminosités différentes.

Pour les étoiles plus brillantes que $m_v = 10.8$, c'est à dire les supergéantes pour lesquelles $M_v < -8$ l'indice de couleur est plus grand de 0.07 que pour les étoiles plus faibles.

$$(B - V) = (B - V) + 0.07$$
$$M < -8 \quad M > -8 \, .$$

Les différences $\Delta(B-V)$ ont été déterminées pour l'ensemble des valeurs du catalogue Ardeberg *et al.* (1972) (Tableau II). Le raccord avec les étoiles III et V sera publié prochainement.

Pour les indices $(U-B)$, la situation est moins nette.

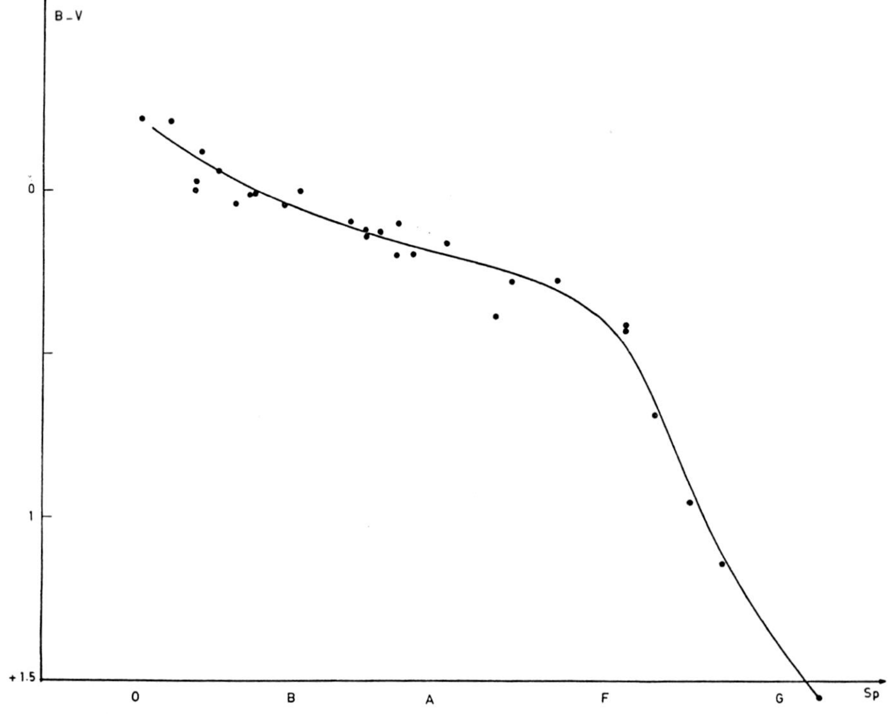

Fig. 2.   Position des étoiles standards dans le diagramme spectre-couleur$(B-V)$.

(a) L'écart est dans le sens opposé et varie avec le type spectral, il est de l'ordre de -0.35 pour les étoiles A5. Mais la dispersion des valeurs est considérable.

(b) Il existe des étoiles F6 ayant sensiblement la même valeur de $(B-V)$ et pour lesquelles $(U-B)$ varie d'une étoile à l'autre de $+0.53$ à -0.10! Il ne s'agit pas d'erreurs de mesures car une étude spectrophotométrique de L. Divan donne des résultats pratiquement identiques. Il ne nous semble pas que les étoiles bleues soient situées à l'intérieur de nébuleuses.

La Figure 3 donne la position des étoiles standards dans le diagramme $(U-B)$ $-(B-V)$. Les différences $\Delta(U-B)$ ont été déterminées pour l'ensemble du catalogue Ardeberg *et al.* (Tableau III).

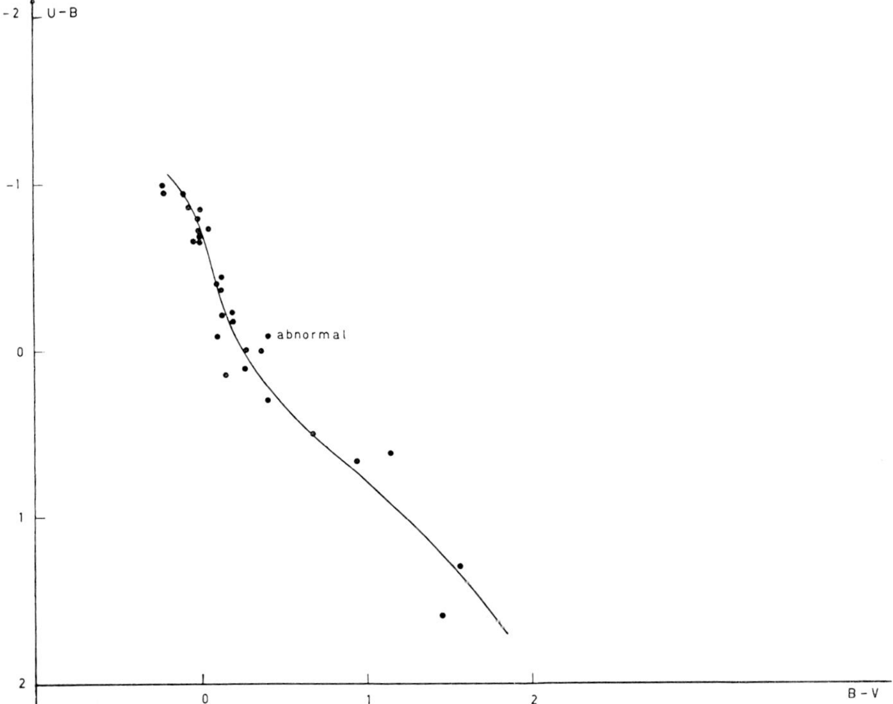

Fig. 3.   Position des étoiles standards dans le diagramme couleur-couleur.

TABLEAU II

Valeur de $(B-V)_0$ pour les supergéantes du GNM

Sp	$(B-V)_0$		$\triangle(B-V)_0$
	(1) $m < 10.8$	(2) $m > 10.8$	(1)–(2)
B0	-0.15	-0.22	$+0.07$
B5	-0.05	-0.13	0.08
A0	0.03	-0.03	0.06
A5	0.15	$+0.08$	0.07
F0	0.27	0.19	0.08
F5	0.42	0.37	0.05
G0	0.80		
G5	1.26		
			$+0.07$

Conclusion

(1) Une mesure de $W\lambda$ de Hγ et Hδ pour les étoiles de spectre $< F2$ et la mesure d'un autre indice spectral pour les étoiles $> F2$ s'impose pour compléter ou remplacer les classes de luminosité MK et ces indices permettront la détermination des magnitudes absolues M_v.

TABLEAU III

Valeur de $(U–B)_0$ pour les supergéantes du GNM

Sp	$(U - B)_0$		$\Delta(U - B)$
	(1)	(2)	(1)–(2)
	$m < 10.8$	$m > 10.8$	
B0	− 0.99	− 0.93	− 0.06
B5	− 0.77	− 0.71	− 0.06
A0	− 0.53	− 0.38	− 0.15
A5	− 0.27	0.08	− 0.35
F0	− 0.01	0.18	− 0.19
F5	0.24	0.21	− 0.03
G0	0.56		
G5	0.98		
G8	1.22		

(2) Les valeurs de $(B-V)\,(U-V)_0$ sont bien définies et permettent la détermination des excès de couleurs E. Lorsque les mesures de $W\lambda$ seront faites, cet ensemble homogène permettra la détermination des distances galactiques.

(3) Les indices $(U-B)_0$ montrent une forte dispersion, néanmoins le diagramme couleur-couleur est utile pour l'étude d'une population stellaire.

Bibliographie

Ardeberg, A., Brunet, J.P., Maurice, E., et Prévot, L.: 1972, *Astron. Astrophys. Suppl. Ser.* **6**, 249.
Fehrenbach, Ch. et Duflot, M.: 1970, *Astron. Astrophys.*, Special Supplement serie No. 1.
Fehrenbach, Ch. et Duflot, M.: 1973, preprint.
Feast, M. W., Thackeray, A. D., et Wesselink, A. J.: 1960, *Monthly Notices Roy. Astron. Soc.* **121**, 337.
Maurice, E.: 1972, Communication faite au cours de ce symposium.

QUANTITATIVE SPECTRAL CLASSIFICATION IN THE BCD SYSTEM FOR LMC SUPERGIANTS

L. DIVAN

Institut d'Astrophysique de Paris and European Southern Observatory, La Silla, Chile

Abstract. In December 1970 and January 1971, spectra of 13 LMC supergiants have been obtained with the Chalonge spectrograph attached to the Cassegrain focus of the 152 cm ESO reflector at La Silla. The spectrophotometric parameters λ_1, D, φ_b and φ_{uv} were measured and the first results deal with (a) spectral classification and (b) the distance modulus of the LMC.

1. Spectral Classification

The $\lambda_1 D$ diagram has been calibrated in absolute magnitudes by means of galactic supergiants belonging to clusters. The result (curves of equal absolute magnitude from $M = 0$ to $M = -8$) was presented at the IAU meeting in Brighton (1970).

If the LMC distance modulus is as large as generally thought, the Cloud contains supergiants 1 or 2 magnitudes brighter than the brightest ones known in our Galaxy. Thus, the program stars in the LMC have been divided into two groups:

(I) the brightest B and A type stars,

(II) a few B and A type stars about 1.5 mag. fainter than those of the first group.

The results are:

(1) The two groups of stars are well separated in the $\lambda_1 D$ plane: the second group lies in the region of the brightest Ia galactic supergiants, between the curves $M = -7$ and $M = -8$, and the first group well outside the curve $M = -8$ in the direction of highter luminosities.

(2) The spectral types of all these stars, deduced from their position in the $\lambda_1 D$ diagram, are generally in very good agreement with the spectral types determined by Ardeberg *et al.* (1972).

2. Distance Modulus of the LMC

Making the assumption that the LMC stars do not differ fundamentally from the galactic ones and that the curves of equal absolute magnitudes in the $\lambda_1 D$ plane are still valid for them, the M values for stars of the second group (stars between the curves, $M = -7$ and $M = -8$) result from the position on the $\lambda_1 D$ diagram. The V_0 magnitudes were obtained from the V magnitudes corrected for the interstellar absorption A_v; the value of A_v, which is always small, was calculated from the excess in blue gradient, $\varphi_b - \varphi_{0b}$, and assuming a normal reddening law. The mean distance modulus $V_0 - M$ given by five different stars (with only one spectrum for each) is 18.1 ± 0.5. This value is smaller than those generally derived and more observations are desirable. The absolute magnitudes of the first group of stars (the brightest LMC supergiants) calculated from this distance modulus, lie between -9 and -9.5, and the curve of equal absolute magnitude $M = -9$ could be added to the $\lambda_1 D$ diagram.

Finally the case of the F type object G 296 has been discussed: with a line spectrum F6Ia, it has the blue gradient φ_b of a F0 or F2 star and the $\lambda_1 D$ type is about F9Ia; up to now, nothing similar has been observed in our Galaxy.

Reference

Ardeberg, A., Brunet, J.-P., Maurice, E., and Prévot, L.: 1972, *Astron. Astrophys. Suppl. Ser.* **6**, 249.

PHOTOMETRIC INVESTIGATION OF IC 2944

A. ARDEBERG, E. MAURICE, and J. RICKARD

European Southern Observatory, Santiago, Chile

Abstract. An investigation has been made of the central region of the cluster IC 2944. For about 70 stars *UBV* photometry has been made. For the brighter of those stars also Hβ has been measured. Slit spectra of intermediate dispersion (73 Å mm^{-1}) have been made for the 40 brightest stars. Coudé spectra (12 Å mm^{-1}) have been obtained for 8 stars. Fabry-Pérot measurements have been made in 5 points within the nebulosity. The number of blue stars is shown to be extremely high. Down to visual magnitude 11.5 the cluster seems to contain only O and B stars, the majority of spectral class earlier than B3. The Fabry-Pérot measurements give evidence of high internal gas motion. Evidently the gas is thin, visible only because of the great number of hot stars.

Details will be published elsewhere.

Ch. Fehrenbach and B.E. Westerlund (eds.), Spectral Classification and Multicolour Photometry, 29. All Rights Reserved.
Copyright © 1973 by the IAU.

SPECTRAL CLASSIFICATION STUDIES OF CEPHEIDS USING A SMALL TELESCOPE WITH AN IMAGE TUBE SPECTROGRAPH*

M. F. McCARTHY

Specola Vaticana, Castel Gandolfo, Vatican City State

Abstract. Most of the results reported thus far of image tube spectroscopy as applied in astronomy have been concerned with emission features and have been the result of a combination of the largest available light gathering power of large telescopes with the improved sensitivity of image intensifiers. We consider here the application of image tube techniques to moderate sized telescopes; concretely, we describe the first program of spectral classification using absorption features as carried out with the image tube spectrograph of the Vatican Observatory attached to the Zeiss 60-cm reflector at Castel Gandolfo. Treanor (1970) has described the optical design, construction and the first tests of this instrument. The receiver is an RCA cascaded image tube loaned to the Vatican Observatory by the Carnegie Image Tube Committee.

The principal conclusions from the present study are the following:

(1) With sufficiently large dispersions in the grating selected, one can overcome most of the limitations of resolution inherent in the electronically formed spectral images; existing criteria for determining temperatures and luminosities can be used successfully. The dispersion used here is 95 Å mm^{-1}; the resolution is estimated at 10 Å.

(2) The intensity gains of the recorded signal allow reasonable exposures to be made even with telescopes of moderate diameters; this is of special importance in the study of the spectra of variables.

(3) The absence of 'dips' in the spectral range available to the S 20 surface used in the Carnegie image tube provides many additional features for observation especially in the green-yellow-red region of the spectrum. New criteria here can complement existing criteria most of which by reasons of limited emulsion sensitivity have been limited to the spectral interval between 4000 Å and 5000 Å.

(4) The combination of increased speed of observation and extended spectral range also facilitates one's observations of sufficient standard spectra. These can be observed very rapidly and do not constitute a block to the observation of programmed variable stars.

The objects studied here were eight Cepheids observed near maximum phase plus twenty two standard supergiant stars of types F, G and K. The following are the Cepheids observed, SZ Tau, S Sge, SU Cyg, DT Cyg, SU Cas, FF Aql, U Aql and

---

* The paper in full will be published in *Ric. Astron. Spec. Vat.*

η Aql; they were chosen because they were relatively bright Cepheids and because abundant data for the correlation of their light and colour curves with spectral variations were available.

The limitation of observations to phases near maximum was made because both Struve (1944) and Kraft (1960) had noted that the spectral differences between variable and non-variable supergiants of late types were more pronounced near maximum phases whereas Code's suggestion (1947) was that the colours of Cepheids would be best determined at maximum phase when Cepheids of a wide range of periods had a nearly identical colour. Our observations confirm the enhancement of hydrogen and the strengthening of Ti as noted by the authors cited above for Cepheids at maximum as compared with non-variable standard supergiant stars of similar spectral type. The enhancement on spectra observed in this program does not seem to be of sufficient strength to cause serious blanketing problems for broad band photometry. It would be interesting on successive short exposures of the same duration to see if the enhancement of hydrogen near maximum undergoes any flickering or if it maintains a monotonic variation with phase. Another interesting point raised by the present series of observations is this: how does the enhancement of the different hydrogen lines correlate with the maximum phase which is observed to be retarded at longer wavelengths. These questions cannot be resolved from the present plate material. When the proper spectrophotometric standards can be imposed on image tube plates along with the comparison spectra and the stellar spectrum, then the use of fast image tube spectral studies can provide answers to these questions.

The preliminary attribution of features noted in spectra of late type supergiants in the interval between 4861 Å and 6530 Å is summarized in Table I; here the numbered feature, as observed from tracings made of the image tube spectra studied here, is given, then the approximate wavelength of the feature, its description and finally some possible sources are suggested.

TABLE I

Features observed in spectra of late-type supergiants

Feature	Wavelength	Description	Suggested possible sources
1	4861	Hβ	
2	5180	Break begins	MgI (5167, 5173, 5184)
3	5300	Line	CrII (5306); FeII (5316); FeI (5324, 5328)
4	5505	Break midpoint	CaI (5505); CrII (5503, 5508)
5	5680	Break begins	ScII (5667); ScI (5672)
6	5893	NaI	
7	6110	Break begins	CaI (6102, 6122)
8	6260	Line	FeI (6231, 6254)
9	6350	Break begins	FeI (6336)
10	6450	Break begins	CaI (6439, 6462, 6493); FeII (6456)
11	6563	Hα	

References

Code, A. D.: 1947, *Astrophys. J.* **106**, 309.
Kraft, R.: 1960, *Stars and Stellar Systems* **6**, 370.
Struve, O.: 1944, *Observatory* **65**, 257.
Treanor, S. J., P. J.: 1970, *Ric. Astron. Spec. Vat.* **8**, 61.

SPECTRAL CLASSIFICATION OF SOME LONG-PERIOD AND SEMIREGULAR VARIABLES NEAR TIMES OF MAXIMUM

L. W. SIMON and W. BUSCOMBE

Department of Astronomy, Lindheimer Astronomical Research Center, Northwestern University, Evanston, Ill., U.S.A.

(Read by P. Rybski)

Abstract. Spectra of 60 M-type long-period and semiregular variables, obtained near the time of maximum at Siding Spring Observatory in Australia, from 1965–1967, have been classified on the Keenan system.

Between September 1965 and November 1967, spectra of standard stars and long-period and semiregular variables near their times of maximum were obtained on the Meinel Spectrograph at Siding Spring Observatory in Australia. They cover the wavelength region 3600–5100 Å with a dispersion of 118 Å mm^{-1} on baked IIa–O emulsion. The spectra are widened 0.25 mm. Sensitometer spots were exposed for each night of observation. The material includes 89 spectra of long-period and semiregular variables of types K and M, and 26 (mostly irregular) of types R, N and S.

The 60 variables listed below are classified on the system presented by Keenan (1966). Classifications for 41 of these stars are listed in Bidelman's catalogue of emission-line stars (1954). As seen in the diagram, there is a very good correlation between these classes and the Siding Spring classes. The radial velocities have been measured, and will be presented with further details of the spectra in Simon's doctoral dissertation.

TABLE I

Spectral classes of long-period and semiregular variables near times of maximum[a]

HD number	Name	Spectral class	Type[b]	Julian date of observation, 2439000 +
151	SW Scl	M2e	Sr	452
151	SW Scl	M3	Sr	814
409	V Scl	M6e	M	811
1115	S Scl	M6.5e	M	808
1760	T Cet	M4	SRb	015
1760	T Cet	M4	SRb	017
1760	T Cet	M4	SRb	451
1925	S Tuc	M4e	M	781
5774	U Tuc	M4e	M	365
6592	Z Cet	M6e	M	810
6592	Z Cet	M6.5e	M	812
17491	Z Eri	M4e	SRb	015
17491	Z Eri	M4	SRb	017
17491	Z Eri	M4	SRb	451
17895	RR Eri	M5	SRb	451

Table I (continued)

HD number	Name	Spectral class	Type [b]	Julian date of observation, 2439000 +
18242	R Hor	M7	M	365 [c]
18242	R Hor	M8	M	808 [d]
18949	T Hor	M4e	M	780
18949	T Hor	M4e	M	814
20646	X Cet	M6.5	M	346
24754	T Eri	M4e	M	810
25725	V Eri	M6.5	SRc	163
25725	V Eri	M6.5	SRc	451
29383	R Ret	M4e	M	164
29383	R Ret	M5e	M	780
30551	R Pic	M0e	SRa	452
33894	S Pic	M6.5e	M	165
40913	V 352 Ori	M6	Lb	165
41698	S Lep	M5	SRb	450
41698	S Lep	M4	SRb	808
71793	R Cha	M4e	M	636
73766	RV Hya	M4	SRc	165
73766	RV Hya	M4	SRc	223
81137	WY Vel	M3e P		165
81137	WY Vel	M3e P		225
84474	RR Hya	M4e	M	224
−21°2931	SU Hya	M4	SRb	223
105266	RW Vir	M5	Lb	165
105266	RW Vir	M4	Lb	224
109372	BO Mus	M4	Lb	165
118767	V 744 Cen	M5	Lb	227
118767	V 744 Cen	M5	Lb	634
120285	W Hya	M7e	SRa	224
120460	VX Cen	M4	SR	225
121518	V 412 Cen	M4	Lb	635
138547	RU Lib	MO	M	633
149234	X Ara	M5	M	365
329889	RX Lup	M4e	M	226
152476	RS Sco	M6e	M	636
172301	U CrA	M2e	M	364
192702	RT Sgr	M6.5e	M	364
199003	S Ind	M6e	M	370
199003	S Ind	M4e	M	780
201866	W Ind	M4e	SRc	786
202306	RR Aqr	M3e	M	752
207192	R Gru	M6.5e	M	364
212537	T Gru	M1e	M	365
212539	S Gru	M5e	M	780
216907	S Aqr	M6e	M	810
218541	Y Scl	M6.5	SRb	814
221433	V Phe	M6.5e	M	780
224269	R Phe	M4e	M	752
224269	R Phe	M4e	M	782

[a] Including 9 observations of irregular variables
[b] *General Catalogue of Variable Stars,* 3rd edition
[c] Phase: − 106 days.
[d] Phase: − 72 days.

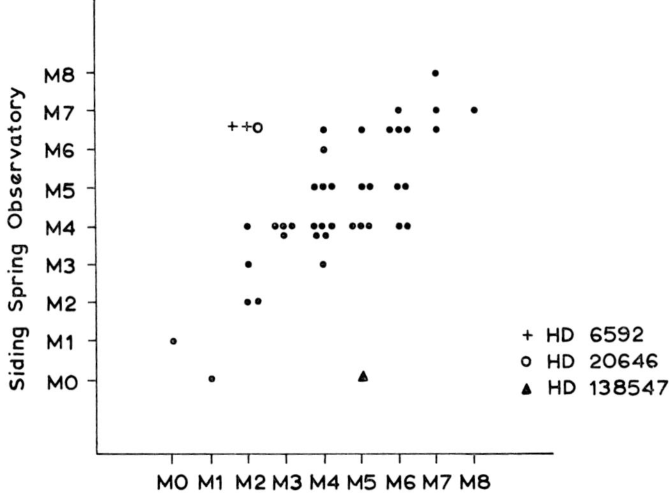

Fig. 1.   Classification correlation.

References

Bidelman, W. P.: 1954, *Astrophys. J. Suppl. Ser.* **1**, 175.
Keenan, P. C.: 1966, *Astrophys. J. Suppl. Ser.* **13**, 333.

SOME NOTES ON STARS NEAR K0

N. G. ROMAN

NASA Headquarters, Washington, D.C., U.S.A.

Abstract. Two hundred giant and subgiant stars near K0 were classified twice by somewhat different techniques. The resulting types were in excellent agreement but a comparison of the assigned types with *U, B, V* colors indicated both an intrinsic scatter in the colors for stars of similar spectroscopic appearance and a problem in either the luminosity classifications or the standard colors or both. Several peculiar stars and a luminous supergiant were also detected.

Photoelectric *UBV* photometry and slit spectra with dispersions and resolutions near that of the MK system have been obtained for more than 700 stars brighter than twelfth magnitude, photographic, in Kapteyn Selected Areas. The majority of the areas studied are at high galactic latitudes since the intent of the program was to provide homogeneous data for a large random sample of moderately faint stars for studies of population effects. These data may also serve as standards for photographic photometry and, particularly, objective prism spectroscopy.

In general, the observed spectral types agree well with those predicted from the photometric colors, but near K0 the scatter is larger than would be expected from the internal consistency of either the photometric measurements or the spectral classification. Individual outstanding cases of disagreement were reclassified with no significant improvement in the agreement and, usually, no change in the assigned types. The 200 stars of luminosity classes III and IV and spectral classes G8, K0, and K1 were then arranged in groups of about 25 in a two-dimensional array on the basis of intercomparisons only among the stars being classified. Since the original types were determined by the direct comparison of each spectrogram with those of standard stars, the arrangements in a two-dimensional array, was independent of my earlier classifications except for the use of the same classification criteria.

More than 85% of the newly derived types were identical to those I had assigned previously. The scatter in the colors for stars of the same type was unaffected. The 15% of the stars for which the new types differed by one classification interval from the earlier ones indicates a reasonable uncertainty, considering that the spectral classes are quantized. Finally, all of the G8 stars were again intercompared and arranged in order of apparent luminosity and the class III stars were again arranged in order of spectral type. These intercomparisons yielded no deviations from the assigned classifications.

The mean colors of the stars at latitudes more than 45° from the plane were then examined. Table Ia lists the mean colors for each class of stars. It is clear that, contrary to expectations, there is no difference in mean color as a function of luminosity. Table Ib shows the same thing in a different way. The G8 stars have been divided into groups of two or three, and the groups paired by $(B-V)$ color. Again, there is no difference between the brighter and the fainter members in each color grouping.

Ch. Fehrenbach and B. E. Westerlund (eds.), Spectral Classification and Multicolour Photometry, 36–41. All Rights Reserved.
Copyright © 1973 by the IAU.

TABLE I

Average colors for the faint stars

(a)					(b)		
Type	No. of stars	$B-V$	$U-B$		Luminosity	G8 Stars $B-V$	$U-B$
G8III	4	0.94	0.57		BR	0.85	0.44
G8IV	12	0.92	0.61		FT	0.88	0.52
K0III	32	1.00	0.75		BR	0.93	0.56
K0IV	13	0.98	0.74		FT	0.94	0.57
K1III	32	1.09	0.99		BR	0.99	0.72
K1IV	2	1.18	1.09		FT	1.03	0.76

Average $(U-B)$ excess 0.05.

Thus, the observations show two effects which are larger than the accidental errors of measurement. There is a spread in colors for the same spectral type and the expected correlation of luminosity class and color does not appear. The colors of the stars assigned to G8IV can be explained if the stars are actually G9III–IV stars. This is not entirely unreasonable since the hydrogen lines decrease in intensity with both decreasing luminosity and increasing spectral type and the CN and $\lambda 4077$ of Sr II become slightly weaker with advancing type as well as much weaker with decreasing luminosity. However, the strontium line should be enough weaker at G8IV than at G9III–IV to make it unlikely that a systematic classification error of this magnitude would have survived the arrangement of the spectra in a two-dimensional array. Moreover, the effect at K0 cannot be explained as simply.

TABLE II

Mean colors of bright stars

Type	No. of stars	$B-V$	$U-B$
G8II – III	6	1.01	0.79
G8III	47	0.94	0.67
G8III – IV	15	0.97	0.74
G8IV	4	0.86	0.52
K0II – III	5	1.15	1.06
K0III	80	1.02	0.86
K0III – IV	6	1.02	0.86
K0IV	2	0.99	0.85
K1II – III	1	1.12	1.02
K1III	28	1.00	0.99
K1III – IV	2	1.12	1.08
K1IV	2	1.04	0.98

The mean colors are listed in Table II for the stars brighter than 5.5 and north of declination $-20°$, which are in the *Naval Observatory Catalogue* (Blanco *et al.*, 1968). The mean $B-V$ colors for each spectral type agree well with those given in Table I for the fainter stars. The $U-B$ colors for the G8 and K0 giants are about 0.1 mag. redder than those for the faint high-latitude stars although the colors for the K1 giants agree perfectly. The brighter U magnitude for the high-latitude giants is interesting but hardly surprising in view of earlier evidence that high-velocity stars appear brighter in the ultraviolet than low-velocity stars. Although, particularly at G8, Table II gives an indication of the expected luminosity effect in the colors, it is far

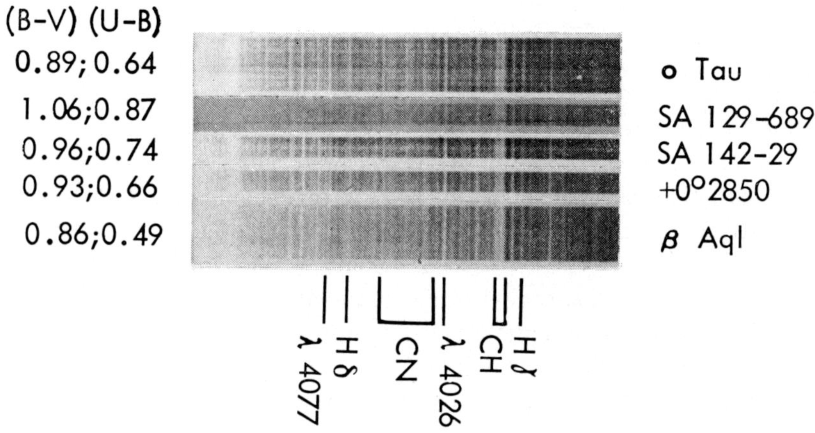

Fig. 1.   The G8IV stars SA 129 $-$ 689 (BD $-$ 14°3683), SA 142 $-$ 29 (CPD $-$ 29°233) and BD $+$ 0°2850 are arranged in order of $B-V$ colors between the G8III and G8IV standards, o Tau and $β$ Aql respectively.

less marked than one would expect (see, e.g., Fitzgerald, 1970). In addition, as for the fainter stars, the scatter in each spectral type is significantly larger than would be expected.

It appears probable that the mean colors of stars near K0 should be revised. It is also possible that most of the stars observed in surveys to a limiting apparent magnitude are basically evolved stars and, hence, that the subgiants are closer to the giants than to the dwarfs. Nevertheless, it is also possible that the large individual residuals reflect a problem with the classification criteria. Near K0, the classification criteria in the MK system are based on the strength of the hydrogen lines, the strength of Ca I, $λ4226$, and the appearance of the G band of CH; the luminosity criteria are based on the strength of Sr II, $λ4077$ and of CN. It is well known that CN is often weak in high-velocity stars and it was given relatively little weight in the present classifications.

Strontium is an element which often appears in anomalous strength in earlier stars as well as in such stars as the BaII stars. I suspected some time ago that there are variations in the strength of the strontium line among the high-velocity stars which are uncorrelated with luminosity and the present study appears to confirm this. It also is well known that, among the high-velocity stars, the hydrogen lines are often strong compared to the strength of the metallic lines and the CH strength is frequently anomalous. Thus, all of the classification criteria are suspect, at least for stars which are not bonafide members of the spiral-arm population.

Figure 1 illustrates three stars near G8, arranged in order of decreasing redness

K STARS

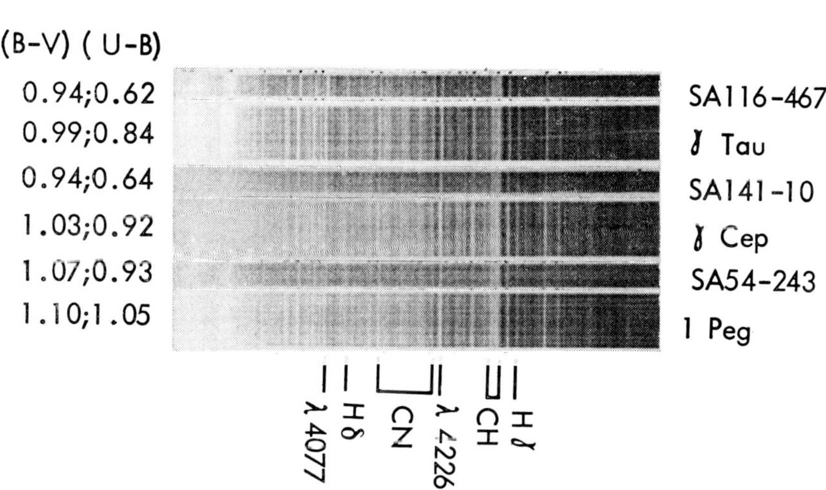

Fig. 2. The standard stars γ Tau, γ Cep and 1 Peg are K0III, K1IV and K1III respectively. SA 141 − 10 (CPD − 29°108) and SA 54 − 243 (BD + 30°2022) are apparently normal K0III stars although CN is weak in SA 141 − 10. Sr II λ4077, is abnormally strong in SA 116 − 467 (BD − 15°43).

between a G8III and a G8IV standard. It is clear that the ratio of Ca I/H indicates that SA 129 − 689 (14°3683) is no later than o Tau in spite of its substantially redder color. The spectrum bears little resemblance to a star as late as K1. Similarly, BD +0°2850 has weaker Sr II than β Aql in spite of its redder color. Both the hydrogen-line strength and the appearance of the G band would be hard to reconcile with a type much later than G8 for any of these stars. Figure 2 illustrates these discrepancies for three stars at K0 and K1, compared with standard stars at K0III, K1IV and K1III. Although SA 54 − 243 (+ 30°2022) is nearly as red as 1 Peg, neither the strontium nor the CN is significantly stronger than in γ Cep. By comparison, SA 141 − 10 (CPD − 29°108) appears both later and brighter than SA 54 − 243 but is substantially bluer. In SA 116 − 467 (− 15°43), the strontium line is obviously too strong. If this were really a luminosity class II star, it would be more than three kiloparsecs below the galactic

TABLE III

Peculiar giants near K0

$-15°564$	CH strong, Ba II star
$-15°43$	Sr strong, CH normal
$-15°3440$	CH strong, remaining spectrum normal
$+45°1951$	CH strong, remaining spectrum normal
$+0°\ 2971$	CH normal, all other features weak
$-0°\ 294$	Lines washed out, CN probably weak
$+45°1764$	CN weak
$-14°6438$	Sr normal, no CN, other lines weak (G8IV)
$+30°3866$	4150 star (K1IV)

SUPERGIANTS

Fig. 3.   SA 64 − 382 (BD + 29°3865) is compared with three other supergiants.

plane. There are no other indications of high luminosity. The strength of the strontium line resembles that in the Barium II stars, but there is no trace of the Ba II line and CH is normal in intensity. Table III lists nine other obviously peculiar stars observed in this study. Again, the tendency for CH, CN, and Sr II to be abnormal is clear.

These results indicate that both the standard classification criteria and the predicted colors for stars near K0 must be re-examined. Until these problems are understood, narrow-band colors must be used cautiously. It may be possible to test the luminosity discrimination in the present material through the computation of secular parallaxes but the small proper motions and tendency for these stars to have large space-velocity dispersions will make this difficult.

Figure 3 illustrates another unusual star near K0, a Ia supergiant. While SA 64 − 382 (+ 29°3865) is slightly later than RW Cep, the strength of the hydrogen lines indicates that it cannot be much later than K0. It appears much more luminous spectroscopically than either ε Gem (G8Ib) or ζ Cep (K1Ib). Thus it adds another star to a very sparsely settled region of the HR diagram.

References

Blanco, V. M., Demers, S., Douglass, G. G., and Fitzgerald, M. P.: 1968, *U.S. Naval Obs. Publ.*, *2nd Ser.* **21**, Washington.
Fitzgerald, M. P.: 1970, *Astron. Astrophys.* **4**, 234.

stars fainter than 6^m5 and stars which had companions closer than $d = 15''$ and/or $\Delta m < 4^m$. In the case of the A-type dwarfs we have examined the existence of reddening corrections by forming average colors for dwarfs both brighter and fainter than 5^m0. If reddening were statistically significant, one would expect both averages to differ systematically, the average for the fainter group being redder than the average of the brighter stars. Since such trend did not show up, we omitted reddening corrections for the A-type dwarfs. For B-stars this is clearly not permissible and it was decided therefore to correct the $U - B$ index by means of

$$E(U - B) = 0.71 \cdot E(B - V).$$

Because of this correction, only the dispersion in $U - B$ can be used in the B dwarfs.

The *UBV* colors of the A-type dwarfs were taken from Cowley *et al.* (1969), whilst those of the B-type dwarfs were taken from Blanco *et al.* (1968).

Tables I and II summarize conveniently the results. In conclusion it can be said that the average value of the dispersion for

B-type dwarfs is $\pm\ 0^m07$

A-type dwarfs is $\pm\ 0^m04$.

Both values are very large and imply that the relation is a band rather than a curve. This is a very obvious fact, which is, however, very seldomly mentioned.

TABLE I

Dispersion of $U - B$ indices in B-type dwarfs

	Hiltner (1956)		Slettebak (1954, 1955)		Hiltner *et al.* (1969)		Cowley *et al.* (1969)		Adopted
B0	0^m08	13							0^m08
B0.5	07	30							07
B1	07	60							07
B2	08	21	0^m11	14	0^m05	25			07
B2.5					04	15			
B3	07	11	10	22	04	18			07
B4					03	12			
B5			04	13	05	16			05
B8			06	15			0^m09	18	07
B9			06	9			08	36	08
B9.5							05	37	05

The first column gives the dispersion, the second column gives the number of data.

The easiest objection one can make to these high dispersions is that they contain the dispersion of the spectral types. The next step is therefore to calculate the dispersions without intervention of the spectral types. This can be done by calculating the dispersion in $U - B$ colors from all stars having $B - V$ indices within prefixed intervals. From the data assembled in Table III it can be seen that the dispersion is practically

TABLE II

Dispersion of $U-B$ and $B-V$ indices in A-type dwarfs

	N	$\sigma(U-B)$	$\sigma(B-V)$
A0	101	0^m07	0^m03
A1	101	5	3
A2	96	4	4
A3	73	3	3
A4	20	2	4
A5	20	3	3
A7	17	4	2
A8	10	7	5
F0	18	6	3

$N=$ number of stars. The colors were taken from a variety
of sources. See Cowley *et al.* (1969).

TABLE III

Dispersion of $U-B$ indices in A type stars
for prefixed values of $B-V$

$B-V$		N	$\sigma(U-B)$
0^m00 to 0^m05		141	$\pm 0^m054$
05	10	100	046
10	15	46	034
15	20	35	034
20	25	23	036
25	30	13	044

constant and of the order of $\pm 0^m04$. In conclusion, the dispersion in spectral type does not influence the dispersion in $U-B$. This can be explained through the slope of the relation spectral type vs $U-B$, which is very small for the range A0–A9.

One possible contribution to the dispersion could originate from systematic errors between observers. Since our colors were averaged from series of different observers, there exists the possibility that this heterogeneity increases considerably the dispersion. To check this possibility we derived the dispersions anew from the material gathered by Johnson *et al.* (1966). The observations contained in this paper were all made by Johnson and collaborators, in the most homogeneous fashion possible. Since the list provides not only UBV but also RI measurements, we calculated the dispersions also for these additional indices. The results are given in Table IV. Since also here the average dispersion $\sigma(U-B)= \pm 0^m04$, it seems clear that systematic differences between observers do not constitute a large source of error.

A second conclusion from this table is that in no color the dispersions are less than 0^m03. Therefore one can say quite generally that broad band two color diagrams give 'bands' and not 'curves'. The dispersion around the average curve is thus an essential feature of the two color diagrams. This dispersion is caused by a number of sources,

C. JASCHEK AND M. JASCHEK

TABLE IV

Dispersion of multicolor indices in A-type stars

Group	N	$\sigma(U-V)$	$\sigma(B-V)$	$\sigma(V-R)$	$\sigma(V-I)$
B9.5 V	12	$0^{\rm m}06$	$0^{\rm m}03$	$0^{\rm m}03$	$0^{\rm m}05$
A0	18	9	4	3	5
A1	19	6	3	2	4
A2	31	5	3	3	5
A3	22	4	2	3	5
A5	9	2	2	4	6
A7	7	(2)	(2)	(2)	(2)
F0	5	(2)	(4)	(3)	(5)
$\langle\sigma\rangle$		$0^{\rm m}05$	$0^{\rm m}03$	$0^{\rm m}03$	$0^{\rm m}05$

The colors were taken exclusively from Johnson *et al.* (1966).

as for instance observational errors, interstellar reddening, chemical composition effects and rotational effects. We will not deal here with the disentanglement of these different effects, but we will rather show some consequences of the existence of a scatter in the two color diagrams.

Let us take the case of the interstellar extinction corrections. Usually stars are 'de-reddened' by moving them back to the 'mean curve'. But if this relation is a 'band' the reddening is indeterminate by a amount of the order of

$$\Delta A_v = \pm \, 3 \cdot \sigma(B - V).$$

Since the dispersion is independent of the number of objects, the ΔA_v cannot be diminished by enlarging the number of stars. It should be added that this criticism was first raised by Becker (1966).

A somewhat similar situation arises if spectral types are used to determine intrinsic colors. Such a method only works if the scatter around the average relation is negligible although this basic assumption is never explicitly mentioned. In order to see what happens, we have given in Table V the correlation between $U-B$ colors and spectral types for bright southern stars. The spectral types were taken from Hiltner *et al.* (1969) and the colors from Blanco *et al.* (1968). It can be seen that no unique intrinsic color

TABLE V

Distribution of $U-B$ indices and spectral types in early nearby B-type dwarfs

$U-B$:	$-0^{\rm m}85$	80	75	70	65	60	55	50	45	40	35	N
B2	6	10	4	5	2	2						29
B2.5	2	–	3	7	2	1						15
B3			1	4	8	4	–	1	1	1		19
B4					3	3	2	1				9
B5					1	8	3	2	1	1		16
B6						1	3	1	2			7

exists for each spectral type, the range being of the order of 0^m2 to 0^m3. This is in line with dispersions in $U-B$ of the order of 0^m07, but shows that intrinsic colors derived via spectral types are imprecise. In general one should avoid such a procedure, and use spectral classification only for selecting normal dwarfs.

One objection which can be raised to the previous procedures is that the scatter is large because broad band photometry is used, which is less precise than narrow band photometry. To examine this argument we have taken several intermediate band and narrow band indices and have analysed the scatter in these indices. Of the many available indices we have selected three, namely the $H\beta$ index of Crawford *et al.* (1966, 1970) the K-index of Henry (1969, 1971) and the '$b-y$' index of Strömgren (Strömgren and Perry, unpubl.). The first two were chosen because Balmer lines and the K-line are also used in spectral classification of A-type stars, which permits an easy comparison of the different techniques. The '$b-y$' index on the other side was included essentially for comparison with the broad band index '$B-V$'.

We have calculated the scatter of these indices in a sample of or A-types dwarfs, the sample being selected according to principles described in more detail below. The dispersions were calculated from the relation of each parameter with $B-V$, and the results are given in Table VI. As one can see, the dispersions are very large even with narrow band indices. The exception is '$b-y$' which correlates very strongly with '$B-V$', and has accordingly a very small scatter.

TABLE VI

Dispersion of photometric parameters

Parameter	Dispersion
K index	$\pm 0^m06$
$H\beta$ index	04
$b-y$	01

Let us now examine the question of the relative precision of these indices. To the three parameters quoted above we will add the '$B-V$' index and the spectral type. Luminosity classes can be left out because we deal only with dwarfs, defined as such by the usual spectroscopic criteria. We have thus five different parameters at our disposal and we would like to examine which parameter gives the least dispersion.

This problem becomes mathematically treatable if it can be assumed that the relation between any pair of the five parameters is linear. This hypothesis can be checked easily and is valid probably in all cases if one uses only small parts of the main sequence.

If X and Z are two of the indices, we have

$$Z = \alpha X + \alpha_0 .$$

Now Z and X have both errors which we will denote by ε and δ. We assume further that $\langle \varepsilon \rangle = 0$ and $\langle \delta \rangle = 0$. The dispersion of the errors of Z is $\sigma(\varepsilon)$ and that of X,

$\sigma(\delta)$. The ε and δ combine both the observational errors and the intrinsic errors of each variable, because at the moment it does not seen profitable to split them into components. Statistical theory (Kendall and Stuart, 1967; Deeming, 1968) shows that it is possible to formulate a system of three equations involving the second moments of X and Z and the first mixed moment, with four unknowns. This system is obviously not solvable. But if instead of having two linearly related variables X and Z, one has three – X, Z and Y – then the problem can be solved. The appropiate formulae are given in the mathematical appendix.

Proceeding along these lines we have studied a sample of A-type dwarfs selected according to the following criteria.

(1) All stars should be dwarfs between spectral types B9 and F0.

(2) All stars should be nearby, to avoid reddening corrections, $m < 6^{m}5$.

(3) All stars should have measures of the K index, the Hβ index, $B-V$, $b-y$ and estimates of the spectral type in the MK system.

We found finally a sample of 87 stars which satisfy all these criteria. The indices were taken from Henry (1971), Crawford *et al.* (1966, 1970), Blanco *et al.* (1968), and Cowley *et al.* (1969). We applied then the technique described above and in more detail in the appendix. Since with five indices there exist five possible combinations of the three indices needed to apply our technique, we decided to combine in turn $b-y$ and $B-V$ with each one of the remaining indices. From the results given in Table VII it can be seen that some indices have a very large scatter, both in absolute value as when compared with its range of variability.

TABLE VII

Intrinsic dispersion of different parameters, from a sample of 87 A-type dwarfs

Index	Dispersion
$b-y$	$\pm 0^{m}011$
$B-V$	$\pm 0^{m}013$
K	$\pm 0^{m}073$
Hβ	$\pm 0^{m}041$
S	0.62

Attention is also called to the value of 0.6 obtained for the spectral type, which agrees very well with the one found by Gliese (1971).

A sample of 76 B type stars was analysed in the same way. Here we used the parameters $U-B$, c_1, Hβ and S. It turned however out that S is not linearly related to the other variables, a fact shown by the imaginary values for $\sigma(S)$ obtained when solving the equations. For the other variables the results are those given in Table VIII.

We can compare now the dispersions obtained for each parameter with its range of variation. The results are given in Table IX. In A-type dwarfs it is certainly no surprise that $B-V$ and $b-y$ do have the same efficiency, defined as the ratio of range and dispersion. Spectral class turns out to be approximately as precise as $b-y$ or

TABLE VIII

Intrinsic dispersion of different parameters, from a sample of 76 B-type dwarfs

Index	Dispersion
c_1	$\pm 0^m020$
$U-B$	$\pm$ 032
Hβ	$\pm$ 017

TABLE IX

Dispersion and range of classification parameters

	Parameter	Range	σ	R/σ
A-type stars:	$b-y$	0^m24	0.011	22
	$B-V$	0^m37	0.015	24
	S	11	0.62	18
	K	0^m97	0.083	12
	Hβ	0^m20	0.041	5
B-type stars:	c_1	1^m08	0.020	54
	$U-B$	0^m97	0.032	30
	Hβ	0.25	0.017	15
	S	12	0.7	17

$B-V$ and to be definitely more precise than both the K and Hβ indices. This is understandable because of the large dispersions found in Table VI. It implies that for A-type dwarfs spectral classification is about as accurate as photometry.

For B-type stars the situation is different. Here Table IX shows beyond doubt that photometry is at least twice as precise as spectral classification. Spectral classification is not competitive here, but remains essential even in this case because of the need to eliminate non-dwarfs.

Acknowledgements

Thanks are due to Mr B. Kucewicz who carried out the calculations.

Mathematical Appendix

Let X_i and Y_i be two mathematical variables affected by observational errors so that

$$x_i = X_i + \delta_i \tag{1}$$

$$y_i = Y_i + \varepsilon_i \tag{2}$$

and assume that X_i and Y_i are related linearly

$$Y_i = \alpha X_i + \alpha_1 . \tag{3}$$

Assume further that $\langle\varepsilon_i\rangle=\langle\delta_i\rangle=0$ and that the dispersions of ε_i and δ_i are σ_ε and σ_δ. Then by substitution

$$y_i - \varepsilon_i = \alpha(x_i - \delta_i) + \alpha_1 \tag{4}$$

and averaging

$$\langle y\rangle = \alpha\langle x\rangle + \alpha_1. \tag{5}$$

By calculating dispersions of (1) one gets

$$\sigma_x^2 = \sigma_X^2 + \sigma_\delta^2 \tag{6}$$

and by calculating the dispersion of (4), after some algebra

$$\sigma_y^2 = \alpha^2\sigma_x^2 + \sigma_\varepsilon^2 \tag{7}$$

provided

$$\langle X_i\delta_i\rangle = \langle Y_i\varepsilon_i\rangle = 0.$$

One can also calculate the covariance of x and y and obtains

$$\mathrm{cov}(x, y) = \frac{1}{n}\sum(x_i - \langle x\rangle)(y_i - \langle y\rangle) = \alpha\cdot\sigma_X^2 \tag{8}$$

provided $\langle\varepsilon_i\delta_i\rangle=\langle\varepsilon_iX_i\rangle=0$. The condition on $\langle\varepsilon_i\delta_i\rangle$ implies that the errors are uncorrelated, the same as $\langle\varepsilon_iX_i\rangle=0$.

Under these hypothesis, equations (6), (7) and (8) form a system of three equations with the four unknowns $\sigma_X\ \sigma_\varepsilon\ \sigma_\delta\ \alpha$

$$\begin{aligned}
\sigma_x^2 &= \sigma_X^2 + \sigma_\delta^2 \\
\sigma_y^2 &= \alpha^2\sigma_X^2 + \sigma_\varepsilon^2 \\
\mathrm{cov}(x, y) &= \alpha\cdot\sigma_X^2.
\end{aligned} \tag{9}$$

A small reflection shows that if one wants $\sigma_\delta>0$ and $\sigma_\varepsilon>0$ then

$$\begin{aligned}
\sigma_x^2 &> \sigma_X^2 \\
\sigma_y^2 &> \alpha^2\sigma_X^2 \\
\mathrm{cov}(x, y) &= \alpha\cdot\sigma_X^2
\end{aligned} \tag{10}$$

implying that one must have

$$\mathrm{cov}(x, y) < \sigma_x\cdot\sigma_y. \tag{11}$$

This condition proves to be very helpful.

Obviously the system (8) cannot be solved. Assume now that a third variable Z_i exists with

$$Z_i = z_i + \zeta_i \tag{12}$$

connected with X_i and Y_i linearly, like

$$Z_i = \beta Y_i + \beta_1$$
$$X_i = \gamma Z_i + \gamma_1 . \tag{13}$$

Then one can write down the system (8) for each pair of variables, yielding nine equations with twelve unknowns. By dropping those equations and unknowns which repeat themselves because of the cyclic arrangement (13) one gets a system of seven equations with seven unknowns, which becomes solvable.

It was found convenient in practice not to solve directly the system of seven equations, but to solve for instance the systems (X, Y) and (X, Z) by introducing assumed values of σ_X and then to select that triple of values σ_X, σ_Y and σ_Z which satisfies simultaneously all equations.

References

Becker, W.: 1966, *Z. Astrophys.* **64**. 77.

Blanco, V., Demers, S., Douglass, G. G., and Fitzgerald, M. P.: 1968, *Publ. U.S. Naval Obs., 2nd Ser.* **XXI**.

Cowley, A., Cowley, C., Jaschek, M., and Jaschek, C.: 1969, *Astron. J.* **74**, 375.

Crawford, D. L., Barnes, J. V., Faure, B. Q., Golson, J. C., and Perry, C. L.: 1966, *Astron. J.* **71**, 709.

Crawford, D. L., Barnes, J. V., and Golson, J. C.: 1970, *Astron. J.* **75**, 624.

Deeming, T. J.: 1968, *Vistas in Astronomy* **10**, 125.

Gliese, W.: 1971, *Veröffentl. Astron. Rechen-Inst. Heidelberg*, No. 24.

Henry, R. C.: 1969, *Astrophys. J. Suppl.* **18**, 47.

Henry, R. C.: 1971, *Astrophys. J. Suppl.* (in press).

Hiltner, W. A.: 1956, *Astrophys. J. Suppl.* **2**, 389.

Hiltner, W. A., Garrison, R. F., and Schild, R. C.: 1969, *Astrophys. J.* **157**, 313.

Jaschek, C. and Jaschek, M.: 1966, in K. Lodén, L. O. Lodén, and U. Sinnerstad (eds.), 'Spectral Classification and Multicolor Photometry', *IAU Symp.* **24**, 6.

Johnson, H. L., Mitchell, R. I., Iriarte, B., and Wisniewski, W. Z.: 1966, *Commun. Lunar Planetary Lab.*, No. 63.

Johnson, H. L.: 1966, *Ann. Rev. Astron. Astrophys.* **4**, 193.

Kendall, M. G. and Stuart, A.: 1967, *The Advanced Theory of Statistics* **II**, Charles Griffin and Co., London.

Slettebak, A.: 1954, *Astrophys. J.* **119**, 146.

Slettebak, A.: 1955, *Astrophys. J.* **121**, 102.

ON THE ACCURACY OF SPECTRAL CLASSIFICATIONS
OF MAIN-SEQUENCE STARS

W. GLIESE

Astronomisches Rechen-Institut, Heidelberg, F.R.G.

Abstract. By examining the observed dispersion in (colour, spectral type) relations, classification errors have been derived from the data of nearby stars. The comparisons of the colour deviations observed in spectral regions of large variations of colour with type with the deviations in regions of small variations give the following standard errors in units of a tenth of a spectral class: For K dwarfs ± 0.6 (MK), ± 1.2 (Mt. Wilson), ± 0.7 (Kuiper); for early M dwarfs ± 0.9: (MK), ± 0.7 (Mt. Wilson), ± 0.5: (Kuiper); and for late M dwarfs ± 0.7 (Kuiper).

In the preceding paper, Jaschek has mentioned an estimated standard error of MK classifications of ± 0.6 subclasses which has been given in a publication on mean relations between spectral types and photoelectric colours of nearby main-sequence stars (Gliese, 1971). The preliminary values of the accuracy of spectral classifications published there have been superseded by the results of a more refined investigation shown in the following. C. Jaschek has also demonstrated the method and the difficulties encountered when using the variation of colour with spectral type and the observed dispersions for deriving accidental errors in the classifications.

In the above cited publication 12 mean relations between MK, Mt. Wilson, and Kuiper types and the colours $B-V$, $U-B$, $(U-B)_{\mathrm{Cape}}$, and $R-I$ have been derived using the data of the 'Catalogue of Nearby Stars' (Gliese, 1969) which have been supplemented by further data collected in card catalogue form:

$B-V$, MK type	$B-V$, Mt. Wilson type	$B-V$, Kuiper type
$U-B$, MK type	$U-B$, Mt. Wilson type	$U-B$, Kuiper type
$(U-B)_{\mathrm{C}}$, MK type	$(U-B)_{\mathrm{C}}$, Mt. Wilson type	$(U-B)_{\mathrm{C}}$, Kuiper type
$R-I$, MK type	$R-I$, Mt. Wilson type	$R-I$, Kuiper type

As an example Figure 1 shows the smoothed curve of the $(U-B$, Kuiper type) relation. It is obvious that classification errors produce an additional dispersion in the colours which increases with increasing slope of the curve.

The principle of this method consists in comparisons of observed colour deviations in spectral regions of large variations of colour with type with the deviations in regions of small variations. In the $(U-B$, Kuiper type) relation deviations in the K-dwarf region should be compared with those in the G or F or A classes.

But before investigating how this method works a few remarks should be made about the spectral data used. Figure 2 shows the frequency of types from F0 to the late M dwarfs among the stars nearer than 22 pc. For K and M stars Kuiper has given numerous intermediates as K4+, K5+, and so on. The original MK system (the MKK) had a framework of only some special types; but in the mean time nearly

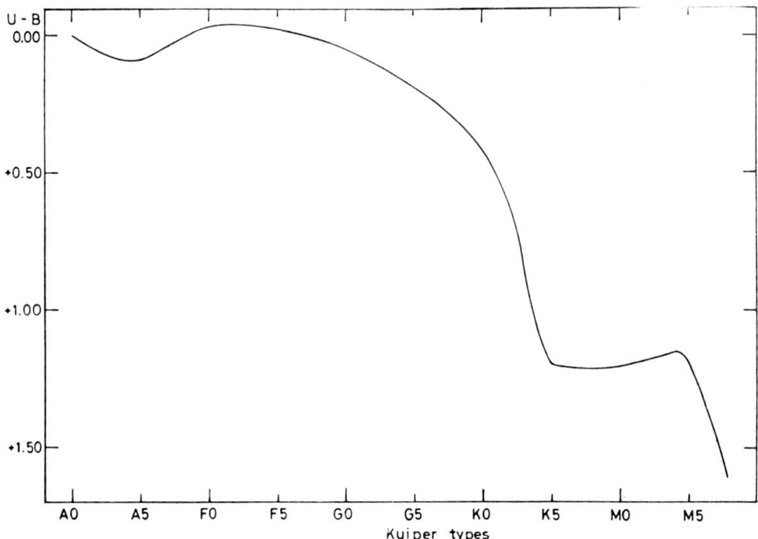

Fig. 1. $(U-B$, Kuiper type) relation of nearby main-sequence stars.

every tenth of a class is found in the literature. But even now we see the preference for the original types.

The main source for MK types has been the 'Catalogue of Stellar Spectra Classified in the Morgan-Keenan System' by Jaschek *et al.* (1964). Objects with dubious luminosity classifications have been excluded. The material used here consists only of stars lying on the main sequence. If different classifications by various observers were available the one type seeming best fitted has been used. Therefore it may be expected that the standard errors derived here will be somewhat smaller than those found by Jaschek and Jaschek (1966).

Nearly all Mt. Wilson types have been taken from the *General Catalogue of Stellar Radial Velocities* (Wilson, 1953). Mostly these types agree with the classification of 1935 (Adams *et al.*) investigated by Butler and Thackeray (1940).

Kuiper's types are found in various publications, mainly in his list of stars nearer than 10 parsecs (1942) and in the *General Catalogue of Trigonometric Stellar Parallaxes* (Jenkins, 1952, 1963). Further red dwarf data were made available by private communication.

In the 12 (colour, spectral type) relations the observed standard deviations (s.d.) of the colours have been derived in various spectral regions; the s.d. result from observational errors in colour ε_c, from observational errors in spectral type ε_t, and from a 'cosmic dispersion' σ. As we are dealing with nearby stars only no further dispersion in colour by different effects of the interstellar matter will occur.

Many of the colour data used here are mean values of various measurements. Their errors are small and ε_c is nearly insignificant, namely of the order ±0.01 or ±0.02.

The classification error, ε_t, produces an additional contribution, ε_s, to the observed

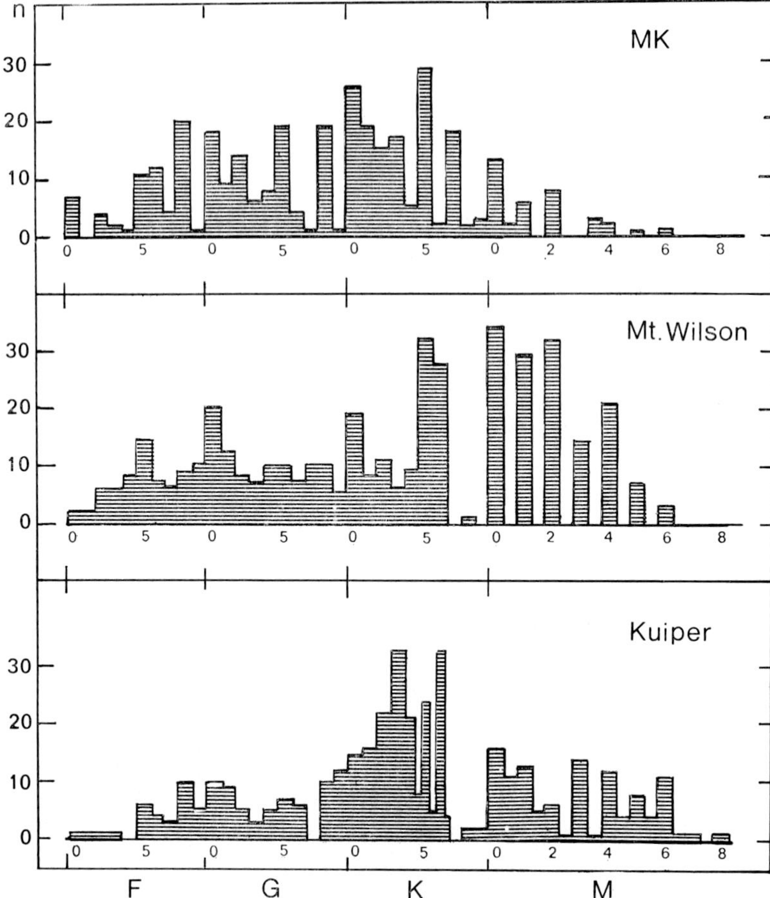

Fig. 2.   Frequencies of the various spectral types of nearby stars.

standard deviations, which is $\varepsilon_s = \varepsilon_t \, v_c$, where ε_t is given in units of a tenth of a spectral type and v_c is the variation of colour with a tenth of a spectral class: $v_c = \Delta$ colour / 0.1 spectral class.

Since there is no one to one correlation between spectral type and colour, we define σ as the dispersion which occurs even after elimination of all observational errors; that is, the true colour dispersion in a (colour, spectral type) relation.

For each spectral region $(s.d.)^2 = \varepsilon_c^2 + \varepsilon_s^2 + \sigma^2$ has been derived from the observations. Substracting an estimated value of the square of the observational error, ε_c, a quantity l is introduced by $l^2 = (s.d.)^2 - \varepsilon_c^2$, or $l^2 = \sigma^2 + \varepsilon_t^2 \, v_c^2$. σ and ε_t are unknown; probably both quantities vary with spectral type.

Figures 3–5 show in the 12 relations the observed l^2 (solid lines) together with the squares of the colour variation v_c^2 (dashed lines). Obviously in all diagrams there is a strong correlation.

From A to G the variation of colour with type is small. Normally, the solid lines

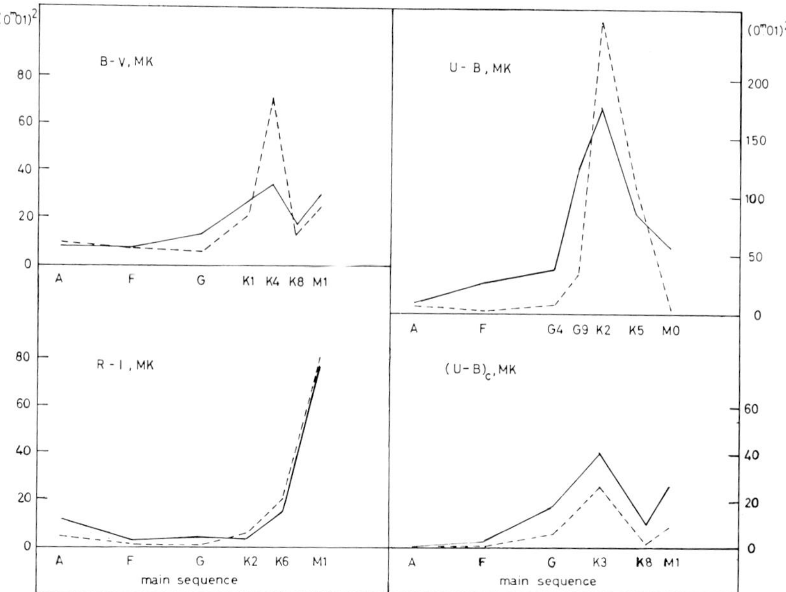

Fig. 3. Comparison between the colour dispersions observed in (colour, MK type) relations with the variation of colour with type for nearby main-sequence stars. Solid line: Squares of the dispersions observed in different spectral regions. Dashed line: Squares of the variation of colour for 0.1 spectral class. Unit for colours: 0.01.

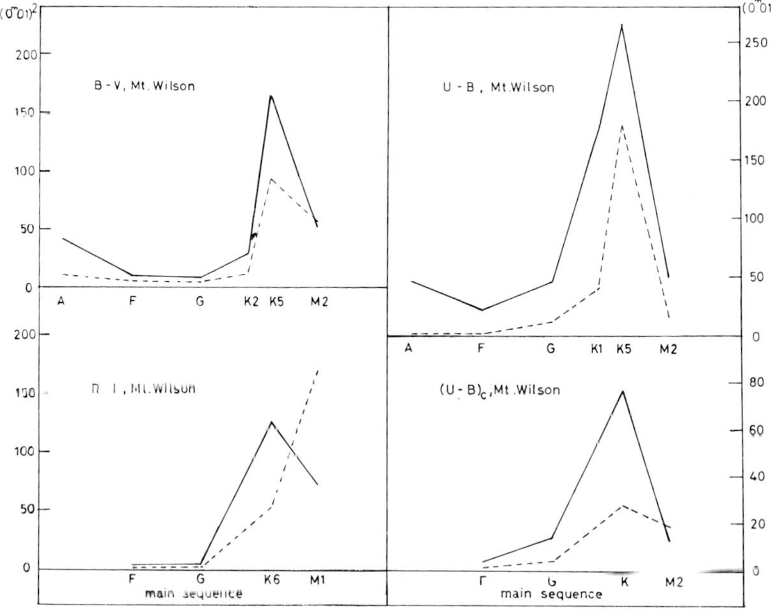

Fig. 4. Comparison between the colour dispersions observed in (colour, Mt. Wilson type) relations with the variation of colour with type for nearby main-sequence stars. Solid line: Squares of the dispersions observed in different spectral regions. Dashed line: Squares of the variation of colour for 0.1 spectral class. Unit for colours: 0.01.

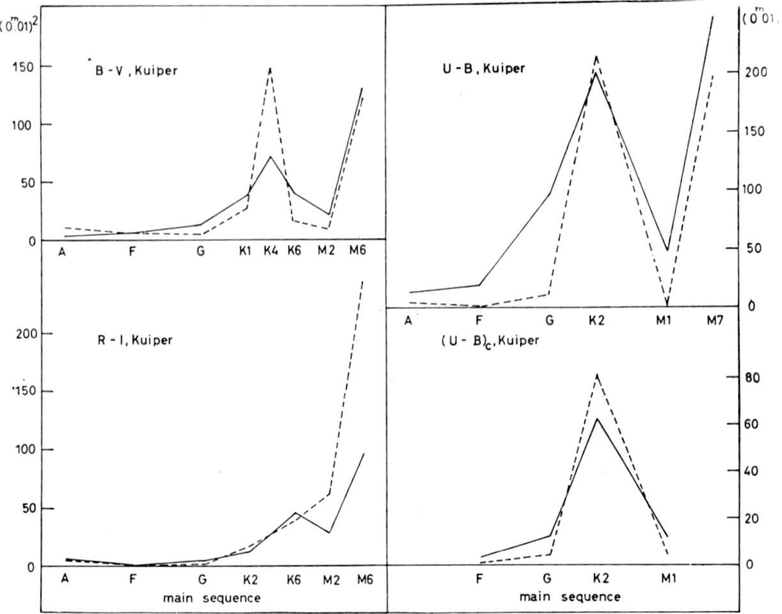

Fig. 5.   Comparison between the colour dispersions observed in (colour, Kuiper type) relations with the variation of colour with type for nearby main-sequence stars. Solid line: Squares of the dispersions observed in different spectral regions. Dashed line: Squares of the variation of colour for 0.1 spectral class. Unit for colours: 0.01.

are somewhat above the dashed values–a result which seems to be caused mainly by the σ^2. This region is not suitable for deriving the errors ε_t. Large coefficients v_c^2 are in the K and M star regions.

The limited number of objects does not allow an exact separation of the various quantities contributing to the observed dispersions. A preliminary view of these diagrams shows

(1) in the MK classification, ε_t of the K types is smaller than 0.1 spectral class – except in the $[(U-B)_C,\ \text{MK type}]$ diagram which has only low weight. For the early M dwarfs ε_t is near a tenth of a spectral class.

(2) in the Mount Wilson classification, ε_t is larger than one in class K but decreases for the early M dwarfs.

(3) Kuiper's classification shows an ε_t somewhat smaller than 0.1 spectral type.

To get precise values of the errors, differences between the quantities in regions of large variation v_c of colour with type and v_c in regions with low variations have been computed; for example

$$(l^2)_K - (l^2)_G = (\sigma^2)_K - (\sigma^2)_G + (\varepsilon_t^2)_K\,(v_c^2)_K - (\varepsilon_t^2)_G\,(v_c^2)_G\,.$$

In these adjacent regions $(\sigma^2)_K - (\sigma^2)_G$ will be negligible and $(\varepsilon_t)_G$ will be approximately equal to $(\varepsilon_t)_K$:

$$(\varepsilon_t)_G = (\varepsilon_t)_K + y \quad \text{where } y \text{ is a small quantity: } |y| \ll 1$$
$$(\varepsilon_t^2)_G \sim (\varepsilon_t^2)_K + 2\,(\varepsilon_t)_K\, y.$$

In the first approximation $\Delta\sigma^2 = [(\sigma^2)_K - (\sigma^2)_G]$ and y are supposed equal to zero:

$$(\varepsilon_t^2)_K = \frac{(l^2)_K - (l^2)_G}{(v_c^2)_K - (v_c^2)_G}.$$

For example, the results by Butler and Thackeray (1940) show that the mean errors in the Mt. Wilson classification do not vary from G to K dwarfs. In this way approximate $(\varepsilon_t)_K$ can be derived from a comparison with the dispersions in the G region and also in other spectral regions (A, F, ...).

As an example Table I shows the approximate values of ε_t in the classification of K dwarfs (mean type K4 or K5) as derived from the dispersions in the $(B - V,$ spectral type) relations. The last columns give relative weights in an arbitrary system. The weight of such a determination depends on the number of stars in the compared regions and on the value of the denominator which should not be too small. In the MK and in the Kuiper classifications the approximate values are slightly decreasing from A to G; the same trend is observed in the $(U - B,$ MK type) and $(U - B,$ Kuiper type) relations but not in the Mt. Wilson system. Such variations can be explained by an increase of the cosmic dispersion σ from A to G whereas the true mean errors ε_t probably do not vary appreciably in this spectral region from A to K $(y \sim 0)$. In the $(R - I,$ spectral type) relations no trend in the approximate ε_t is observed.

TABLE I

Approximate values of the classification errors ε'_t of K dwarfs
(mean type: K4 or K5; unit: 0.1 spectral type; p: relative weight)

Comparison region	MK types		Mt. Wilson types		Kuiper types	
	ε'_t	p	ε'_t	p	ε'_t	p
A	±0.66	0.5	±1.21	0.6	±0.71	1.0
F	0.65	2.1	1.32	4.4	0.68	4.1
G	0.57	3.5	1.32	6.1	0.64	8.3
K1	0.37	1.6	1.28	2.5	0.54	5.4
K7	0.55	0.7			0.50	4.8
M1	0.32	0.7	1.73	3.3	0.60	10.0

Corresponding series of approximate values of ε_t have been computed from the dispersions in all 12 relations and also for other spectral regions with the mean types K1, K7, M1...M6. From the trends of these ε_t the true values of the mean errors are estimated making allowance for the different weights. The combination of all results gives the values in Table II.

But these results need some qualifying remarks; the complications in this method arise from the details. Spectral types are discrete points along a continuously filled

TABLE II

Mean errors ε'_t of spectral classifications, still affected by the 'slope effect'
(Unit: 0.1 spectral type; p: relative weights)

Spectral class	Colour	MK types		Mt. Wilson types		Kuiper types	
		ε'_t	p	ε'_t	p	ε'_t	p
K	$B-V$	± 0.6	11	± 1.4	17	± 0.65	36
	$U-B$	0.7	22	1.2	35	0.85	56
	$(U-B)_C$	1.1	1	1.6	0.4	0.8	2
	$R-I$	0.8	0.5	1.6	1	1.05	2
K	mean	± 0.7	35	± 1.25	54	± 0.8	96
early M	$B-V$	± 1.0	1	± 0.8	14		
	$U-B$			1.2	1		
	$(U-B)_C$			0.7	0.1		
	$R-I$	1.0	2.5	0.7	14	± 0.6	2.5
early M	mean	± 1.0	4	± 0.8	29	± 0.6	2.5
late M	$B-V$					± 1.0	6
	$U-B$					1.0	4
	$R-I$					0.6	14
late M	mean					± 0.8	24

sequence of spectra. Type t_i is assigned to all spectra from $t_{i-1/2}$ to $t_{i+1/2}$ with the mean colours between $c_{i-1/2}=\frac{1}{2}(c_{i-1}+c_i)$ and $c_{i+1/2}=\frac{1}{2}(c_i+c_{i+1})$. Therefore, even if all spectra are classed correctly (and σ is equal to zero) a dispersion in colour will be observed. In a homogeneous filled sequence the average square of the colour deviations originating in this effect is one sixth of the square of the colour variation with a tenth of a spectral class: $\frac{1}{6}\,v_c^2$. This percentage is included in the ε_t in Tables I and II. Its exact elimination seems to be difficult as this dispersion has no Gaussian distribution and only a limited number of objects is available. But it may be estimated that the values of the classification errors ε_t in Table II have to be diminished by at most 0.1.

The elimination of this 'slope effect' from the results in Table II gives the true values of the standard errors in classification which are listed in Table III.

TABLE III

Standard errors ε_t of spectral classifications
(Unit: 0.1 spectral type)

Spectral class	MK	Mt. W.	Kuiper
K	± 0.6	± 1.2	± 0.7
early M	0.9:	0.7	0.5: :
late M			0.7

A comparison of the data in the three systems shows the superiority of the MK classification in the range of the K dwarfs. The value ± 0.6 agrees fairly well with the accuracy derived by C. Jaschek and M. Jaschek (1966). $\varepsilon_t = \pm 1.2$ for the Mt. Wilson system confirms the result by Butler and Thackeray (1940). It is very interesting that the early M dwarfs have been classified with smaller errors than the Mt. Wilson K dwarfs even though their apparent magnitudes normally are fainter than those of the dK stars. The relatively high accuracy of the late M types classified by Kuiper should be emphasized.

An application of this method to other luminosity classes such as subgiants or giants will be complicated by the reddening effects.

Errors in the luminosity classification cannot be investigated. C. Jaschek and M. Jaschek (1966) have given estimates for the MK classes. The nearby-star data indicate that among F and G dwarfs and subgiants the percentage of dubious luminosity classifications in the Mt. Wilson system is nearly twice that of the MK system.

Concluding this report on the accuracy of classifications derived from nearby-star data, it should be pointed out that the errors given here are not valid for MK classifications on objective prism plates.

Acknowledgements

The author is indebted to Dr R. Laubscher for his assistance in the preparation of the English version of the text.

References

Adams, W. S., Joy, A. H., Humason, M. L., and Brayton, A. M.: 1935, *Astrophys. J.* **81**, 187; *Contr. Mt. Wilson Obs.*, No. 511.

Butler, H. E. and Thackeray, A. D.: 1940, *Monthly Notices Roy. Astron. Soc.* **100**, 450.

Gliese, W.: 1969, *Veröffentl. Astron. Rechen-Inst. Heidelberg*, No. 22.

Gliese, W.: 1971, *Veröffentl. Astron. Rechen-Inst. Heidelberg*, No. 24.

Jaschek, C., Conde, H., and de Sierra, A. C.: 1964, *Obs. Astron. Univ. Nac. La Plata Ser. Astron.* **28** (2).

Jaschek, C. and Jaschek, M.: 1966, in K. Lodén, L. O. Lodén, and U. Sinnerstad (eds.), 'Spectral Classification and Multicolour Photometry', *IAU Symp.* **24**, 6.

Jenkins, L. F.: 1952, *General Catalogue of Trigonometric Stellar Parallaxes*, Yale Univ. Obs., New Haven, Conn.

Jenkins, L. F.: 1963, *Supplement to the General Catalogue of Trigonometric Stellar Parallaxes*, Yale Univ. Obs., New Haven, Conn.

Kuiper, G. P.: 1942, *Astrophys. J.* **95**, 201.

Wilson, R. E.: 1953, *General Catalogue of Stellar Radial Velocities*, Carnegie Inst. Washington, D. C., Publ. 601.

ADVANTAGES AND LIMITATIONS OF
QUANTITATIVE SPECTRAL CLASSIFICATION

F. SPITE

Observatoire de Paris, 92190 Meudon, France

(Read by E. Maurice)

Abstract. The measurements of spectral lines of moderate dispersion slit spectrograms for F, G and K type stars give a large quantity of information. The measurements are time-consuming in comparison with visual classification, but the use of digitized microphotometers reduces the task, and the method is now promising. The interpretation of the measurements raises some problems, which are briefly discussed.

The great success achieved by the spectral classification has not to be emphasized here. The one-dimensional classification is an easy and quick procedure. A two-dimensional classification, such as the MK system, requires considerably more care. Deciding for a type and a class is not always easy, even for Population I stars. Population II stars do not fit in the MK system, defined by a set of standards of Population I; a special classification, such as that proposed by Landi-Dessy (1964), should be used, supposing that a complete set of Population II standards could be found. A three-dimensional classification by visual inspection seems very hard to achieve and, except for the case of more or less extreme Population II stars, the attempts for such a visual classification have not proved to be entirely successful. A precise and detailed three-dimensional classification falls in the field of quantitative classification.

The simplest idea is to measure the intensity of lines in the spectra, instead of evaluating intensities (or intensity ratios) by visual inspection. This means a heavy task, calibrating the plates with scaled photometric sources, registering spectra and calibrations through a microphotometer, determining the characteristic curve of the emulsion, measuring the tracings and reducing the measurements. This has been done in the past by several astronomers and the use of digitized microphotometers and computers reduces now considerably this task.

Let us suppose that the astronomer has a set of measures of lines. Multivariate analysis seems to be an adequate tool for using this set of data. However, under careful examination, a number of problems arises.

For example, as the temperature decreases, the intensities of most of the easily measurable lines grow more or less at the same rate, so that these lines carry essentially the same information. A rather great accuracy is necessary to measure the small differential variations which display a variation of luminosity or abundance. Such an accuracy can only be obtained by exercising great care in every step: observation, measurements, reduction. The observations have to be homogeneous, and a rather *rigid* technique has to be used.

Another difficulty arises with the meaning of the measured intensities of lines.

On the empirical side, the measurements do not usually fit exactly the visual estimates. On the theoretical side, they are not necessarily simply related to equivalent widths, since only small dispersion spectra are used; a theoretical interpretation is not always straightforward. For instance, let us consider the ratio of the lines $\lambda 4077$ Sr II / $\lambda 4071$ Fe I (a classical luminosity criterion); from the statistical analysis of a sample of 129 stars, the measures of these two lines do not lead to a good luminosity criterion (Spite, 1968).

Finally, when measurements are made, it is generally expected more from these measurements than from visual inspection; quantitative estimation of T_{eff}, g or M_v and [Fe/H] are expected instead of a mere qualitative (even three-dimensional) classification; a fit with theory is required.

The best way to relate measurements of lines and theory, is to compute a synthetic spectrogram from a model (Cayrel, 1968). Good models are now found in the literature. The main difficulty is the lack of suitable physical constants (or more generally: physical knowledge) of atomic and molecular processes, so that the calculation of the synthetic spectra of late-type stars is now quite uncertain. Some progress has to be made in this field. The theory of heavy lines (chromospheric lines) has to be improved. Then, a three dimensional quantitative classification could be built on sound basis from measurements of spectrograms. This would be very valuable, especially for the study of parts of the Galaxy consisting of a mixing of Population I and II.

References

Cayrel, R.: 1968, in O. Gingerich (ed.), *Third Harvard Smithsonian Conference on Stellar Atmosphere*, MIT Press, 1969.
Landi Dessy, J.: 1966, in K. Lodén, L. O. Lodén, and U. Sinnerstad (eds.), 'Spectral Classification and Multicolour Photometry', *IAU Symp.* **24**, 33.
Spite, F.: 1968, in O. Gingerich (ed.), *Third Harvard Smithsonian Conference on Stellar Atmosphere*, MIT Press, 1969.

PART II

CLASSIFICATION OF OBJECTIVE-PRISM SPECTRA

INTRODUCTORY TALK FOR SESSION ON OBJECTIVE PRISM SPECTRAL CLASSIFICATION

C. B. STEPHENSON

Warner and Swasey Observatory, E. Cleveland, Ohio, U.S.A.

Traditionally, introductory talks usually have either the function of justifying the meeting at which they are given, or of attempting to present some sort of overview of the whole subject, or sometimes both or neither. A meeting such as this one is generally in order when the field in question has for any reason been very active since the last meeting about it; and by that standard I should say that this symposium is indeed appropriate inasmuch as the field of spectral classification has surely been very active during the seven years since the last symposium on spectral classification and multicolor photometry. In the area of objective prism spectral classification, the last several years have not seen very much really new in the way of observing techniques or methods of analysis that I know of, at least in comparison with some other fields; but they have certainly seen a great deal of astronomy done.

The vast majority of objective prism spectral classification has been and will continue to be done by visual inspection. This is mainly because the most common object of objective prism spectral classification is the formation of finding lists of special classes of stars. For this purpose the human eye, with its ability to examine large numbers of spectra as it were simultaneously, in order to select without detailed individual analysis those meriting further individual attention, is very well suited. This is doubly true because relatively few classes of stars have differences of major importance between them without also showing fairly strong spectral differences, though the exceptions are large in absolute number and their definition as important is a matter of taste to some extent. On the other hand, if the object is the positive detection of a weak, narrow line – which will, of course, require individual and careful inspection – then the human eye is actually superior to, for example, a tracing (or set of numbers) derived from a slit scanner, because the latter loses the information inherent in the two-dimensionality of the photographed spectrum; the tracing cannot distinguish between a pinhole and a weak and narrow line. For weak features that are broad (the wings of a strong line, for example) the situation is entirely otherwise, but these are seldom the object of an objective-prism search. As an aside, these remarks evidently bear also on the fairly frequent proposal that visual examination of widened spectra be replaced by plate-transmission measurements of unwidened ones.

Considering the span of time during which procedures for performing objective prism surveys has remained but little altered, it might seem remarkable that so much work should be in progress, or remain to be done, or to have been but recently completed. Recently completed projects of appreciable size that I am aware of and that might be mentioned here are Westerlund's infrared survey of the southern Milky Way for M, S, and C stars (cf. Westerlund, 1971); the south galactic pole survey by Slettebak and Brundage (1971); and the survey of the southern Milky Way for luminous stars of early spectral type by Sanduleak and myself. Three major surveys now in progress, the first two far advanced and the third just begun, are Stock's blue-region survey of the sky between the Sanduleak-Stephenson and Slettebak-Brundage surveys just mentioned, the search for southern peculiar stars by Bidelman and his colleagues on ~ 110 Å mm^{-1} plates taken with the Curtis Schmidt telescope and its 10° objective prism, and the reclassification of the stars of the HD catalogue on the MK system by Miss Houk using the same plates as Bidelman's group. This last project, if carried to completion, will falsify at least one prophesy that I made some years ago. Writing in *Vistas in Astronomy* in 1964 about objective prism astronomy, I said: "Higher dispersion work will probably be done by objective prisms in the years ahead, but sheer considerations of manpower will probably dictate for many years to come that most of this work will be done in regions of the sky of special interest...". I trust that Miss Houk will have the persistence to make me wrong. I omit a number of other important recent objective prism surveys only because I am not trying here to give an exhaustive summary.

All of the projects that I have just mentioned were started since our last symposium on spectral classification and multicolor photometry, and the main reason that so many things remained then to do is that essentially every new project that I have mentioned pertains to the skies of the far south; even Miss Houk's plate collection is at the moment more or less complete only for the southern hemisphere. Nevertheless the state of work in the southern hemisphere is becoming, surprisingly, more complete than that in the north; complete surveys in the north have yet to be carried out that would correspond to the ones just mentioned by Westerlund, Bidelman, Stock and, to some extent, by Houk.

Some of the astrophysically most interesting work that gets done via objective-prism spectral classifications is not necessarily part of such large surveys as I have been mentioning; or it may be only a very important by-product of those surveys, though perhaps the main source of zest on the part of the people doing the work. For example, in our southern luminous stars survey we found, and published (Hiltner *et al.*, 1968), an HD star with a spectrum very like a fast nova near maximum; a star which optically is very like the X-ray source Sco X-1 (Stephenson *et al.*, 1968; Hiltner and Gordon, 1971); and a previously unrecognised white dwarf of unusually small proper motion apparently at a distance of only 20 pc (Stephenson *et al.*, 1968). All of the examples just mentioned were incidentally confirmed independently by other means.

I should like to comment briefly upon two other accomplishments of recent times that bear on our subject. One is the atlas of objective prism spectra of normal stars that Dr Seitter has produced at Bonn. No doubt everyone here has seen this long since; but if anyone has not, I commend it to your attention. The other work is Wackerling's catalogue of early-type stars that have shown Hα in emission (Wackerling, 1970). In my own work I have found Wackerling's catalogue immensely valuable, and there is no doubt that objective prism workers in particular owe him a great debt of gratitude.

Mentioning Wackerling's catalogue brings me to a point bearing upon the question of how to facilitate our work, and that after all is one of the central questions addressed by the IAU. There is no doubt in my mind that one of the serious needs of observational astronomers today is for a greater number of general catalogues of astronomical data, and this is particularly true of the people who look at objective prism plates. By catalogues, I mean published catalogues lodged in observatory libraries. I am aware that thought is being given to establishing astronomical data centers, where everything but the local telephone directory will be stored upon magnetic tape. I am entirely in favor of such centres, but it is unfortunately true that information that is stored in a distant center is a bit like a dictionary two rooms away – it simply does not get used as often as if it were right at hand, and when it does the information too often does not come forth at the ideal moment at which it would have been most useful. A number of such general catalogues have recently appeared; Wackerling's is one example, the MK classification catalogue of Jaschek *et al.* another; and I hope myself within a year or two to publish a catalogue of all stars that have been published as carbon stars plus some new ones. But we need many more, which brings me to my point. We who do objective prism astronomy are responsible for no small portion of the great flood of astronomical data that is forever pouring into our libraries. Surely we would be wise, as an aid to compilers of general catalogues, always to include in our publications star positions with respect to a single standard equinox and equator (in addition to positions for observers), or at least to include one position referred to an equinox that is one of the two most commonly used for documentation? I am aware that an IAU group is supposed to be preparing a recommendation on this right now; meanwhile I can recall offhand having had to work with papers published in the last 25 yr that used equinoxes of 1855, 1875, 1900, 1945, 1950, 1955, and several others; and now, increasingly, we see 1975 used as the sole equinox for a given paper. This sort of thing is guaranteed to discourage compilers of general catalogues! Personally, for catalogues of spectral types I think the only logical standard equinox is 1900; it is the equinox of the HD, of nearly all the Warner and Swasey surveys except for the northern Luminous Stars, and incidentally our southern Luminous Stars catalogue uses 1900; it is the equinox of Bidelman's late-type emission-line catalogue, of his unpublished spectral bibliography, of the GCVS, of the Jaschek MK-type catalogue, the Wackerling Hα-emission catalogue, of my unpublished carbon star catalogue, and so on. Surely, however, if we don't use 1900, then the only other equinox to refer positions to for compilers is 1950.

From time to time the subject of automation is introduced in connection with the searching and classifying of objective prism plates. There is, of course, no doubt whatever that the growth of computer technology in recent years has been a great boon to objective prism workers – enabling us, for instance, to make very rapid identifications of stars on a plate from their equatorial coordinates by means of overlays, to cite only one simple example. But the automation I refer to is the virtual elimination of human manipulation of the plate. I have no doubt that this can be done, although I do not think that a successful computer program for classifying a random spectrum about which nothing is known in advance will be a very simple one, and I would certainly prefer the scanner to do a two-dimensional analysis of the image for the reasons given earlier. However, like the supersonic transport the question is, is it worth it? Having just made a very long air journey myself, perhaps I should have phrased that question as, To whom is it worth it? In an age of dwindling funding, I do not think that question is very trivial. Although experiments along these lines are bound to be interesting, for my part I would prefer to see any massive infusions of automated labor in the immediate future directed to a few alternate tasks. High on the list of priorities I would put: Preparation of charts for the parts of the C.P.D. catalogue that were not done by Ristenpart; a new edition of charts for the astrographic catalogue; preparation of charts for the Hamburg zones of Luminous Stars in the Northern Milky Way; and publication of a collation of C.D. vs C.P.D. numbers. I am aware, and grateful, that the same computer science that would make these things more feasible nowadays also makes their lack often less painful than formerly; but still it would be nice to have them done. Although the items mentioned would in some measure benefit practically all observational astronomers at some time or other, I believe that their lack is more troublesome to objective prism workers than to most people. Another item that we sorely need is the publication of Dr Bidelman's bibliography of spectroscopic data, or something very like it; but I do not see any prospect of automation contributing very much here.

Lest the foregoing remarks seem too disrespectful of the march of technological progress, let me say that I have become sufficiently impressed by the great improvements brought about in recent years at my own institution by our modest acquisitions of new equipment and materials that I have more than once given serious thought to postponing indefinitely some new project until more expeditious means of carrying it out come into being. I can illustrate this with a few examples, all drawn from the OB surveys done at the Warner and Swasey Observatory. Several years ago, when we did the winter Milky Way survey that was published as Luminous Stars VI, we were plagued by a long-standing problem: the Schmidt mirror would be pinched in cold weather, or if relieved would slip around and spoil the collimation during a warming trend; many of the plates that we had to use for that survey were not very good. A few years later we got funded to re-mount the mirror, and now for its radial support it floats in mercury and we can get perfectly good plates all the time in cold weather; but the Luminous Stars VI survey was finished years before that happened. When we took our objective prism plates at Cerro Tololo for the southern luminous stars survey they were having trouble with the spectral widening, with the result that some of the plates have the wrong widening or the wrong exposure times; today I understand that there is a new and very satisfactory system for widening the spectra – but our plates are all taken. When we started using the Smithsonian Astrophysical Observatory Star Catalogue to mark astrometric reference stars on these plates, we were plagued by the fact that the southernmost zones give mainly C.P.D. numbers, and we had no satisfactory C.P.D. charts. Pursuing the matter, we learned of the existence of the Ristenpart charts, and eventually obtained some – after we had finished working in the regions covered by those charts. Still later, of course, the SAO charts themselves arrived – after we were through marking reference stars in *all* the zones. And so it went – eventually we acquired a plotter for our desk computer, which makes it very easy and convenient to find out what is wrong when the astrometric plate reduction program rejects a plate – but all 450-plus plates for the survey had been reduced by the time we had the plotter. The moral of this could be that we should have greatly postponed starting the southern luminous stars program – in that case we would still be postponing it instead of being already finished, for ideally I would certainly wait until we had a digitised measuring engine, which even now is not in sight. I think the moral in fact ought to be that one should nearly always go ahead, and just call it the way the cookie crumbles when the sorts of things happen that I have been mentioning. The alternative is to risk the dilemma posed by Bondi (1970) in the talk that he gave for the sesquicentennial cermonies of the Royal Astronomical Society. He proposed that there is a serious, perhaps insurmountable psychological barrier to launching inter-stellar space probes. The barrier works this way: Why should I launch this probe now, seeing it will require 300 yr to return its results to us, when by waiting 5 yr I may be able to launch one that will get the data back in 250 yr? Bondi suggested that by asking such questions continually the result will be that the probe will never be launched. I would add, however, that 300 yr after the first decision not to launch, somebody is going to start being awfully annoyed with the people who gave in to that psychology! And I think that is just the way it is with our science.

References

Bondi, H.: 1970, *Quart. J. Roy. Astron. Soc.* **11**, 443.
Hiltner, W. A. and Gordon, M. F : 1971, *Astrophys. Letters* **8**, 3.
Hiltner, W. A., Stephenson, C. B., and Sanduleak, N.: 1968, *Astrophys. Letters* **2**, 153.
Slettebak, A. and Brundage, K.: 1971, *Astron. J.* **76**, 338.
Stephenson, C. B., Sanduleak, N., and Hoffleit, D.: 1968, *Publ. Astron. Soc. Pacific* **80**, 92.
Stephenson, C. B., Sanduleak, N., and Schild, R.: 1968, *Astrophys. Letters* **1**, 247.
Wackerling, L. R.: 1970, *Mem. Roy. Astron. Soc.* **73**, 153.
Westerlund, B. E.: 1971, *Astron. Astrophys. Suppl.* **4**, 51.

TWO-DIMENSIONAL CLASSIFICATION OF THE HD STARS

N. HOUK and A. COWLEY

University of Michigan, U.S.A.

Abstract. The major project of assigning spectral and luminosity classes to the Henry Draper stars south of $+30°$ is underway. The high-quality objective-prism plates have been taken with the Michigan Curtis Schmidt telescope at Cerro Tololo Inter-American Observatory. The spectra, widened to .8 mm, have a dispersion of 108 Å mm^{-1} at $H\gamma$. Plates exposed for 20^m and for $4^m + 1^m$ yield classifiable spectra between 4 and 10 m_{pg}.

Star identification is completely automatic, using a computer-generated plot to the Schmidt plate scale for each plate center. This plot, having the HD stars identified by number, is copied onto transparent material and placed under the Schmidt plate while the stars are visually classified.

Classification is being carried out by one person (N.H.) to maintain a uniform system. MK spectral standards of similar quality and density, also taken with the Schmidt, are continually referred to. Intercomparison of the new types with existing spectral types shows no systematic differences. However, the HD types themselves are systematically earlier in the range A to F.

Stars between the south celestial pole and $\delta = -55°$ are being classified first, with the first volume of results to be published late in 1973. It is expected that the catalogue of spectral types and remarks will be published in 6 volumes containing all HD stars south of $+30°$.

This research is being supported by the National Science Foundation.

During the past four years $10°$ objective prism plates covering the southern sky have been taken with the Curtis Schmidt telescope on loan at Cerro Tololo Inter-American Observatory. Most of the plates have already been scanned for peculiar spectra by Bidelman, MacConnell *et al.* The goal of the present Michigan spectral classification program is to assign two-dimensional spectral types to all stars in the Henry Draper Catalogue south of $+30°$ (the northern limit of the survey planned to be completed from Chile). It is hoped that this material will greatly add to our understanding of galactic structure, percentages of giants and dwarfs in the solar neighborhood, frequency and galactic distribution of both normal and peculiar stars to mention only a few applications. It will provide a moderately accurate spectral class for any individual HD star as well as being useful for statistical purposes.

The high quality plates have a dispersion of 108 Å mm^{-1} at $H\gamma$ with spectra widened to 0.8 mm. Because of the overlapping of plates about 25% of all stars appear on 2 or more plates. The 20-min exposures give classifiable images between 7 and 10 mag. $4^m + 1^m$ exposure plates are now also being taken to cover stars in the range 4–7 mag. Because of the excellent seeing in Chile the resolution of the spectrograms is comparable to that of the MK atlas. Multiple exposures of widely differing densities have been taken for the standard stars so that an unknown can be compared to a standard of similar density. 260 standard plates are presently available covering 145 MK types. Figure 1 illustrates some of the spectra so that the reader can get a visual impression of the high quality of the plate material.

An identification chart to the Schmidt scale marking the HD stars on each plate is generated by the Cal-Comp plotter and then copied on transparent Xerox material

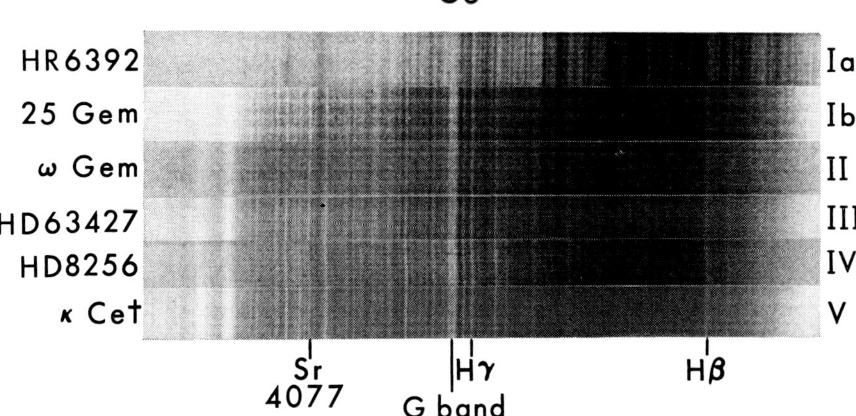

Fig. 1. Some G5 stars. The Sr ɪɪ4077/Fe ɪ4045 and 4063 ratios enable each of these luminosity classes to be differentiated. HR 6392 is a G5 Ia standard, but on our plates it is a K0. It may have a somewhat variable spectrum.

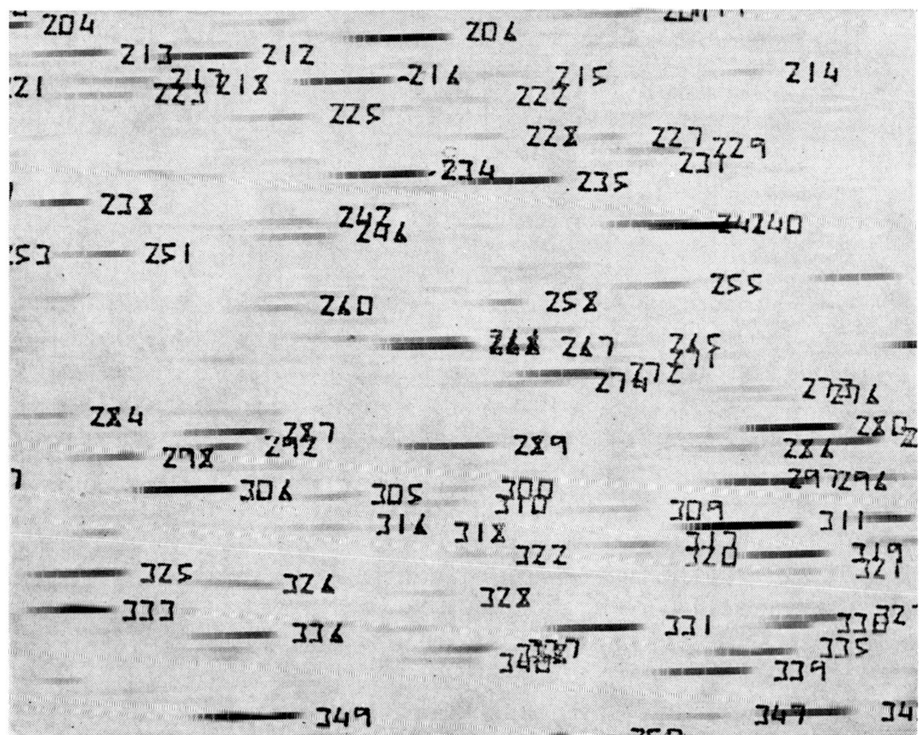

Fig. 2. Sample region ∼ 2° × 2° showing spectra and superimposed identification chart. For each plate the computer generates both an overlay and an accompanying list giving the HD number, position, and magnitude corresponding to the identification number on the overlay.

to form an overlay. An example of a stellar field with a superimposed identification plot is shown in Figure 2.

The classifications of all stars are now being made by one person (N. H.) to maintain a uniform system. A quality estimate ranging from 1 to 4 is assigned to each star. A description of each quality group is given in Table I. The table also lists the percentage of stars in each quality group for a representative sample. Although there are problems with overlapping spectra, especially in the galactic plane, this does not greatly affect the percentages of stars for which good types can be obtained. The main plate to plate variations arise because the limiting magnitude of the HD varies substantially from one region to another; on some plates there are many faint HD stars while on others almost none. In summary, good spectral and luminosity types should be obtained for 80% to 90% of the HD stars.

TABLE I

In plane (~500 stars)		Out of plane (~1000 stars)	
(1) 50% (2) 30	} 80%	(1) 55% (2) 31	} 86%
(3) 17 (4) 3	} 20%	(3) 11⁻ (4) 4⁻	} 14%

(1) Highest confidence; approximately equivalent to slit spectrum of similar dispersion and resolution.
(2) High quality but spectrum somewhat faint or overlapped; in many cases as accurate as group 1.
(3) Faint or overlapped to extent that spectral or luminosity type or both uncertain; range of possible types is often given.
(4) Poor type but better than nothing; usually no luminosity – 'early A', 'K', etc.

Several preliminary estimates of the internal and external accuracy of classification have been made. The first is from stars classified independently on overlapping plates by N.H. In all the comparisons only stars having quality ratings of 1 or 2 were included. Secondly a comparison between A.P.C. and N.H. on the same plates was made. Our agreement is good especially for spectral types (see Table II). This sample includes a large number of A stars for which luminosities are more difficult than for later spectral types. Finally, Morgan, Bidelman and Keenan have given us much valuable advice and stars which have been classified in common with them also show good agreement. Table II also shows comparison with HD types. The standard deviations are large, of course, and in some spectral regions are systematic. The types F2–G0 show relatively small scatter and no systematic differences. There is a large systematic difference between the Michigan types and the HD types for the A stars in the sense that the HD types are *earlier*. However, the scatter is such that no effective corrections to the HD types can be made (see Figure 3). A star classified as A2 in the HD is about equally likely to really be of type A2, A3, A5, A7 or F0. The difference is not magnitude dependent as is shown by Figure 4. There are also systematic differences in the B

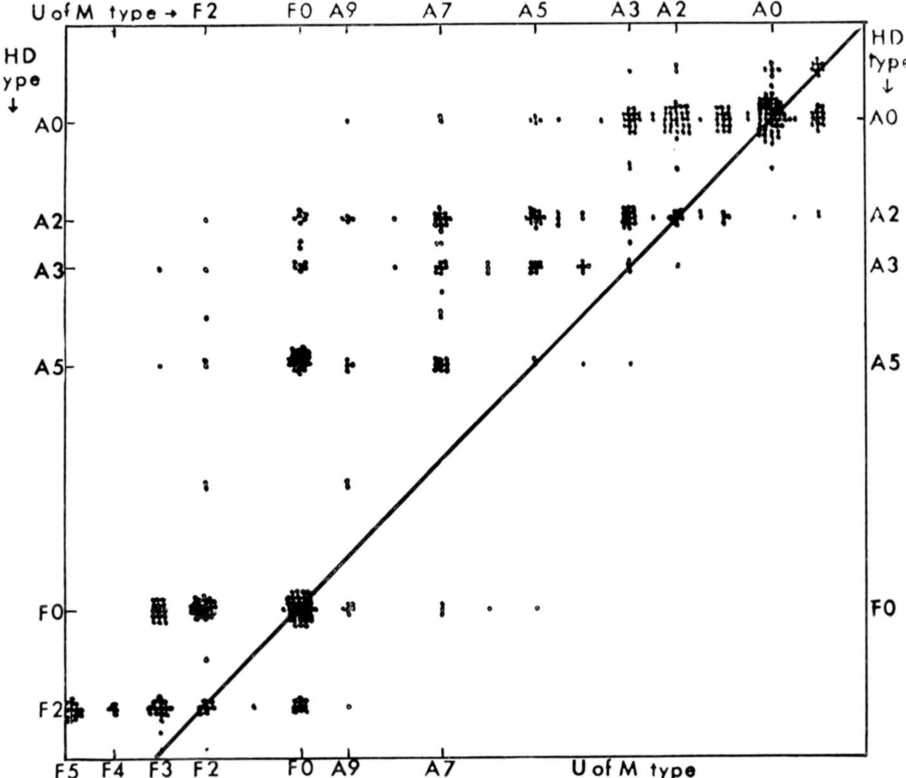

Fig. 3. Comparison of Michigan revised types with HD types for the A stars. Note the systematic difference in the middle A's such that the HD types are earlier.

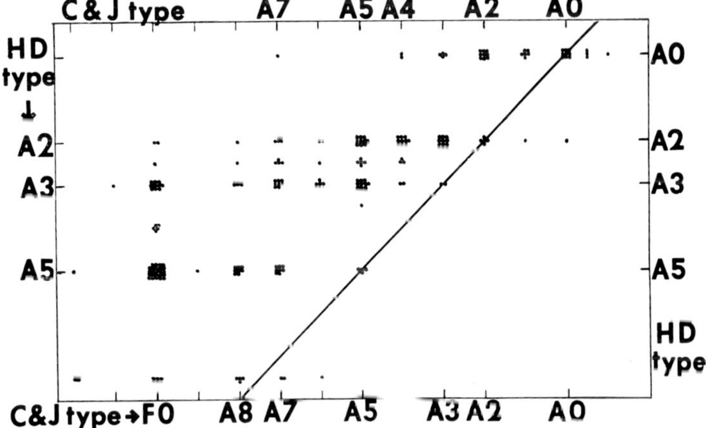

Fig. 4. Comparison of the types for the bright A stars as classified by Cowley et al. (1969) compared to HD types. Note the similarity of the systematic difference to that shown in Figure 2.

Acknowledgements

This work is supported by a grant from the National Science Foundation.

References

Cowley, A., Cowley, C., Jaschek, M., and Jaschek, C.: 1969, *Astron. J.* **74**, 375.
Feast, M. W., Thackeray, A. D., and Wesselink, A. J.: 1955, *Mem. Roy. Astron. Soc.* **67**, part II, 51.

EARLY RESULTS FROM THE MICHIGAN
SOUTHERN-HEMISPHERE SPECTRAL SURVEY

W. P. BIDELMAN

Warner and Swasey Observatory, Case Western Reserve University, Cleveland, Ohio, U.S.A.

and

D. J. MacCONNELL and R. L. FRYE

The University of Michigan Observatory, Ann Arbor, Mich., U.S.A.

(Read by C. B. Stephenson)

Abstract. This paper discusses the present status of a preliminary inspection of the plates obtained in the Michigan objective-prism survey of the entire southern sky. A very large number of peculiar stars, and of normal supergiants as well, have been noted and lists of these are being prepared for publication.

When the Curtis Schmidt-type reflector began operations at Cerro Tololo in the spring of 1967 the main program undertaken by University of Michigan personnel with the telescope was a systematic objective-prism coverage of the entire southern sky in the blue spectral region with the highest available dispersion, approximately 108 Å mm^{-1} at Hγ. This program is now some 90% complete, and observations are being extended to the northern sky accessible from Chile. Most of the plates were taken by MacConnell, and the entire project has continually had his immediate supervision.

The complete exploitation of the beautiful spectrographic material obtained in Chile will undoubtedly take many years. One phase of this work is the ambitious, but feasible, determination of accurate spectral types and luminosities for the southern stars of the *Henry Draper Catalogue*, so capably being attacked by Miss Nancy Houk, which you have heard described already. But even in the early stages of the plate-taking it was clear that it would be comparatively easy to pick out on this material such interesting and important objects as Ib or brighter supergiant stars, late-type dwarfs, emission-line and other multitudinous peculiar stars, to a considerably fainter limiting magnitude than that of the H.D. catalogue. Thus we began as soon as the plates became available to systematically scan them for such objects. The modus operandi was the following:

(1) The plates were first scanned by eye as carefully as possible and all objects thought to be of interest were marked and their classifications or peculiarities noted. The supergiants were classified with the aid of prints from the MKK and other atlases, slit spectral material and occasional Schmidt standard-star spectra previously obtained at Michigan. The peculiar stars were noted in most cases ab initio. In the course of time, of course, various peculiar stars served as standards for their respective types. This preliminary scanning was done largely by several skilled Michigan graduate students and to some extent by MacConnell. Those participating in this phase of the

work were Bond, Schmitt, Humphreys, and finally, Frye, who has been responsible for the bulk of this work.

(2) The second stage of the work consisted of an inspection by Bidelman of all of those objects marked in stage 1, in the course of which a definitive classification of the various objects was decided upon. In this, appreciable changes were often made in the preliminary classifications and indications of abnormality and many doubtfully peculiar objects were rejected. It should be pointed out that this work was largely done without use of spectral standards, and it is inevitable that substantial errors exist in the results. Some of the presumed supergiants will not prove of high luminosity, and some of the supposed peculiar stars will not prove peculiar, but it is hoped that the number of errors is low. It is also worth specifically noting that if an interesting object were missed in stage 1 it would almost invariably also not have been picked up in stage 2. Thus one may well expect there to be many interesting objects yet left on the plates. This is especially true of marginally peculiar stars that can only be noted by careful use of standards.

(3) The final stage is the determination of coordinates for the objects retained in stage 2, their identification in various catalogues, and their publication.

The present status of this so-called 'early result' program is as follows:

Preliminary scanning (stage 1) has been completed for nearly all of the plates so far obtained, so that exceptionally interesting objects can be noted without undue delay. Bidelman's inspection (stage 2) has been completed for the plates taken through August 15, 1969, which included a coverage of approximately 81% of the southern sky (1003 plates of 922 separate fields). According to present plans his participation in this program will cease at this point; thus every effort is being made at the moment to complete the identifications (which have proved the most time-consuming part of the project, since many important objects are not in programmed catalogues), for those stars that have been checked by him. Carrying on the program for the remainder of the southern plates and for the northern plates now being obtained will be the responsibility of the University of Michigan personnel.

In the preparation of the material for publication emphasis has been placed on publishing only new discoveries. There seems little point in including in our lists southern supergiants already classified (presumably more accurately) on slit spectograms, or already known cepheids, peculiar objects or emission-line stars. Thus known unusual objects have in general been deleted from our final catalogues, which are now in course of preparation. Preliminary lists of various types of objects discovered have already been distributed to many interested astronomers, and a considerable number of stars of special interest have already been noted in the literature (*IAU Circ.* Nos. 2089, 2120, 2130, *Astrophys. J. Suppl.* **22**, 117, *Publ. Astron. Soc. Pacific* **82**, 730 and 1360; **83**, 98 and 485).

Final figures for the numbers of *new* objects of various types that have been found in the "early result" program are not available since the identifications are still incomplete, and thus many at the moment undetermined duplications exist among the various plates as a result of intentional plate overlap and the necessary re-taking of occasional poor plates. Very rough figures which in a few cases are probably fairly close to the final totals but which in others are far from final are:

supergiants	251	weak-line stars	151	Ba II stars	175
Be stars	52	hor.-branch stars	10	br Ca II giants	59
Ap stars	326	carbon stars	14	no-Gbd stars	29
Me (LPV)	32	CH stars	3		
shell stars	10	S stars	10		
He-rich	7	white dwarfs	1		
H-poor	3	dMe	4		
H + Ca II emission	7	H + λ4686 em.	8		

The numerous composite, metallic-line and δ Delphini stars and late type dwarfs noted on the preliminary scanning of the plates will not be included in our lists but will be discussed by others. It is our hope that our lists will provide astrophysicists and galactic-structure astronomers with many profitable objects for further study, and we earnestly apologize in advance for any errors on our part that may cause them to lose invaluable observing time.

THE BONN SPECTRAL ATLAS: PART II

W. C. SEITTER

Five College Astronomy Dept., U.S.A., and Bonn University Observatory, Germany

Abstract and Summary. When work commenced on the Bonn Atlas for Objective Prism Spectra some years ago it was our intention to supply a tool for astronomers working with such widely different dispersions and resolutions as are used in the field of objective prism spectroscopy. This was to be accomplished through an atlas which contained and compared spectra of considerable difference in dispersion.

The Bonn Schmidt telescope which is used to obtain the observational material is equipped with three prisms giving linear reciprocal dispersions of 240, 645 and 1280 Å mm^{-1} at Hγ, thus covering a good part of the range generally used.

The first part of the atlas, containing spectra of the largest dispersion only, was published as soon as it was finished, while work on the second part with the two lower dispersions was still in progress. Thus, one of the main intended features of the complete publication, a comparison of criteria displayed at different dispersions, with a special interest in the appearance and disappearance of certain criteria as one goes from higher to lower dispersions, was not yet possible.

Now, work on the second part of the atlas has progressed far enough for sample pages of the lower dispersion plates to be distributed for inspection and discussion. Plate I gives examples of spectra that will be used on the S-plates (sequences of different spectral types for a given luminosity class) showing both spectra of 645 Å mm^{-1} (right hand side) and 1280 Å mm^{-1} (left hand side).

While the first part of the atlas tried to point out as much detail in the spectra as possible to make the atlas applicable for work with even higher dispersions, the second part concentrates only on those features which are important in the medium- to low-resolution range.

Originally, it was intended to use greater enlargements on the L-plates (sequences of different luminosity classes for a given spectral type). Yet, general agreement upon inspection of the sample plates, which are not shown here, was that the smaller enlargements are better and thus they will be used in the final version of the atlas.

A detailed explanation of the different spectral features useful in classification from medium- to low-dispersion plates will be possible only after completion of all observational work. So far it is indicated, as was to be expected from the work of other authors, that some of the most useful luminosity criteria overlap seriously with population criteria, e.g. the CN bands.

The discussion at the symposium revealed an interest in the sampling of peculiar spectra. Following this suggestion work has began on preparing for a third part of the atlas which is planned to contain spectra of about 70 peculiar stars taken with all three above-mentioned dispersions.

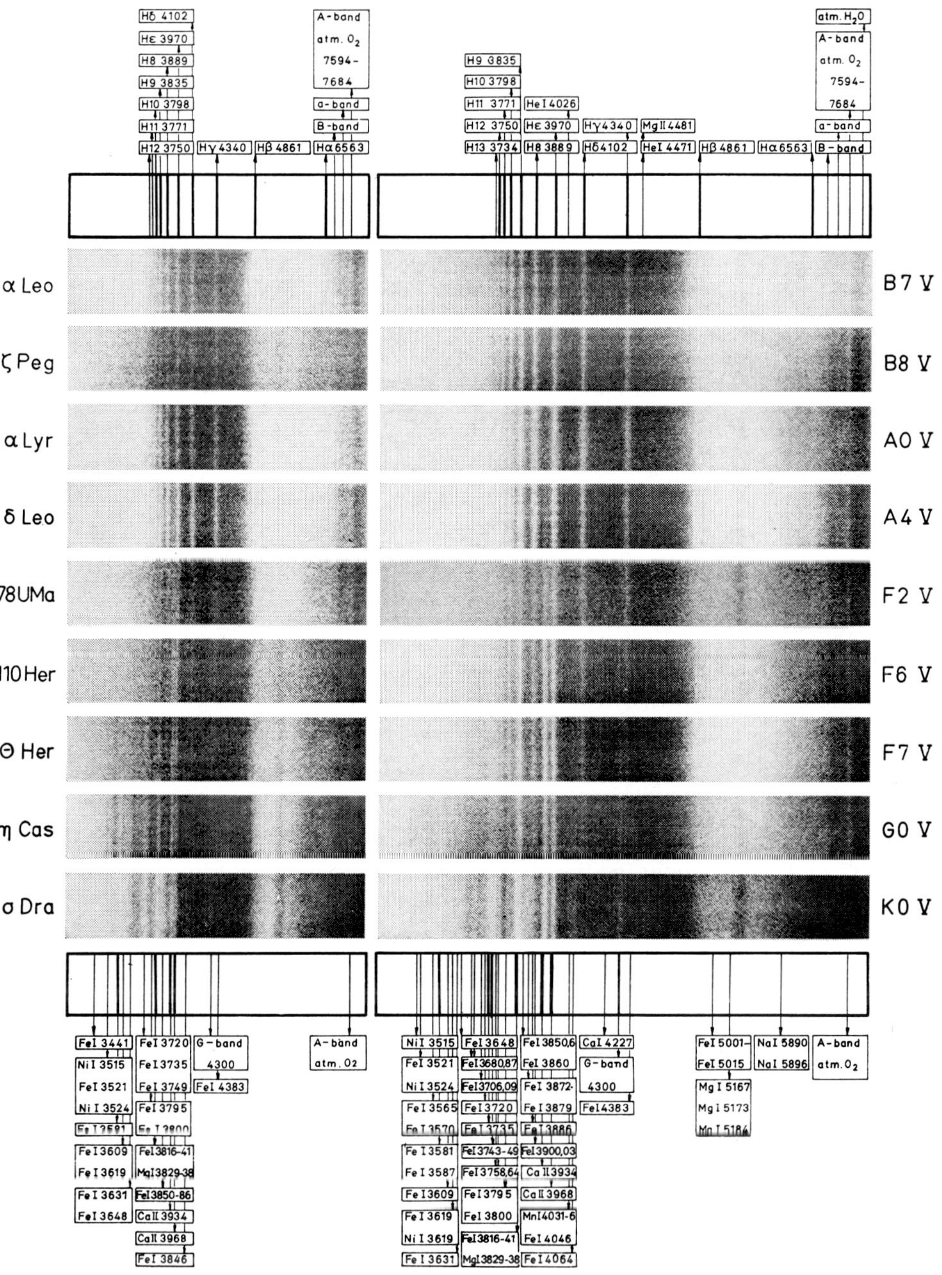

Plate 1.

THE PRESENT STATE OF A SPECTRAL SURVEY OF
THE SOUTHERN MILKY WAY FOR STARS EARLIER THAN A5

E. H. GEYER

Astron. Institut der Universität Bonn – Observatorium Hoher List, Germany

Abstract. The present state of a spectral survey of the southern Milky Way for stars earlier than A5 is described.

In 1962 a duplicate of the original Schmidt-camera, owned by the Hamburg-Bergedorf Observatory, was installed at the Boyden Observatory Bloemfontein/South Africa. This instrument is also equipped with an objective prism, and it was at the author's disposal from April 1962 until November 1963, who obtained the plates for an extensive spectral survey of the Southern Milky Way.

(I) Instrumental and observational data are given by Geyer (1966).

(II) The survey for stars earlier than A5 on the plates.

This observational material had already been surveyed for OB and OB+ stars by Klare and Szeidl (1966). Yet the spectra are of such high quality that better classification can be achieved for early type stars. Therefore in 1968 the author started the thorough classification on a homogenous system of stars earlier than A5 in all this fields in the magnitude range $6.0 \leqslant m_{ph} \leqslant 10.5$. It was intended that the spectral classification should be close to the Henry Draper system. Therefore a classification system adapted for the given reciprocal linear dispersion was established making use of criteria given by Becker (1929) for the blue spectral region, and by Seitter (1970) for the ultra-violet region down to the Balmer limit. As a stand-by served the *Atlas for Objective Prism Spectra* by Seitter (1970) and *An Atlas of Low-Dispersion Grating Stellar Spectra* by Abt *et al.* (1968). The HD-pre- and suffixes were used as usual.

On transparent enlargements from the original plates the stars are marked together with their classification, and afterwards they are identified on star charts (Cordoba-Durchmusterung, Santiago charts, Vehrenberg's Atlas Stellarum), and also in the relevant star catalogues, mainly the Cape Photographic Durchmusterung, from where the photographic magnitudes are taken. For fainter stars not contained in catalogues rough positions with an accuracy of $1'$ are obtained, and they will be marked on finding charts. Finally all data for a star are punched on cards.

Up to now all fields in the galactic longitude interval $230° \leqslant 1^{II} \leqslant 300°$ have been surveyed and some 22 000 stars registered. On the average 500 stars per plate are marked. From these 90% are contained in the Cape photographic Durchmusterung, but only about 50% are HD-stars.

A first volume of this spectral survey catalogue (galactic longitude 230°–300°) will be published in 1972.

Nearly on every plate several star pairs with distances not exceeding $1°$ are found

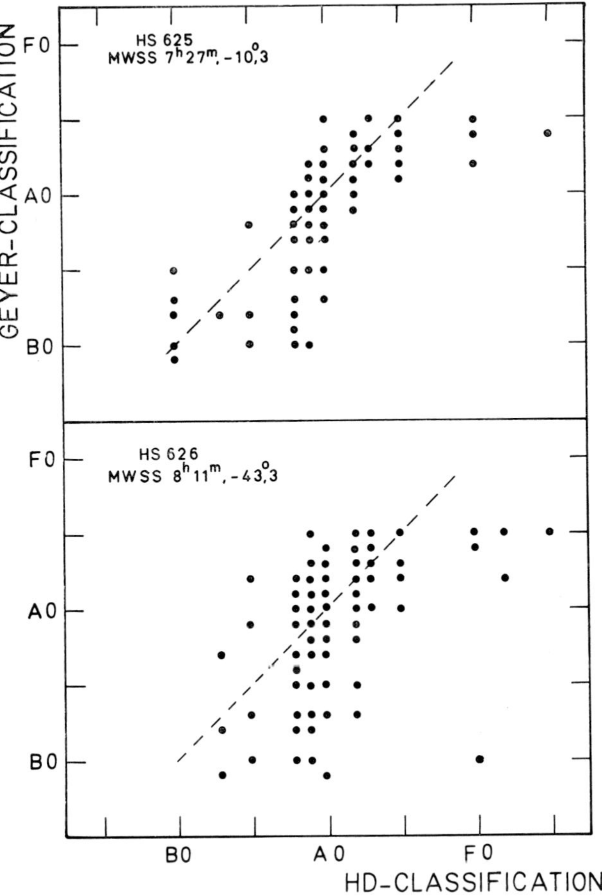

Fig. 1.   Comparison of the HD-spectral classification with that of Geyer for two fields.

TABLE I

Star No.	Spectral type		
Cir.	Photom.	HD	Geyer
2	B0	B5	B0, B3
6	B0	B8	B0
10	B2	B8	B2
12	B3	B8	B5
14	B2	B8	B5
20	B3	B9	B3
25	B7	B9	B8
29	B2	–	B7
37	B3	B8	B5

having identical spectral types and magnitudes. To this also Lodén (1969) called attention.

For two Milky Way fields a comparison of the HD-classification with the author's is presented in Figure 1. Beside the well known fact that some fainter HD-stars are classified too early or too late, the general tendency reflected by the diagrams is that the author classifies the stars earlier. Finally in Table I a comparison of some stars classified photometrically by Haug *et al.* (1966) in the Circinus region is given which supports this result.

Acknowledgements

This investigation is supported by the Deutsche Forschungsgemeinschaft, grants Ge 209/1–4.

References

Abt, H. A., Meinel, A. B., Morgan, W. W., and Tapscott, J. W.: 1968, *An Atlas of Low-Dispersion Grating Stellar Spectra.*
Becker, F.: 1929, *Publ. Astrophys. Obs. Potsdam* **27**, Heft 1.
Geyer, E. H.: 1966, *Veröffentl. Heidelberg-Königstuhl,* **18**.
Haug, U., Pfleiderer, J., and Dachs, J.: 1966, *Z. Astrophys.* **64**, 140.
Klare, G. and Szeidl, B.: 1966, *Veröffentl. Heidelberg-Königstuhl,* **18**.
Lodén, L. O.: 1969, *Vistas in Astronomy* **11**, 161.
Seitter, W. C.: 1970, *Atlas for Objective Prism Spectra*, Dümmlers Verlag Bonn.

OBJECTIVE PRISM SPECTRAL CLASSIFICATION AT THE STOCKHOLM OBSERVATORY

B. NORDSTRÖM and A. SUNDMAN

Stockholm Observatory, Saltsjöbaden, Sweden

Abstract. As a result of the spectral survey at the Stockholm Observatory finding lists are prepared for early and late type stars in the Southern Milky Way. In order to make the lists more useful we present the principles of the stellar classification.

1. Introduction

The spectral survey of the Southern Milky Way which is continuously carried out at the Stockholm Observatory, is based on observations performed by L. O. and K. Lodén and others in the years 1956–1966 at the Boyden Observatory in South Africa. The complete observational material covers the Milky Way from $l = 237°$ to $l = 7°$ approximately between $b = +3°$ and $b = -3°$.

Spectral plates have been obtained with the ADH Baker-Schmidt telescope and with the Hamburg Schmidt telescope at the Boyden Observatory. The spectral plates used until now are ADH-plates with a dispersion of about 200 Å mm^{-1} at Hδ. The widening is between 0.2 and 0.5 mm. The plates cover completely a 6° broad galactic band, but a considerable number of stars listed are found outside this 6° band. A big portion of this survey is now finished and catalogues with spectral classification and approximative magnitudes are prepared for the regions from Puppis to Centaurus ($l = 237°$–$318°$).

These catalogues contain lists of early- and late-type stars and other objects which are considered as interesting, for example carbon and emission line stars. At the moment the catalogues exist in print (Lodén and Sundman, 1966) and preprint (Nordström, 1970; Sundman, 1970). They can be obtained from the Stockholm Observatory.

The principal aim of our classification system is to secure as much information as possible about each star from the inspection of the spectrum.

2. The Classification Principle

2.1. EARLY-TYPE SPECTRA

The early-type spectra which have been recorded at the inspection are roughly those without visible K-line. An important exception from this rule concerns spectra with unusually narrow Balmer lines, indicating giant or supergiant A stars, interstellar K-line, emission features, or spectral peculiarities. The main problem is the uneven quality of the material. Only on optimal exposures both spectral and luminosity classification may be performed with reasonable chance of success. We have therefore accepted the following main principles as the indication of spectral type:

Ch. Fehrenbach and B. E. Westerlund (eds.), Spectral Classification and Multicolour Photometry, 85–90. All Rights Reserved.
Copyright © 1973 by the IAU.

concerned. The scheme presented in Table II will indicate the reasonable borders of uncertainty for each spectral type designation in the present system. Each spectrum has been classified by two persons one of which always has been L. O. Lodén.

2.2. M SPECTRA

The M spectra are identified and classified only from the TiO bands. For the faintest stars of this type, the continuum vestiges between the bands are the only traces visible on the plate. Spectra of high quality with hardly visible TiO bands are classi-

TABLE II

Reasonable misclassification limits for the spectral types given
in the Stockholm catalogues

Spectral type designation	Earliest MK type	Latest MK type
O5, O6	O5	B0V or B2I
O7, O8	O5	B0V or B3I
O9	O6	B1V or B5I
B0	O9	B2V or B6I
B1	B0V	B3V or B6I
B2	B1V	B5V or B8I
B3	B2V	B5V or B8II
B4	B3V	B6V or B8II
B5	B5V	B7V or B8 III
B6	B5V	B8V–III
B7	B6V	B9V–III
B8	B7V	A0V
B9	B8V	A1V
A0	B9V	A2V

fied as M0 and those with very well pronounced bands as M5. Other subtypes are obtained by means of interpolation and extrapolation. A considerable fraction of the M0 spectra will disappear into the late K ones and escape detection. For spectra later than M0 there is generally no detection problem, but it has neither been possible to establish any connection with MK–class on an absolute basis nor to distinguish between main sequence and above-main-sequence M stars.

2.3. OTHER SPECTRAL TYPES

No attempt has been made to subclassify the carbon or S stars because of their rarity which prevents comparison between spectra on the same plate. Spectra of early or late type with evident emission lines superimposed are classified as accurately as possible although the emission lines will generally increase the uncertainty. Therefore, the designations OBe, Be, and Me are rather common in the catalogue. In a number of cases, only the emission lines are visible on the plates. These objects are indicated with an *E* in the catalogue. The spectra of this type might represent any type of emission line stars or nebulous objects. However, certain Be stars with Balmer lines completely filled in by emission may have been mistaken for O stars.

2.4. THE LIMITING MAGNITUDES

There are at least three exposures of each partial region of the Milky Way with different exposure times, generally 1 h, 15 min and 4 min. The corresponding ultimate photographic magnitude limits for a star of spectral type A should roughly be of the order of 15^m, 14^m and 13^m respectively. However, the practical limit is set earlier and mainly conditioned by other circumstances as for instance the state of overlap on the plate and the distortion of the spectra under extreme exposure conditions. Concerning the risk for overlap, it is important to note that the whole project is devoted to extremely star-rich regions of the Milky Way. Therefore, over a certain exposure time that will differ slightly from one region to another, one does not obtain any additional information. The faintest spectra will interfere with those over and well-exposed preventing successful classification. However, as mentioned above, the registration is restricted to spectra with outstanding appearance on the plate, and when severely underexposed most of these spectra will lose their typical features. An underexposed spectrum of type B5 will look more or less like an A0 judged from the Balmer lines and generally escape detection. As a reasonable mean value of the practical limiting magnitude for the O–B9 stars we may indicate $m_B = 12.0$. In the case of late-type stars the situation is quite different. The molecular band features are easy to detect even if the spectrum is extremely underexposed or overlapped. Thus about 85% of the late type spectra can be detected down to $m_B = 15.5$. For natural reasons no limiting magnitude can be indicated for the emission line objects.

2.5. OTHER CONCEIVABLE SYSTEMS

It might be considered whether the classification system which has been applied might be favourably replaced by a less detailed one in order to provide less ambiguous information. The system nearest at hand would then be the rather common one that includes only three principal types, OB^+, OB, and OB^-. Careful comparisons have shown, however, that a *direct* classification in this system will give very ambiguous and indistinct border regions between the types and make the catalogue less useful. There are also appreciable deviations between various catalogues applying the 'OB system'. In spite of that, the 'OB system' or 'system of natural groups' is preferred by several observers because of its simplicity. It directly tells the probable actual luminosity of the star, i.e. the OB^+ stars are either of very early spectral type or of intermediate type and high luminosity. In many cases it is of greatest interest just to select the luminous stars for further investigations. Therefore, we have a programme for translation from the present system to the 'OB' one. Such a translation is easy to perform without introduction of too much uncertainty in the border regions. However it is not reversible which is the main reason for us to maintain our detailed system unchanged.

3. The Card Catalogue

All information about the stars is stored originally on punch cards and transferred to magnetic tape and a computer disc memory. Each record (card) gives identification number, coordinates (rectangular, equatorial, galactic), estimated magnitudes (generally m_B, occasionally complete UBV photometry), spectral type and sometimes

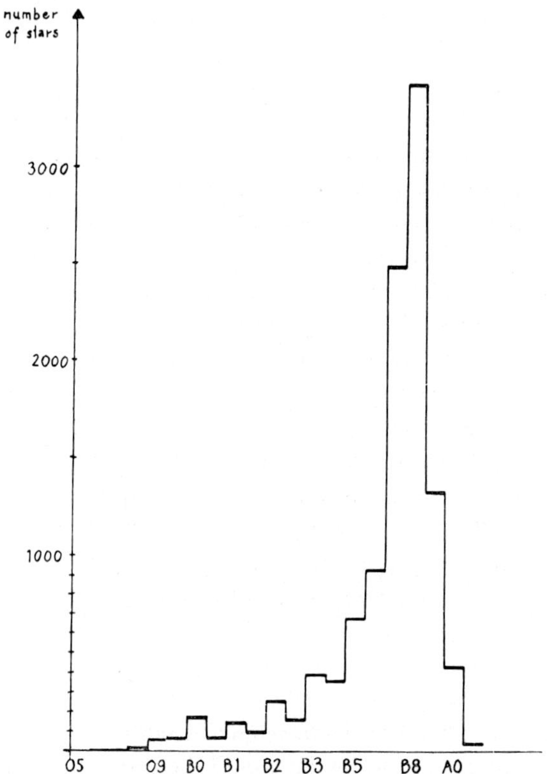

Fig. 2.   The distribution of early spectral types appearing in the catalogue covering the Carina-Centaurus region ($l = 280° - 318°$).

additional information concerning classification – quality of the spectrum, overlap, or other remarks.

Altogether we have registered 13 000 stars of which 2000 are late type stars (Figure 2).

References

Lodén, L. O. and Sundman, A.: 1966, *Stockholms Obs. Ann.* **22**, No. 10.
Morgan, W. W., Kennan, P. C., and Kellman, E.: 1943, *An Atlas of Stellar Spectra with an Outline of Spectral Classification*, Chicago.
Nordström, B.: 1970, *List of 3000 Stars of Spectral Type O–B5 in the Galactic Region $l^{II} = 236°$ to $l^{II} = 318°$*, unpublished.
Sundman, A., *A Catalogue of 3000 OB, B and M Stars in Centaurus*, unpublished.

THE PECULIAR STARS AND SPECTRAL CLASSIFICATION

E. K. KHARADZE and R. A. BARTAYA

Abastumani Astrophysical Observatory, U.S.S.R.

(Read by B. E. Westerlund)

Abstract. The spectral classification experiments with the 70-cm meniscus telescope and 8° objective prism of Abastumani Observatory (dispersion 166 Å mm^{-1}) have shown that if we apply the Kodak IIa-O plates at the 0.4 mm widening of spectra and the 20 minutes exposure time, we can successfully classify with relatively high accuracy in two dimensional MK system the stars down to the 11th magnitude, with simultaneous discrimination of peculiars.

We have undertaken such determinations first of all in Kapteyn Selected Areas aiming to fill the gaps in existing data for these standard areas.

Finally we could mention the high quality of the meniscus telescope optics together with good seeing in Abastumani as favourable factors making the Abastumani spectra of 10–11th magn. stars comparable with the slit spectra of the same dispersion.

The Abastumani spectra allow to apply the Morgan and Kitt-Peak criteria of MK classification only with a slight modification.

We have two-dimensionally classified in MK system about ten thousand stars down to the 11th magnitude in KA Nos. 2–43 (4°.5 × 4°.5 each). Among them 137 peculiar stars have been revealed. They are mainly Ap, Am and BaII (7 stars) and the stars with composed spectra (5). 97 stars have been first identified as peculiars.

It may be stated that there is a general growth of interest now in the peculiar stars, i.e. the stars having some sorts of anomalies in their spectra.

Not only the probable connection of peculiarities with the physical state and evolutionary processes in stars raises the interest. Simply it seems impossible now to make sufficiently accurate spectral classification if the stars belonging to a definite spectral group are not preliminary discriminated.

As a matter of fact the accuracy of the two-dimensional classification is the higher the more uniform the group of classified stars is. But if this is not the case we have to deal with the multi-dimensional classification which is a more difficult task indeed.

Principally we may involve as many parameters as we find it necessary to have manifested all the peculiarities present in the spectra of stars. But the quantitative methods for such aim have not been so widely applied. Till now the aim of these methods consisted in two or three-dimensional classification applicable to the majority of stars.

If we remember also the difficulties attending the realization even of the three-dimensional classification we easily imagine the importance and the relative privilege of the qualitative classification, allowing the discrimination of the peculiars, i.e. the grouping the stars into uniform groups.

Only after such groups of stars have been composed they may be classified quantitatively and the rate of peculiarity may be determined together with the two common parameters.

Peculiarities are represented by very slight effects in spectra. Only the recent development of observational technique made it possible to study the peculiar stars on a large scale. The objective prism spectra serve well for the discoveries of peculiar stars. But the dispersion and the quality of such spectra are restricted. This requires that we consider and attack the problem individually in each case of the given technique and the given dispersion.

The spectral classification experiments with the 70-cm meniscus telescope and the $8°$ objective prism of Abastumani Observatory (dispersion 166 Å mm^{-1}; the edge of the spectrum on the short wave side at about 3500 Å) have shown that if we use the Kodak IIa-O plates, the 0.4 mm widening of spectra and the 20-min exposure time, we can successfully classify with relatively high accuracy in two-dimensional MK system the stars down to the 11th mag., with simultaneous discrimination of peculiars. This procedure is rather effective for peculiars of the types Ap, Am and BaII and for stars having composed spectra.

It scarcely requires to stress the importance of such work from the point of view of stellar astronomy problems.

We have undertaken such determinations first of all in Kapteyn Selected Areas aiming to fill the gaps in existing data for these standard areas. No doubt we remember well the widely known work by Fehrenbach, but the limiting star in his work is much brighter (about 9.5 mg). The definite advantage of Fehrenbach's determinations is the higher dispersion (80–100 Å). Their uniformity should also be recognized. But Fehrenbach had been doing his work when the peculiars had not been attracting much attention. This fact inevitably introduced some errors of classification. We emphasize the necessity of preliminary discrimination of peculiar stars if we wish to secure high accuracy of the MK classification.

We would like to stress the full comparability of our results with those by Fehrenbach in spite of lower dispersion of our spectra. This may be ascribed to two facts.

Firstly we have used a shorter exposure time: 20 min vs Fehrenbach's 2 h.

Secondly, the length of our spectra reaching 3500 Å exceeds on the short-wave side that of Fehrenbach's (3900 Å). This difference arguing in our favour is of essential importance for early type stars and such as the peculiar stars Ap and Am.

Finally we mention the high quality of the meniscus telescope optics together with good seeing in Abastumani as favourable factors making the Abastumani slitless spectra of 10–11th mg stars comparable with the slit spectra of the same dispersion (West, 1970).

In the specified magnitude interval the Abastumani spectra allow to apply the Morgan and Kitt-Peak criteria of MK classification only with a slight modification.

As for the Kitt-Peak spectra, their dispersion (128 Å mm^{-1}) does not differ from that of ours. Moreover: starting from Hδ down to the short-wave region, where the majority of criteria for the early types are present, the Abastumani spectral dispersion is even a little higher.

We have two-dimensionally classified in MK system about ten thousand stars down to the 11th mag. in KA Nos. 2–43 ($4°5 \times 4°5$ each). Among them 137 peculiar stars have been revealed. They are mainly Ap, Am and BaII (7 stars) and the stars with composed spectra (5). Ninetyseven stars have been first identified as peculiars (*Astron. Circ. Acad. Sci. U.S.S.R.* No. 583, 1970). Let us now describe in short the criteria we have used.

Ap stars. These are stars from B5 to F0, but for the most part B9–A2, and of

IV, V class luminosity. Si II 4128–30 is always present. But sometimes Sr II 4077, Mn I 4030–33, Cr II 4171, Eu II 4205 (all or some of them) are visible too. As it is known, the 4128–30 Å region contains, besides Si II, the Eu II, Cr II, Fe I lines. It is difficult to state which of them prevail in any particular case. However, the integral effect is distinct enough to reveal the peculiarity, one never misses it.

In low dispersion spectra we do not see always all individual lines of the mentioned elements. Therefore we restrict ourselves to identifying the peculiars Ap, listing only the lines which are more or less distinctly visible without naming their elements.

One must remember that such a strong characteristics of Ap stars as the 4128–30 Å is also characteristic for supergiants. Therefore if one leaves out of account the hydrogen lines intensity and other characteristic lines of supergiants (especially distinguishable Ti II 3760, 3815), one might erroneously classify a peculiar as a supergiant.

In case of objective prism spectra of the so-called hot Ap stars we encounter some difficulties. As is known they are the early stars up to B5 in whose spectra the He lines are weakened. For this special reason one classifies them erroneously as B8, B9 stars and they are missed as peculiars. Therefore when one has to deal with stellar associations where the probability of B0–B5 stars presence is high by itself, one must be very careful classifying the stars. Generally speaking the suggestion is needed of some quantitative or at least qualitative characteristics to avoid such mistakes.

Am stars. Early type stars A0–A5 if determined according to Ca II K; hydrogen lines are weakened if compared with common IV, V class stars, while the metal lines are strengthened.

In case of A5 and later subclasses it becomes difficult to discriminate peculiars from supergiants. This is evident from the Kitt-Peak spectra reproductions. As for the stars earlier than A5 along with other properties Sr II 4077 line favours the matter very much, which is quite an essential and strong feature for Am stars ,while in our spectra this line is absent even in supergiants up to A5.

It must be mentioned, that both Ap and Am stars are divided into subtypes according to the ratio K/H. It is clear that to take Ca II K line for the Ap stars is not justified. But on the other hand this ratio somehow points to what this Ca II K line was like at the moment of our observation.

Ba II stars. G8–K0 Giants. Sr II 4077 and 4216 lines attract one's attention with their intensity. 4077 (Sr II) $\gg$ 4063 (Fe I) and 4216 (Sr II) $\geqslant$ 4226 (Ca I). G band (CH) is strengthened as a rule. The Ba II 4554 line itself mostly characteristic to these stars is less distinct in our spectra on account of low dispersion.

It must be borne in mind that 4077 Å and 4216 Å are also strong in supergiants together with other characteristics, but they never attain the strength as they do in Ba II stars. Particularly, 4077 surpasses 4063 in intensity slightly and 4216 approaches but never equals 4226.

It is noteworthy that we stress the resemblance of the properties of the Ap, Am, Ba II stars with those of the supergiants, because it is this resemblance and its ignorance that have conditioned those inherent errors, which as mentioned above, are encountered in Fehrenbach Catalogue.

THE SPECTRAL CLASSIFICATION OF SOUTHERN CARBON STARS

P. M. RYBSKI

Lindheimer Astronomical Research Center, Northwestern University,
Evanston, Ill., U.S.A.

Abstract. Well-widened objective prism spectra of carbon stars south of $-25°$ declination have been examined to determine their suitability for classification. These spectra, taken by Henize in the early 1950's from South Africa as part of the Michigan-Mt. Wilson $H\alpha$ survey of the southern sky, have a reciprocal dispersion at Na I 5890 Å of 300 Å mm^{-1} and are in good focus in the range between 5100 and 6600 Å.

Compared with spectra of northern carbon stars taken by the writer at the Lindheimer Astronomical Research Center with the one meter reflecting telescope and Cassegrain spectrograph at a grating reciprocal dispersion of 263 Å mm^{-1}, and compared with spectra taken by Sanford of carbon stars overlapping the Henize sample, the objective prism spectra have been found of sufficient quality to support their classification into five groups following the scheme as proposed by Keenan and Morgan in 1941 and as refined by Gordon in 1967.

The first group exhibit weak C_2 and CN features and $H\alpha$ in absorption; the second, features attributable to a low C^{12}/C^{13} ratio; the third, strong C_2 and CN features and only moderately strong Na I in absorption; the fourth, very strong Na I in absorption; and the fifth, $H\alpha$ in emission. Examples are given of each group, stars not fitting well into any of these groups are discussed, and the significance of each group is mentioned in light of work by Bouigue, Gordon, Peery, and Richer.

1. Observational Material

Between September, 1949 and August, 1951, Henize, working at the Lamont-Hussey Observatory in Bloemfontein, South Africa, completed the Michigan-Mt. Wilson $H\alpha$ survey of the sky south of declination $-30°$. Plates were taken also of the Milky Way up to declination $-8°$, and several plates were taken of a region in Cygnus extending as high as $+42°$ and as low as $+15°$. The instrument employed in the survey was the Mt. Wilson 10-in. camera corrected for the red and used with a 15-deg prism. This system yielded a plate scale of 159″ per mm and a reciprocal dispersion at plate center of 300 Å mm^{-1} at Na I 5890 Å. Under conditions of good seeing, the resolution element at 6000 Å near plate center approached 10 Å. Plate size $= 14'' \times 14''$ $\approx 7° \times 7°$ usable.

Several different filters were employed in photographing any one region to reduce the overlap between spectra while providing the maximum amount of information about any one object. Plate overlap was chosen so that any one object would fall at least once within the central one-half of a plate, and all plates were taken within 2.5 h of the meridian to minimize the effect of atmospheric refraction on the widening of the spectra. Most plates were calibrated with a spot sensitometer to enable later spectrophotometric reduction. The emulsion employed in the survey was Kodak 103a-E. and all plates were developed in Kodak D-19 for 4 min with continuous agitation.

Though intended for the discovery of objects exhibiting $H\alpha$ in emission, the survey

Ch. Fehrenbach and B.E. Westerlund (eds.), Spectral Classification and Multicolour Photometry, 96–108. All Rights Reserved.
Copyright © 1973 by the IAU.

also provided spectra of objects exhibiting spectral peculiarities between 5000 and 6600 Å. Because the moderate dispersion in the red resulted in the spectra near plate center being 2.5 mm in length from 5635–6600 Å, and owing to the careful and extensive widening of the spectra – being 0.5 mm on the average –, the spectra of M, S and C stars appeared of sufficient quality to support some form of classification. The accuracy any such classifications could approach would be limited by the small number of exposures of different lengths that were made of any one field – specifically, 8, 120, and 240 min – and by the fact that plates of less than excellent quality were not retaken. Simon (1964) has investigated the suitability of the M star spectra for a rough temperature and luminosity classification, and Henize (unpublished) has classified the S stars on a scheme proposed by Keenan (1954). The purpose of this paper is to present the results of the writer's attempts since the spring of 1970 to place the carbon stars into some scheme of classification.

2. History of Classification Schemes

The historical background for the scheme proposed in this paper begins with Shane's (1928) reworking of the Harvard R–N scheme. Shane's system was based upon estimates in slit spectra, taken in the blue, of C_2 and CN molecular band strengths. Despite the difference between the Harvard criteria and Shane's, he found it convenient to retain the R–N distinction wherein, for an equal $C_2(1, 0)$ 4737 Å absorption intensity, the N star would always be much fainter in the ultraviolet than the R star. When Shane examined the N sequence separately and attempted to relate the differences in spectral intensity maxima to differences in blackbody temperatures, often he derived temperatures which seemed too low to believe. Then current and later radiometric measures of four such stars by Pettit and Nicholson (1928, 1933) suggested the range in carbon star temperatures to be from 2000–2400 K. The problem of how to relate the R to the N sequence was complicated by Shane's finding that in each sequence the $C_2(1, 0)$ band passed through a maximum while the ultraviolet cyanogen bands appeared to pass through a maximum only in the R sequence. The similarity between the sequences was underscored by Wurm's (1932) showing that by placing the stars in an order determined by a color index defined between the ultraviolet and infrared portions of the spectrum the sequences were intermingled at similar index values.

The first significant improvement in carbon star classification after Shane was that proposed by Keenan and Morgan in 1941. Based primarily on atomic line ratios in blue spectra of reciprocal dispersion 125 Å mm^{-1} at Hγ – line ratios found sensitive to temperature in K and M stars – and secondarily on the $C_2(0,1)/C_2(1, 2)$ bandhead ratio, the absolute strength of the unresolved Na I doublet at 5890 Å, and continuum measures in the green, orange and red portions of the spectrum, this scheme proposed merging the R and N Harvard branches into a single sequence with the early R's placed largely from C0–C3 and with the later R's and all N's placed from C4–C9. They also defined a second dimension – carbon abundance – determined from eye estimates of the absolute strength of the $C_2(0, 1)$ bandhead at 5635 Å. Keenan and

Morgan recognized that this strength would be affected not only by temperature but also possibly by a spread in luminosity and that it would have to be studied in greater detail than was possible for them before its abundance-dependent character would become clear.

The principal weakness of the Keenan-Morgan system as defined was that, for stars defining the system later than C4, blue spectra usually could not be obtained. Consequently, their arrangement by temperature had to be based primarily on the three red criteria mentioned above, and each criterion had its own difficulties. The continuum measures were chosen in places in the red spectra, of reciprocal dispersion 250 Å mm^{-1} at 6100 Å, where it appeared there were few disturbing molecular features. Subsequent research at higher dispersions has shown that only one of these regions – 5670 Å – is undisturbed. Secondly, the absolute strength of the Na I line had to be corrected for the overlying CN(7,2) absorption, a procedure even Keenan and Morgan considered qualitative at best (Keenan, 1971b). Finally, while the use of the ratio of the first two members of the $C_2(0,1)$ band sequence was based on the laboratory work of Wurm (1932) – showing the ratio changing from near unity at 3000 K to 3 at 1500 K –, the whole band sequence was contaminated by the CN(5,0) band. Since the structure of this band, its intensity and complexity is every bit as considerable as the CN(7,2) band,* any variation in CN strength – such as the writer has encountered in at least two CH stars – could introduce systematic errors into a temperature sequence based upon the $C_2(0,1)/(1,2)$ ratio. What resulted, then, was a scheme which at best was heterogeneous in principle and which at worst showed large scatter for stars classified later the C4 (Morgan, 1971).

Three major rediscussions of carbon star classification have appeared since 1941. Bouigue (1954) proposed a scheme based largely on spectra of a dispersion similar to that used by Keenan and Morgan, making use of the absolute strength of Na I corrected for the overlying CN(7,2) band while utilizing ratios of the unresolved band structure of the +5 and +4 CN band sequences and the −1 and −2 band sequences of C_2 as further measures of temperature. Bouigue's results corroborate roughly those of Keenan and Morgan, while at the same time isolating for special consideration a subclass J of carbon stars showing an additional, apparently molecular absorption feature at 6270 Å which Sanford (Keenan and Morgan, 1941) found correlated inversely in strength with increasing apparent C^{12}/C^{13} ratio.

Gordon (1967) re-examined the classification problem from the perspective not only of three times greater dispersion in the red than employed by Keenan and Morgan but also of blue spectra of the brighter stars of all carbon types. By examining blue spectra of K and M giants in the region 4300–4500 Å, she defined additional atomic line ratio criteria for temperature. When these were applied to the carbon star spectra, three conclusions could be drawn about the Keenan-Morgan scheme. First, the order of the stars from C0–C3 was preserved, a result for which one would have hoped, this interval being that for which Keenan and Morgan had adequate blue spectral data.

* Marenin, 1971.

Secondly, those stars classified at C4 – and at least one at C5 – which exhibited features suggestive of a much lower C^{12}/C^{13} ratio than the average for carbon stars, and all of which belonged to Bouigue's subclass J, should have temperatures equivalent to middle K through early M.

Thirdly, the stars between types C5 and C9 divided into three groups with the following characteristics. First, there were those showing only weak to moderately strong Na I with CN and C_2 absorption strengths varying inversely with the Na I strength. Secondly, there were a group of stars exhibiting extremely strong Na I while showing only moderately strong CN and usually weak C_2 features and which had been found by other observers to exhibit features peculiar even for carbon stars. Thirdly, there were the carbon stars exhibiting Hα in emission. From her blue spectra and from the photometry of Mendoza and Johnson (1965), Gordon found that in these cool carbon stars the Na I absorption strength was not always a trustworthy criterion of temperature. Those stars with extremely strong Na I were found to merge into the group in which Na I was only weak to moderately strong. And those stars with Hα in emission appeared to be the coolest of all, despite Na I strengths which would have merged them randomly into the group with weak to moderately strong Na I strengths. All in all, the temperature range in these cool carbon stars appeared to extend from M3.5–M7.

Comprised of all N stars and one R9 star, the three groups of cool carbon stars seemed to have one feature in common: a significantly enhanced Ba II line at 4554 Å. Gordon (1968) took this as evidence for these stars having enhanced abundances of s-process elements and as one of several reasons for associating them generically with the Ba II stars (Bidelman and Keenan, 1951), which are of higher temperature than the cool carbon stars, rather than with the R stars of type 8 and earlier. Not considered carbon stars despite their exhibiting a stronger $C_2(0,0)$ absorption than is considered normal in the oxygen sequence stars (Bidelman, 1956), the Ba II stars are known to be rich in s-process elements (Burbidge and Burbidge, 1957; Warner, 1965). Her suggestion of associating the Ba II with the N sequence has received support from abundance analyses by Utsumi (1967) of five typical N stars, each of which were shown to possess enhanced s-process abundances. Gordon (1967) also found one R9 star – HD 59643 – to have an enhanced Ba II line. More work needs to be done on the very late R stars before it will be clear whether they belong to the R or to the Ba II–N sequence.

Gordon (1968) implies, without stating so specifically, that all N stars should be considered to have enhanced s-process abundances. Her evidence for such a conclusion is that every one of her 14 N stars for which she had a blue spectrum showed the enhanced Ba II line. And since these stars have a unique red spectral appearance which sets them aside from all but the latest R stars and from the subclass J stars, one might be tempted to call all N stars 'late Ba II stars' solely on the basis of this unique red appearance. However, her implication may not be proved conclusively by 14 spectra. Assignment of cool carbon stars to the late Ba II class may require a blue spectrum for each one so assigned – an observational problem of considerable difficulty. Until such a study is completed, the N and late R stars should probably be referred to merely as the 'cool carbon stars', a convention adopted in this paper.

A blue spectral study of these cool carbon stars may well affect the study of the optical spiral structure of our Galaxy. When Peery (1970) sorted the N stars by their light variability; assigned them absolute visual magnitudes by forcing them into the spiral arms defined by young clusters, H II regions and Cepheids more luminous than $M_v = -4.3$; and corrected their mean magnitudes for interstellar absorption; then he found the mean absolute visual magnitude of 26 irregularly varying N stars to be -3.5 with only a small scatter about this mean. Such a small scatter in the mean is acceptable, says Peery, because spectrally and photometrically these stars are so similar. In short, Peery feels the irregular N variables may be used to trace optical spiral structure. One of the results of the classification of the Henize carbon stars will be having brought research one step closer to a large-scale Peery-like treatment of southern N irregular variables, since, first, the cool carbon stars stand out so clearly from the hotter ones, and, secondly, because among the cooler stars those potentially of Mira-type can be sorted out from those which are semiregular and irregular in their light variability.

Two additional reports deserve mention since they relate directly to the scheme of objective prism classification presented below. The first is the work by Yamashita (1967) in which he gathers together plate material on carbon stars accumulated over 20 yr at Victoria by McKellar – blue and red spectra of generally higher dispersion – than that used by Keenan and Morgan. First, from his own study of K and M stars, Yamashita defines blue criteria which he then applies to carbon stars of all varieties, confirming generally Keenan and Morgan's results for all stars save those exhibiting Na I in great strength. Secondly, he extends what is essentially the Keenan-Morgan system to many carbon stars without a previous Keenan-Morgan type.

The second report is that by Richer (1971) on a scheme of temperature classification employing the infrared CN bandheads at 7852, 7876 and 7899 Å and the infrared Ca II triplet at 8498, 8543 and 8662 Å in spectra of reciprocal dispersion 124 Å mm^{-1}. A comparison (see Figure 8) of Richer's types with Mrs Gordon's unpublished blue types shows a surprising lack of correlation for the cool carbon stars for which Richer's scheme could have provided a valuable means of temperature discrimination. Gordon's hotter carbon stars do fall at earlier Richer type, while the cooler ones cluster largely at Richer type C5II. It remains, then, an unsolved problem why cool carbon stars, which vary between Gordon's equivalent blue type M3.5 to M7, show little or no temperature dispersion in the Richer system. (The error bars in Figure 8 were provided by Gordon (1971) and represent 75% confidence limits. Richer makes no star-by-star estimate of classification error, stating his error on the average as plus or minus one subclass.)

Richer's most significant contribution seems his having found an apparent dispersion in luminosity for the carbon stars. This dispersion makes its appearance through a general washing-out of the infrared molecular features and an enhancement of infrared atomic lines isolated by Keenan and Hynek (1945) as sensitive to luminosity in M stars. In the group of stars to which Richer assigns a higher luminosity are found a surprising number of stars which Bouigue either did or would have assigned to his subclass J. If one believes the higher temperatures Gordon assigns to these stars, then the enhancement of the luminosity-sensitive lines in them may only be an effect of their higher temperature. Clearly, further research seems justified to sort out in Richer's class I which effects may be attributed to temperature and which to luminosity. Identifying additional stars of subclass J would aid in the solution of this problem. It is one of the features in the classification scheme proposed below that carbon stars of subclass J form a separate class.

3. The Objective-Prism Classification Scheme

Difficulties in classifying the Henize spectra were encountered immediately. Since these spectra resembled most closely in dispersion the spectra reproduced in Keenan and Morgan's (1941) article, a comparison between the two was made only to reveal that the Henize spectra contained far more information longward of 6000 Å than did the reproductions of Keenan and Morgan's. When Mrs Gordon's reproductions (1967) were compared to the Henize sample, her spectra were found of too high a dispersion for direct comparison. Finally, of the few Henize spectra north of declination $-15°$, most were too faint to have had any previous classification. This made impossible the direct comparison of Henize spectra which might have had some form of classification to Henize spectra without any types.

An attempt was made at using slit spectra obtained by Sanford for some kind of comparison and analysis, but little success was encountered owing to their widely varying resolution, dispersion and degree of exposure. Only Sanford's spectrum of HD 75021 proved useful, for it contained broad features which could also be seen in the Henize spectrum of the same star, features which have since been used to select out subclass J stars from the Henize sample.

Even though individually the Sanford spectra were not of aid in solving the classification problem, they pointed the way toward the use of broad features in spectra similar but not identical to the Henize spectra in their instrumental quality and which could be taken of as many stars in the north as would be necessary to sort out the Henize sample. To this end, the writer has secured some 185 spectra of 103 northern carbon stars brighter than $m_v = 10.0$ at the Lindheimer Astronomical Research Center with the one meter reflector and Cassegrain spectrograph. The reciprocal dispersion of these spectra is 263 Å mm^{-1}, the projected slit width is 6 Å, and all the spectra were widened to 0.62 mm. The resulting length-to-width ratio of these grating spectra is nearly identical to that found in most of the Henize spectra. All spectra were exposed to similar densities at 5670 Å to facilitate intercomparing Na I intensities and were taken on Kodak IIa-F plates to minimize having to take more than one spectrum of a given star to insure proper exposure between 5635 and 6600 Å. All plates were developed in Kodak D-76 at 20° Centigrade for 15 min with continuous, vigorous agitation in a large tray rocked by an International Observatory Instruments tray agitator to insure not only properly but also reproducibly developed spectra.

From this sample, coarse criteria were devised which placed these and the Henize spectra into seven groups, the first five of which were suggested by the major groups in Mrs Gordon's scheme, while the sixth contained peculiar stars with carbon characteristics and the seventh held those stars which might possibly appear like carbon stars at some time during their variability. Examples of each of these groups is given in Figures 1 through 7. With the exception of the stars HD 91708, Co.D. $-62°466$, and those stars with a catalog name followed by a small letter p – standing for prismatic dispersion – only the spectra identified by a simple sequence number were taken by Henize. Those taken by the writer are identified with a catalog name or number which may or may not be followed by a small letter g – standing for grating dispersion. The use of sequence numbers to identify spectra of Henize carbon stars was necessitated by the absence for these stars of reliable position measurements which would have made possible identification and a catalog reference. These stars will be identified in a later article published jointly by Henize and the writer.

In Figures 2 and 4 examples are given of stars for which the writer has in his possession a spectrum taken by Henize and one taken by himself. These have been shown to illustrate how much detail really is retained in the Henize spectra. Where such a direct comparison was not possible, a northern grating spectrum was chosen to match a southern prismatic spectrum to illustrate features in the southern spectrum which can also be seen – and whose source can be identified – in the northern.

3.1. The early R stars. Keenan and Morgan types C0–C3. Figure 1

These stars always exhibit Hα in absorption if Hα is present at all. Na I shows weakly in absorption as does the $C_2(0,1)$ band sequence. If the $C_2(0,2)$ band sequence is present, it is usually very weak. The CN(4,0) band sequence may be present, but it is considerably weaker here than in any other class save the subclass J stars.

The stars chosen to illustrate this group are the northern R3 star HD 5223, for which a grating spectrum is shown, and the southern R0 star HD 91708 shown in a prismatic spectrum. Both are rather featureless except for the $C_2(0,1)$ band sequence and a weak Na I line. Owing to this featurelessness of the early R stars, these were the most difficult to detect and classify on the Henize plates. It is doubtful any stars not already in the Henry Draper survey as early R stars, will be found in the Henize sample.

Please note that marked on the spectrum of HD 5223 are the yellow and green lines of mercury vapour from the street lamps of Chicago. These lines are marked in each grating spectrum in other figures where they are sufficiently intense to be mistaken for part of the stellar spectrum.

3.2. The subclass 'J' stars. Keenan and Morgan type C4. Figure 2

Most stars in this group exhibit not even a trace of the CN(4,0) band sequence. Instead, the region between 6191 and 6600 Å is marked by the absorption band at 6270 Å used by Bouigue to define this class and by a triplet of dark bands – a pair between the expected positions of CN(5,1) and (6,2) and a single dark band just

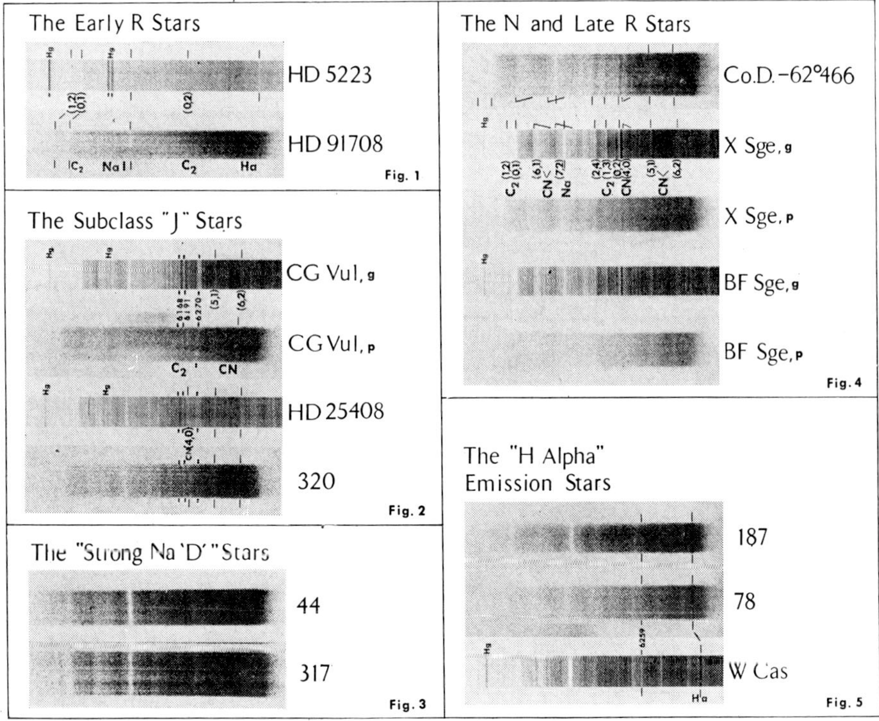

Figs. 1–5.

blueward of the expected position of Hα. The regions immediately redward of $C^{12}C^{12}$ (0,2) 6191 Å and $C^{12}C^{13}$ (0,2) 6168 Å are each marked by dark narrow bands of equal strength. These bands will be seen in stars of later type, but the band at 6168 Å never approaches the intensity of the one at 6191 Å as it does in this class. Similar dark bands each redward of the $C^{12}C^{12}$ and $C^{12}C^{13}$ (1,3) and (2,4) bandheads may appear and only in this class. Na I may be absent, but it may also be weak to moderately strong. But it never approaches the intensity reached in any of the normal cooler stars. When the CN(5,0) band sequence can be seen, (6,1) appears most strongly while (7,2) and (8,3) seem washed out. Finally, on the E plates, the continuum redward of 5635 Å appears steplike. The portion blueward of 6191 Å is always uniformly fainter than the portion redward of that point.

Two extremes of this class are illustrated in Figure 2. The northern subclass J star CG Vul shows very strongly the features characterizing the group. Note the strength of the 6270 Å band, the strengths of the dark bands in the C_2(0,2) band sequence, and the overall weakness of the CN(1,0) band sequence. This should be compared to the northern star HD 25408 and to the southern star Henize 320. In both, the CN(4,0) band sequence is much stronger than in CG Vul, and the 6270 Å is much reduced in strength. In fact, Henize 320 represents a limit detection problem in this class.

3.3. The strong Na 'D' stars. keenan and morgan types C8–C9. figure 3

Stars included in this group were selected almost entirely on the basis of their extra-ordinarily strong Na I absorption. Some exhibited additional peculiarities, such as the dark bands at 6191 and 6168 Å typical only of a subclass J star. More often, they exhibited very weak C_2 features but readily visible CN features. It is not clear these stars should form a group apart from the peculiar stars discussed later. Only a small number of Henize stars fall into this group, yet the strength of their $C_2(0,1)$ bandhead and the presence of an absorption at 6259 Å – characteristic of stars of apparently low temperature – justifies temporarily their separate consideration.

3.4. The N and late R stars. keenan and morgan types C5–C9. figure 4

To be a member of this class, a spectrum had to satisfy three criteria. The first was that all visible features had to be assignable to normal molecular and atomic sources, i.e., only the usual features due to C_2, CN and Na I were allowed. Secondly, Hα could not be in emission. The third criterion required the following behavior of the usual molecular and atomic sources. Since in the grating spectra dark bands each redward of a normal bandhead in the $C_2(0,2)$ band sequence were found to correlate directly in their strength with the absolute strength of the absorption bandhead of the $C_2(0,1)$ band sequence at 5635 Å (originally Keenan and Morgan's abundance index); and because it was found that in the 'normal' northern carbon stars the latter bandhead strength correlated inversely with decreasing temperature, as deter-mined from Gordon's (1967) blue spectra, while correlating directly with the expected temperature-dependent ratio of the (5,0) and (4,0) CN band sequences; then these dark bands – which could always be seen in the Henize spectra when the region of the CN (4,0) band sequence was properly exposed, a condition seldom met by the $C_2(0,1)$ band sequence – had to correlate directly in strength with the CN band sequence ratio and inversely with the absolute strength of Na I. That is, when the former two were weak, the latter had to be strong; when the former two were strong, the latter had to be weak. Only a few stars were found in the Henize sample in which this final condition could not be met, and these spectra are receiving further consi-deration.

The writer was fortunate to have two northern N stars – X and BF Sge – for which he had for direct comparison both Henize spectra of good quality and his own grating spectra. A spectrum of Co.D. $-62°466$ has been included in Figure 4 to show the difference between spectra taken by Henize at small zenith distance – typified by $-62°466$ – and spectra taken at large zenith distance – typified by the stars in Sagitta. Owing most likely to differences in average conditions of seeing, these differences make difficult or impossible direct comparisons of absorption intensities in spectra of stars north and south of the celestial equator in the Henize sample.

The important molecular and atomic absorption features seen in this type of star are identified both in the grating spectrum of X Sge and in the prismatic spectrum of

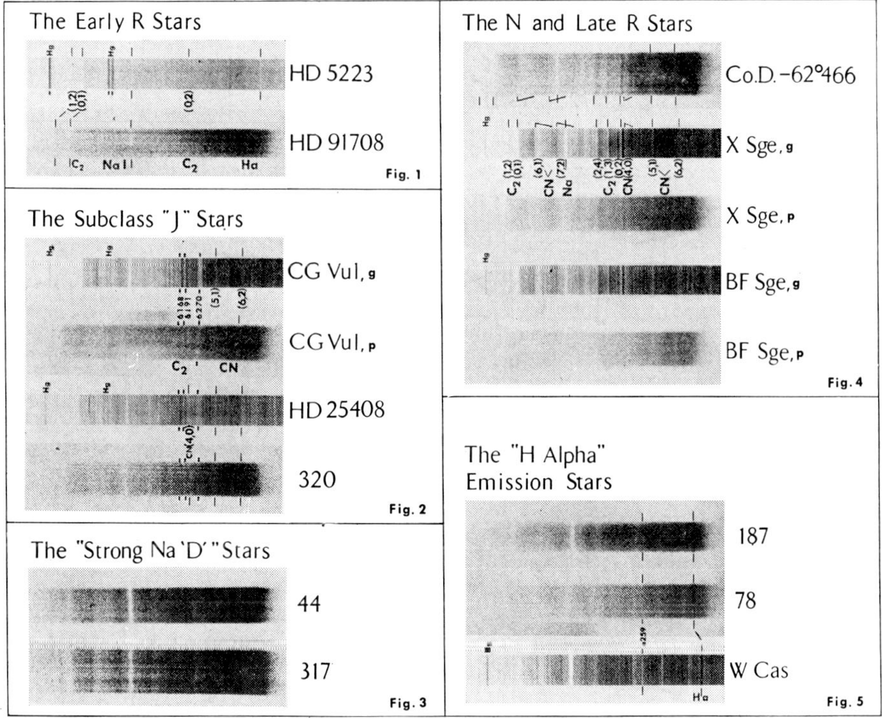

Figs. 1–5.

blueward of the expected position of Hα. The regions immediately redward of $C^{12}C^{12}$ (0,2) 6191 Å and $C^{12}C^{13}$ (0,2) 6168 Å are each marked by dark narrow bands of equal strength. These bands will be seen in stars of later type, but the band at 6168 Å never approaches the intensity of the one at 6191 Å as it does in this class. Similar dark bands each redward of the $C^{12}C^{12}$ and $C^{12}C^{13}$ (1,3) and (2,4) bandheads may appear and only in this class. Na I may be absent, but it may also be weak to moderately strong. But it never approaches the intensity reached in any of the normal cooler stars. When the CN(5,0) band sequence can be seen, (6,1) appears most strongly while (7,2) and (8,3) seem washed out. Finally, on the E plates, the continuum redward of 5635 Å appears steplike. The portion blueward of 6191 Å is always uniformly fainter than the portion redward of that point.

Two extremes of this class are illustrated in Figure 2. The northern subclass J star CG Vul shows very strongly the features characterizing the group. Note the strength of the 6270 Å band, the strengths of the dark bands in the C_2(0,2) band sequence, and the overall weakness of the CN(4,0) band sequence. This should be compared to the northern star HD 25408 and to the southern star Henize 320. In both, the CN(4,0) band sequence is much stronger than in CG Vul, and the 6270 Å is much reduced in strength. In fact, Henize 320 represents a limit detection problem in this class.

3.3. The strong Na 'D' stars. keenan and morgan types C8–C9. figure 3

Stars included in this group were selected almost entirely on the basis of their extraordinarily strong Na I absorption. Some exhibited additional peculiarities, such as the dark bands at 6191 and 6168 Å typical only of a subclass J star. More often, they exhibited very weak C_2 features but readily visible CN features. It is not clear these stars should form a group apart from the peculiar stars discussed later. Only a small number of Henize stars fall into this group, yet the strength of their $C_2(0,1)$ bandhead and the presence of an absorption at 6259 Å – characteristic of stars of apparently low temperature – justifies temporarily their separate consideration.

3.4. The N and late R stars. keenan and morgan types C5–C9. figure 4

To be a member of this class, a spectrum had to satisfy three criteria. The first was that all visible features had to be assignable to normal molecular and atomic sources, i.e., only the usual features due to C_2, CN and Na I were allowed. Secondly, Hα could not be in emission. The third criterion required the following behavior of the usual molecular and atomic sources. Since in the grating spectra dark bands each redward of a normal bandhead in the $C_2(0,2)$ band sequence were found to correlate directly in their strength with the absolute strength of the absorption bandhead of the $C_2(0,1)$ band sequence at 5635Å (originally Keenan and Morgan's abundance index); and because it was found that in the 'normal' northern carbon stars the latter bandhead strength correlated inversely with decreasing temperature, as determined from Gordon's (1967) blue spectra, while correlating directly with the expected temperature-dependent ratio of the (5,0) and (4,0) CN band sequences; then these dark bands – which could always be seen in the Henize spectra when the region of the CN (4,0) band sequence was properly exposed, a condition seldom met by the $C_2(0,1)$ band sequence – had to correlate directly in strength with the CN band sequence ratio and inversely with the absolute strength of Na I. That is, when the former two were weak, the latter had to be strong; when the former two were strong, the latter had to be weak. Only a few stars were found in the Henize sample in which this final condition could not be met, and these spectra are receiving further consideration.

The writer was fortunate to have two northern N stars – X and BF Sge – for which he had for direct comparison both Henize spectra of good quality and his own grating spectra. A spectrum of Co.D. $-62°466$ has been included in Figure 4 to show the difference between spectra taken by Henize at small zenith distance – typified by $-62°466$ – and spectra taken at large zenith distance – typified by the stars in Sagitta. Owing most likely to differences in average conditions of seeing, these differences make difficult or impossible direct comparisons of absorption intensities in spectra of stars north and south of the celestial equator in the Henize sample.

The important molecular and atomic absorption features seen in this type of star are identified both in the grating spectrum of X Sge and in the prismatic spectrum of

$-62°466$. It is easily seen from this example how different these stars are in appearance on the Henize plates from the subclass J and early R stars.

3.5. THE Hα EMISSION STARS. KEENAN AND MORGAN TYPES C5–C9. FIGURE 5

To be included in this group, a star had to exhibit only Hα in emission and either the CN or C_2 band sequences in absorption. In the northern sky such a selection criterion would net not only *bona fide* carbon stars but also stars showing simultaneously features characteristic of carbon and S stars. Owing to Henize's survey favoring objects exhibiting Hα emission, many such carbon stars were detected. But many such stars were so poorly exposed blueward of Hα that it was impossible to say with certainty that they were normal or peculiar. It was felt more consistent to keep all the Hα emission stars together and to comment on peculiarities individually when seen than to divide up these stars into a 'normal' and a peculiar class.

The stars included in Figure 5 would probably fall into the 'normal' group at the dispersion of the survey. Henize stars number 187 and 78 show all the features marked for the N and late R stars in Figure 4, though somewhat less intensely, while showing an additional feature at 6259Å which Gordon (1967) attributes to a zero-volt transition of Sc I but which Keenan (1971b) would prefer to attribute to a confluence of low-excitation Ti I lines. Gordon found this line to strengthen at later equivalent blue spectral types, implying it was a temperature-sensitive line.

3.6. THE PECULIAR STARS. FIGURE 6

All stars in this group have the common property of exhibiting weakly the CN (4,0) band sequence, and most show evidence for the presence of the low-temperature 6259Å line. After these similarities, the stars appear to fall into two groups. The first are similar to CY Cyg in showing apparently no CN(5,0) band sequence and no evidence for the $C_2(0,1)$ band sequence. A prismatic spectrum of Henize 201 compares favorably to a grating spectrum of CY Cyg obtained by the writer this past September. The second group appears similar to a recently obtained grating spectrum of WZ Cas in showing a weak but definitely present $C_2(0,1)$ absorption, the low-temperature 6259Å line somewhat more strongly than in the CY Cyg group, and an additional, apparently molecular feature redward of CN(5,1) which degrades to the red and is most intense at 6379Å. Examination of grating spectra in the writer's collection of $+38°955$, RS Cyg, and R CMi showed that the band appeared in the first two but not in the last.

One encounters confusing evidence when trying to ascribe this band to the (1,1) R_2 band of zirconium oxide. When the illustration in the article on GP Ori by Bidelman (1950) was examined, both GP Ori and FU Mon exhibited the band in greater strength than did either $+38°955$ and WZ Cas also shown in the article. But while both GP Ori and FU Mon seemed to show in great strength the (0,0) R_3 band of ZrO at 6473Å, WZ Cas and $+38°955$ did not. The 6259Å line in WZ Cas (in the grating spectrum) seems peculiarly stronger than in any other star of accepted low temperature – such as Z Psc –, and this enhancement may be due to the (1,1) R_1 band

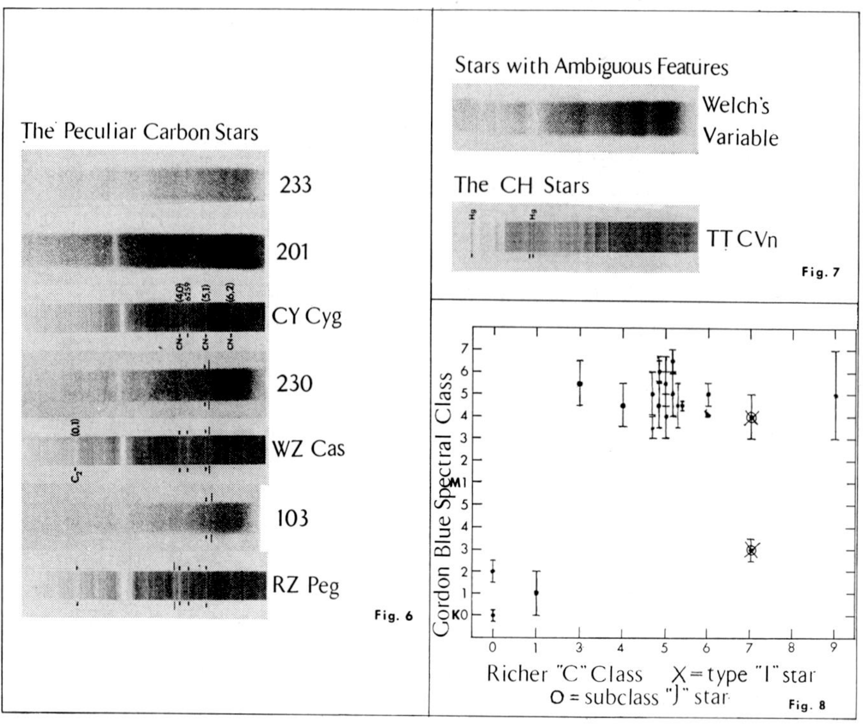

Figs. 6–8.

of ZrO at 6260Å. But when laboratory evidence is consulted, the tentative identification of the new band with a bandhead of ZrO is not supported: bands which should be present along with the (1,1) R_2 band, and which should be stronger than it, are not seen in WZ Cas or $+38°955$. Further complicating the matter is that direct comparison of this spectrum of WZ Cas with a spectrum of the S star HR 8714 taken with the same instrument reveals the band present in the latter star but only weakly so when compared with the expected, properly stronger bands of ZrO.

The purpose of discussing this band at length is not to argue for a particular identification but rather to point to the stars in the south which exhibit the band in the Henize spectra. Numbers 230 and 103 are shown in Figure 6 along with WZ Cas, and large tic marks are used to set off the unidentified band in all three stars as it appears just redward of CN(5,1) marked by smaller tic marks.

Finally, included in Figure 6 is a spectrum of RZ Peg obtained recently by the writer. Examination of the spectrum reveals not only an unexpected enhancement of what heretofore has been the CN(4,0) bandhead but also the presence of an additional band blueward of both CN(4,0) and $C_2(0,2)$. Also, the region surrounding Na I looks peculiarly depressed despite the spectrum's showing hardly any evidence of the CN(5,0) band sequence. There is sufficient evidence from other researchers (Greene, 1971; Keenan, 1971b) that these peculiar absorptions are caused by the red and orange systems of CaCl. No Henize star has been found to date with these features.

3.7. STARS NOT YET INCLUDED IN THE CLASSIFICATION SCHEME. FIGURE 7

Several stars mentioned above have been referred to in the literature (Keenan, 1971a) as stars with features similar both to carbon stars and to S stars. One of these is CY Cyg which in the spectrum illustrated looks rather like a carbon star. Most of these stars are variable in spectral appearance as well as they are variable in light, and several of these stars have been known to look more like S stars than carbon stars at their cooler phases. Such a spectrum variable is Welch's in Crux (Keenan, 1971a). The writer was privileged to see a spectrum of this star taken recently at moderate dispersion by Dr Keenan. At the time the spectrum was obtained, the star was about 8.5 visual magnitude; and the usual features of C_2 and CN could be made out, though they were probably weaker than in the normal N stars. Included in Figure 7 is a Henize spectrum of this variable when the star was about 10 visual magnitude, and it is easily seen to bear little resemblance to any other spectrum shown in this article. This star was not included by Henize in his list of potential carbon stars because it appeared more like an S star than like a carbon star. Had the $C_2(0,1)$ bandhead been visible instead of being obscured by the spectrum of another star, this star might have found its way into the list of carbon stars. Suffice it to say that the writer has not included stars such as Welch's variable in the Henize list unless $C_2(0,1)$ is clearly present.

Finally, spectral peculiarities can often be confused with poor resolution in the Henize spectra. Such would probably be the conclusion if a spectrum like that of the northern CH star TT CVn were seen in the sample. This is because both the CN(4,0) and (5,0) band sequences are so weak as to reveal an almost pure C_2 absorption spectrum, making the star look like an early R star. It is not clear at the moment whether such stars will find their way into a separate class or accidentally be misclassified.

Acknowledgements

The writer would like to express his extreme gratitude to Dr Karl Henize for placing at the writer's disposal his valuable plate collection and for invaluable assistance in approaching the classification problem. Special thanks must be extended to Dr W. Buscombe under whom the writer is using the material discussed for his thesis and to Dr J. Allen Hynek for scheduling the excellent observing facilities of the Lindheimer Astronomical Research Center for obtaining the northern slit spectra. Finally, the writer owes a special debt of gratitude to Drs P. C. Keenan, W. W. Morgan, and C. P. Gordon for their encouragement and private communications concerning their classification schemes. This work was completed while the writer held an NDEA Title IV fellowship.

References

Bidelman, W. P.: 1950, *Astrophys. J.* **112**, 219.
Bidelman, W. P.: 1956, *Vistas in Astronomy* **2**, 1428.

Bidelman, W. P. and Keenan, P. C.: 1951, *Astrophys. J.* **114**, 473.
Bouigue, R.: 1954, *Ann. Astrophys.* **17**, 104.
Burbidge, E. M. and Burbidge, G. R.: 1957, *Astrophys. J.* **126**, 357.
Gordon, C. P.: 1967, Ph. D. Thesis, University of Michigan.
Gordon, C. P.: 1968, *Astrophys. J.* **153**, 915.
Gordon, C. P.: 1971, private communication.
Greene, A. E.: 1971, Ph. D. Thesis, Ohio State University.
Keenan, P. C.: 1954, *Astrophys. J.* **112**, 219.
Keenan, P. C.: 1971a, *Monthly Notices Roy. Astron. Soc.* **153**, short communication.
Keenan, P. C.: 1971b, private communication.
Keenan, P. C. and Hynek, J. A.: 1945, *Astrophys. J.* **101**, 265.
Keenan, P. C. and Morgan, W. W.: 1941, *Astrophys. J.* **94**, 501.
Marenin, I.: 1971, private communication.
Mendoza, E. E. V. and Johnson, H. L.: 1965, *Astrophys. J.* **141**, 161.
Morgan, W. W.: 1971, private communication.
Peery, B.: 1970, *Kitt Peak Symposium of Late-type Stars*.
Pettit, E. and Nicholson, S. B.: 1928, *Astrophys. J.* **68**, 279.
Pettit, E. and Nicholson, S. B.: 1933, *Astrophys. J.* **78**, 320.
Richer, H. B.: 1971, *Astrophys. J.* **167**, 521.
Shane, C. D.: 1928, *Lick Obs. Bull.*, **13**, 123.
Simon, L. W.: 1964, M.S. Thesis, Northwestern University.
Utsumi, K.: 1967, *Publ. Astron. Soc. Japan* **19**, 342.
Warner, B.: 1965, *Monthly Notices Roy. Astron. Soc.* **129**, 263.
Wurm, K.: 1932, *Z. Astrophys.* **5**, 260.
Yamashita, Y.: 1967, *Publ. Dominion Astrophys. Observ.* **13**, no. 5, 67.

AUTOMATIC CLASSIFICATION OF OBJECTIVE
PRISM SPECTRA

R. M. WEST

European Southern Observatory, Hamburg, Germany *

Abstract. Methodological problems and hardware/software requirements of efficient automatic classification of objective prism stellar spectra are described on the basis of experience from 'microphotometer-computer' classification. Significant advances in hardware (two-dimensional scanners, on-line computers) as well as in software (image processing techniques) have now brought high-speed, exhaustive, automatic classification within reach. Some astronomical implications of automatic spectral surveys are discussed.

1. Introduction

There is hardly any field of modern astronomy in which digital equipment has not yet been successfully applied. Many laborious tasks which are unavoidable in astronomical work and which formerly occupied significant amounts of an astronomer's time, are now performed by various kinds of machines. It is quite natural that first of all those straightforward problems have been automated that only to a lesser degree require the astronomer's skill; those problems that demand a greater degree of versatility and personal judgment are still, to a large extent, solved by the astronomer's manual efforts.

The classification of stellar spectra is a fundamental astronomical function that has not yet been automated. Anyone who has tried to classify spectra visually, or who has had the opportunity to acquaint himself with the working methods of the great visual classifiers, must admit that spectral classification requires nothing less than an artist's approach. The whole method of spectral classification, and especially the visual application of spectral criteria to objective prism spectra, is a very delicate operation for which a successful result can only be reached by great skill and patience.

In view of the many inherent difficulties one might therefore ask if it justified to investigate the possibilities of automating spectral classification. However, attempts to increase the objectivity of the classification by means of microphotometer registrations have definitely demonstrated the advantage of *partial automation* in this field.

Automation is always applied most efficiently where large masses of data are treated in a not necessarily simple, but at least monotonous way. Within the field of classification of stellar spectra, objective prism plates with hundreds or even thousands of spectra constitute an obvious object for automation efforts. The usefulness of quantitative, two-dimensional interpretation of objective prism spectra has been demonstrated repeatedly. Moreover, investigations by McCarthy *et al.* (1966), Samson (1969) and West (1970, 1971) have shown that quantitative three-dimensional classification of late-type objective prism spectra is possible, even at rather low dispersion. The magnificent spectra of the Curtis Schmidt telescope on Cerro Tololo prove that objective prism plates that are taken under excellent atmospheric conditions also yield

* Now at ESO Sky Atlas Laboratory, Geneva.

excellent results; thus further adding to the increasing importance of objective prism work in modern astronomy.

With this in mind, we shall therefore now consider the possibility of *complete automation* of spectral classification, applied to objective prism spectral plates. The word 'complete' here signifies the automatical assignment of parameter values to stellar spectra on a photographic plate by means of an apparatus (an 'automatic classifier') into which the plate is inserted.

Let us for a moment forget about the vast difficulties which must be overcome before an automatic classifier can be put into operation and consider the major advantages that would then become available (not listed in order of relative importance):
– Position determination and automatic identification of the stars.
– Magnitude and colour determination by means of appropriate mathematical filters.
– Possibility of determination of coarse radial velocities from plate sets with prism in normal and reversed position.
– Direct estimate of *basic* stellar parameters (e.g. T_e, g, abundances).
– Possibility of classification of higher dimension.
– High speed and great endurance, i.e. many spectra can be classified within a reasonable time.
– Homogeneity, elimination of personal errors.
– Accuracy.
– Detection of spectral variability.
– Possibility of rapid and easy re-classification, e.g. if the calibration of the stellar parameters changes or when new spectral criteria become available.

In what follows, an outline is given of the methods (Section 2) and the means (Section 3) that are now available for high-speed, exhaustive, automatic classification of objective prism plates. The present state of automatic classification and some of the problems that still have to be solved are summarized in Section 4, and possible astronomical applications of future, large-scale spectral surveys are discussed in Section 5.

2. The Methods

Automatic classification of objective prism spectra can be accomplished in several ways within the following general scheme:

2.1. REGISTRATION

The spectral plate is scanned and the transmission is measured in a network of points.

2.2. RECOGNITION

A spectrum is identified as a plate area with reduced transmission. The measured transmission values within the spectrum are arranged in a predetermined way, e.g. according to wavelength.

2.3. TRANSFORMATION

The spectrum is transformed in some way or another in order to have it in a convenient form for further reduction, e.g. to intensity units by means of the photographic calibration curve; Fourier- or Hadamard-transformation; convolution with suitable function, etc. In what follows the transformed spectrum will be referred to as the *image* of the spectrum.

Two different procedures are possible at this step:

2.4a. CRITERIA EVALUATION

Certain numerical or logical values (criteria) are computed which characterize the image of the spectrum in an unambiguous way, e.g. line depth ratios, equivalent widths, amplitudes in power spectrum, etc.

2.4b. CROSS CORRELATION

The image of the spectrum is compared with the images of stored standard spectra by cross correlation techniques. By iteration those standard spectra are found that surround the spectrum in the n-dimensional classification space.

2.5. CLASSIFICATION

If step 2.4a was performed, then the values of the stellar parameters are found from the computed values of the criteria and a certain class is assigned to the spectrum. Following step 2.4b, however, an interpolation in the classification space by means of the appropriate correlation measures leads to the spatial position of the image of the spectrum and thus the class. Peculiarities, e.g. unusual features in the spectrum, spectroscopic binaries, etc., are detected here or possibly at an earlier stage.

It follows from this scheme that a large number of standard spectra (or synthetic spectra of the relevant dispersion, if these should become available) for which the stellar parameters are known, must be secured. In case the cross correlation method is used (step 2.4b), the images of these spectra will be stored in the computer, possibly after some preliminary processing to ensure homogeneity.

When the criteria evaluation method (step 2.4a) is used, then the automatic classification must be preceded by the establishment of optimized spectral criteria that are evaluated from standard stars.

The accuracy of the automatic classification is readily estimated from automatic re-classification of standard spectra.

3. The Means

There has been a tremendous progress in image processing techniques during the past few years and the image types that can now be automatically interpreted range from cells in biological tissue to elementary particle tracks and all kinds of objects on aerial survey photographs. An extensive literature exists on this subject, and a series of excellent review papers is found in the Proceedings of a NATO Summer School (Graselli, 1969) and a Conference on Methodologies of Pattern Recognition (Watanabe, 1969). For a very recent, brief survey of the INTER/MICRO Exhibition in London in September 1971 that almost exclusively dealt with the analysis of images by computer (see Fisher, 1971).

The automatic classification of objective prism spectra has much in common with other types of automatic image interpretation. Similar to features on aerial and satellite photographs the spectra may be *easily* identified when they stand out clearly from the plate background, but only *with difficulty* if they are 'camouflaged' on a variable background or by overlapping. The complexity of the software required for the classification is rather *high*, since the spectra are classified according to distribution of surface density and not to the size and shape alone as most other images that are now being automatically interpreted.

We shall now take a look on the hardware and software that will be necessary for the realization of automatic classification. In general, only qualitative facts will be given, since the ultimate speed and accuracy of an automatic classifier will depend critically on the software that controls the scanning device and performs the classification. Although hardware already exists that is very similar to what will be needed for automatic classification, no detailed study of the necessary amount and substructure of the software has so far been completed.

The hardware consists of two closely interacting parts: the scanning device and the on-line computer. If, for practical reasons such as speed and storage limitation, the entire classification procedure can not be carried out in real-time by the on-line computer, then a storage medium (tape- or disc-unit) must be provided for. From this medium the data are later transferred to another (larger) computer.

3.1. HARDWARE: THE SCANNING DEVICE

It is convenient to distinguish between *one-dimensional* and *two-dimensional* scanning on the basis of the freedom of the spot/slit during computer controlled movement. A one-dimensional device can, of course, simulate two-dimensional scanning by consecutive offsettings perpendicular to the scanning direction, but only at a vast expense in time. Considering the general advantages of two-dimensional scanning (that will be further discussed below), there is no doubt that this technique must be preferred in automatic classification of objective prism spectra.

The most important characteristics of the scanning device for the present purpose may be summarized as follows:
- Spot/slit dimensions.
- Spot/slit speed; plateholder speed.
- Positional accuracy and stability.
- Light source stability.
- Ability to keep the plate continuously and optimally focussed.

3.1.1. *One-Dimensional Scanning*

Digitized slit microphotometers are now in use at many observatories. So far they have mainly been used for providing the input material in computer analysis of high-dispersion spectra, but recently studies of automatic classification of objective prism spectra have also been undertaken (cf. Section 4). The conventional microphotometer is a very stable device and the slit can be set to almost any size. A major drawback in this context is, however, the necessity of individual setting on each spectrum, including manual focussing. Moreover, most of the digitized microphotometers that have been put into operation have been somewhat limited in speed, partly because of slow output devices.

3.1.2. *Two-Dimensional Scanning*

Besides a few machines that have been specifically developed for astronomical purposes, a large number of automatic two-dimensional scanning devices for the measurement of particle tracks are now in operation. Among these, reference will here only be given to those that were developed at CERN (the European Organization for Nuclear Research) for scanning of bubble chamber tracks. These machines are, however, very representative for the whole class, as it can also be inferred from the Proceedings of the International Conference on Data Handling Systems in High Energy Physics (Lord and Powell, 1970). A brief survey of two-dimensional scanning devices has been given by Hambleton (1971).

One type of two-dimensional, 'flying spot' scanning device is based on a light spot that is moved across the plate by means of *digitally controlled, mechanically moved mirrors*. The light source may be *conventional* (US Naval Observatory Parallax Measuring Machine (cf. e.g. Strand, 1971); the HPD Machine at CERN (Benot *et al.*, 1968)) or a *laser* (Luyten Proper Motion Machine (Newcomb *et al.*, 1970); the Cambridge Machine (Kibblewhite, 1971)). This scanner type is fairly stable and the obtained speeds of the light spot are high; however, the complete digital control of the light spot is, where present, somewhat slowed by the inertia of the mirror mechanics. The laser devices offer the advantages of high spot intensity and small spot size.

In recent years, *Cathode Ray Tubes (CRT)* have been used to generate flying spots. Typical examples hereof are the GALAXY Machine (cf. e.g. Walker, 1971) and the LUCY Machine at CERN (Anders *et al.*, 1970). The major problems here are the stability of the light source, due to variations in the CRT phosphor screen, and unavoidable image aberrations. Furthermore, multiple reflections in the CRT screen may diffuse the spot and create problems in the astronomical application. On the other hand, the CRT technique allows a very high speed and certainly a great versatility. Comparison of the CERN machines HPD and LUCY seems to indicate that the CRT devices can now be made as stable as the 'moving mirror' devices, if geometric and photometric calibrations are performed carefully and regularly by means of standard grids and sensitometric scales.

The latest technical innovation which might prove to be of great use for scanning apparatus is the *Diode Array*. It consists of several hundred mini-diodes (until now manufactured with areas of 2.5×2.5 to $100 \times 100\ \mu m^2$) in a one-dimensional array on the surface of an integrated block that also contains the circuitry for reading a digitized signal from the diodes. A scanning apparatus that moves a photographic plate per-

One way of overcoming some of the recognition problems is to use predetermined stellar positions that are obtained by scanning a direct plate in advance and storing the positions of the stars for which the spectra are later to be classified. This method is advantageous in several respects, especially for the safe recognition of stellar pairs that might otherwise simulate spectroscopic binaries during the spectral classification. It presupposes, on the other hand, that a direct plate is available and that the measured plate positions can be stored in the computer or in a peripheral unit and later made available at the right moment. It can probably best be judged after practical tests whether or not a direct plate should be registered before the spectral plate.

After registration and possible disentangling from neighbouring spectra, a spectrum is now stored in the computer as a two-dimensional array of transmission registrations. At this point it is possible to correct for some types of plate errors, e.g. lines that extend across the spectrum and thus can be recognized above the surrounding plate background. Smaller errors entirely within the spectrum or just at the border must still await detection.

In making the recognition of a spectrum complete, a rigid wavelength calibration is needed. Since the mean dispersion curve of the plate may be considered as known, it is first of all necessary to establish at least one fixpoint of known wavelength in the spectrum. This fixpoint can be the cut-off of the photographic emulsion or the telescope optics. It may also be some of the major lines in the spectrum, although this implies some kind of rough classification already at this point, which might be somewhat impractical. The aforementioned use of predetermined plate positions from a direct plate may solve this problem if the corrections between the direct and the spectral plate – caused by the addition of the objective prism to the optics – are sufficient accurately determined.

Following the preliminary wavelength calibration by means of one fixpoint and a mean dispersion curve, it now becomes possible to identify major spectral features so that an individual, rigid wavelength calibration can be established for each spectrum. It must be emphasized that the problem of a precise wavelength calibration is a very fundamental one; without a correct knowledge of the wavelengths, the classification criteria may be incorrectly evaluated due to faulty feature identification. Furthermore, the existence of peculiarities at certain wavelengths, e.g. Barium and Silicium lines, can not be convincingly ascertained.

It is in principle possible to perform the wavelength calibration by means of optimal fitting through cross correlation with a standard spectrum, similar to the Griffin (1967) method for radial velocity determination. However, this method implies that the spectral type is fairly well known and therefore necessitates a preliminary classification of the spectrum already at this stage.

By integrating over the spectrum, a measure of the total density is obtained from which the apparent magnitude can be computed if the plate has been photographically calibrated or contains a standard magnitude sequence. With the application of one or more mathematical filters, the magnitudes and colours can be computed in any photometric system within the sensitivity region of the plate emulsion. When the photographic calibration curve is wavelength dependent, as is practically always the case, this must be taken into account during the integration.

The final output of the recognition part of the program consists of a two-dimensional array of plate transmissions that has been wavelength calibrated.

3.3.2. *Transformation*

In almost all cases where images are handled by a computer, it has been found very useful to let an image transformation follow after the recognition phase in order to facilitate the access to the important image features. Several monographs and articles are available on the subject of image transformation; e.g. two excellent introductions have been written by Rosenfeld (1969) and Andrews (1970). There are evidently many levels of complexity in this part of image processing, ranging from very simple to extremely sophisticated statistical-mathematical treatment of the image array. It is difficult to foresee exactly what kind of technique that will be the most advantageous to apply in our case, but in view of the limited time available for each spectrum, the highly advanced and thus very time consuming methods appear to be out of question. But we should not *a priori* discard the possibility of making at least partly use of some of the powerful tools of modern numerical analysis.

It is possible to reduce immediately the two-dimensional transmission array to one dimension by averaging or integrating the transmission readings at the same wavelength. This is what is implicitly done by the slit scanning microphotometer. However, notwithstanding the increased amount of computation involved, there are advantages in a further two-dimensional evaluation. This concerns especially the plate errors which do not extend across the whole width of the spectrum and the noise. The recorded noise in the spectrum depends on the intrinsic graininess of the plate, the spot size and, depending on the scanning principle, on internal noise in the scanning device. In the past, noise has been eliminated either by smoothing (averaging) or cutting of the high frequencies in the one-dimensional Fourier transform.

After high-speed, two-dimensional Fourier transformation has become available (cf. Special Issues of IEEE Trans. Audio Electroacoustics AU-15 (June 1967) and AU-17 (June 1969) on Fast Fourier Transforms), this specific kind of transformation has repeatedly been successfully applied in image processing. Some one-dimensional astronomical applications of the Fast Fourier Transform have been given by Brault and White (1971).

Ideally, the density at a certain wavelength should be the same across the spectrum, but it is seldom so because of noise and plate errors, and sometimes because of incorrect guiding of the telescope. It would therefore be very interesting to see two-dimensional Fourier transformation applied to objective prism spectra for noise smoothing and error detection in the inverse image of the spectrum. Other transformations, e.g. the Hadamard-transformation, are presently in much use in image processing, mostly in their two-dimensional forms. Their usefulness in the present context should also be investigated.

There are other ways of transforming the spectrum and simultaneously make the elimination of noise possible, e.g. convolution or cross correlation with suitable functions (filtering).

A transformation from plate transmission to intensity can take place if the plate has been photographically calibrated. It is advantageous to use individual calibration curves for each wavelength, obtained by interpolation between the curves for each 50 or 100 Å.

After the spectrum has been transformed and possibly cleaned for noise and plate errors, the image of the spectrum can now be compressed to a one-dimensional image. Hopefully the plate errors for which no correction could be made are now known so that the affected spectral regions can be disregarded in the subsequent analysis.

3.3.3. *Criteria Evaluation*

This part of the program must be distinguished from the succeeding part, the classification, because it establishes inherent numerical or logical values of the spectrum. The classification part thereafter interpretes these values in terms of stellar parameters, but this interpretation, i.e. the calibration of the criteria, may change, for instance if the basis for the calibration becomes better.

Well defined spectral criteria exist at present for visual classification at various dispersions and in several wavelength regions. Criteria that have been established in the MK system and which play a role in the classification of objective prism spectra have been published in many places, but are best illustrated in the spectral atlases of Morgan, *et al.* (1943), Abt *et al.* (1968), Seitter (1970), and Landi Dessy *et al.* (1972), However, since automatic classification may not be optimally performed with the already existing criteria, and since new criteria are needed if the criteria values are to be derived from transformed spectra, it is necessary to establish (new) criteria independently. Moreover, the same set of criteria can only be applied to stellar spectra taken through the same optics and on the same photographic emulsion. New, optimal criteria must be determined whenever one of these factors is changed.

Before any automatic classification can be performed, it is therefore necessary that a large number of standard spectra with known values of the stellar parameters are processed until the stage of criteria evaluation. At this point the images of the spectra are thoroughly searched for optimal criteria. This may seem an enormous task in view of the many spectral features, but it can be very efficiently conducted by means of a computer. All possible criteria, such as line depths, line-depth ratios, peak heights, feature areas (e.g. equivalent widths) and absence/presence of specific features in the image of the spectrum can be tested for correlation with the stellar parameters, and those criteria that appear the most promising are selected for refinement. In this process the *accuracy* as well as the *general availability* of the criteria must be considered. There should preferably be two or more criteria for each spectral parameter so that any reasonable number of plate errors does not exclude the evaluation of at least one criterion for each parameter and thus the complete classification of any spectrum.

The possible automatic use of the present spectral criteria in visual classification should be investigated.

When the optimal criteria have been decided on, they are calibrated directly in terms of the stellar parameters.

It is possible only to use major criteria in the beginning of the evaluation of the criteria in order to save time. Depending on a preliminary 'coarse' assignment of spectral class, other, more refined criteria are thereafter activated for the 'fine' classification.

3.3.4. *Cross Correlation with Standard Spectra*

An alternative to the criteria evaluation is the cross correlation with standard spectra. This technique implies that a sufficient number of standard spectra is stored in the computer. The spectrum for which values of the stellar parameters shall be found is brought into the same format as the standard spectra; this involves a very accurate wavelength calibration and some standardizing transformation. As earlier mentioned the wavelength calibration may be part of the cross correlation procedure which in that case begins with a cross correlation with a standard spectrum of similar spectral type to obtain the best fit in wavelength. Thereafter the spectrum is correlated with a number of standard spectra in order to place it in the correct box in the n-dimensional classification space, n being the number of independent classification parameters.

It will be very interesting to see the cross correlation method applied to automatic classification of objective prism spectra. However, notwithstanding its many advantages, it might be feared that a rather large number of correlations must be performed, even in two-dimensional classification, which will be of disadvantage in computing time required. It is quite possible that a combination of the criteria evaluation and the cross correlation technique will be the most useful approach. By alternate use of spectral criteria and cross correlation in narrow spectral regions, the correct assignment of a stellar parameter may be achieved in a miminum of time.

3.3.5. *Classification*

As the last step the classification is now performed by assigning to each spectrum those values of the stellar parameters that correspond to the criteria values or, if the cross correlation method is used, by interpolation in the classification space. Peculiarity features are looked for, and it is noted if there are significant discrepancies between the values of any stellar parameter when determined by means of different criteria.

It is possible that the classification procedure will have the form of an iteration, e.g. first the luminosity class is estimated from certain criteria and, depending on this class, other criteria give the temperature class, the knowledge of which in turn makes possible a more accurate determination of the luminosity class, etc.

The endproduct of all these computations is a list, giving the plate positions of the spectra, the magnitudes and colours and the stellar parameters with their internal errors. External errors may be computed if some of the spectra have been classified by other investigators. The plate positions can be utilized for automatic identification of the stars by number or designation if an astrometric catalogue of the field is available on tape. When other spectral plates of the same field have been processed, mean values of the stellar parameters can be computed and stars with spectral variability detected. The intrinsic accuracy of the automatic classification method is determined through the comparison of the results obtained from plates of the same field.

By analysis of spectral plates with the prism in normal and reversed (180° rotation) positions, coarse radial valocities may be determined from the positions of the lines (cf. Stock and Osborn, 1973).

4. The Present State of Automatic Classification

Apart from a few, isolated experiments with flying spot scanners, all experience with automatic classification of stellar spectra has so far been gained in one-dimensional 'microphotometer-computer' classification. This means that the recognition problem hardly has been touched upon, except for some experiments with automated wavelength calibration (West, 1971). Various aspects of the computer reduction of digitally registered slit spectra have been described by Bonsack (1970), Furenlid (1971), Gratton *et al.* (1971), Hutchings (1971), Hutchison (1971), Latham (1971), Leushin *et al.* (1970), Peat and Pemberton (1970), Thompson (1967, 1971a, 1971b), and Yoss and Lutz (1968). Automatic reduction of objective prism spectra is described by Cassatella *et al.* (1973), Clausen (1973), Gratton *et al.* (1971) and Yoss (1973). Mathematical discussions of the computer analysis of stellar spectra are found in the papers of Rusconi and Sedmark (1971) and Thompson (1971a). Radial velocities were determined by a computer from microphotometer scannings of late-type, slit spectra (Lutz and Yoss, 1969; Yoss, 1966). A search for optimal criteria in the ultraviolet part of objective prism spectra of late-type stars has been described by West (1970). Partially automated classification of early-type objective prism spectra with a digitized microphotometer has been performed by Rossen (1969) and of late-type spectra by West (1971). Unpublished work on automatic reduction of objective prism spectra has been reported by Th. Schmidt-Kaler and collaborators in Bochum, by B. Virdefors in Lund and F. Simien in Marseille.

Obviously quite an amount of work is presently being invested in *partially* automated reduction of stellar spectra by means of digitized microphotometers. With the advantages of two-dimensional scanning in mind, let us now discuss some of the major problems that still have to be solved before *complete* automation be realized.

Some of the scanning devices that are available for automatic processing of circular images have been referred to above. A few of these regularly measure direct astronomical plates and much of the experience can undoubtedly be directly applied to the problem of automatic stellar classification. However, there are still important modifications that must be made.

Among the necessary changes in the *hardware*, one of the most significant is the digitization of the plate transmission, i.e. of the recorded spot intensities. Until now, most devices only discriminate between 'clear' and 'dark' plate areas (1 bit digitization) since this is fully adequate in the registration of circular images for which only the position and size are important. An exception herefrom is the GALAXY Machine in the measuring mode.

The digitization will slow down the scanning speed for two reasons. First, the illumination time of each image element must be increased to reduce correspondingly the statistical noise due to the corpuscular nature of light, and secondly, more computing time is needed for the discrimination of the light levels. To retain the photographic accuracy, a transmission digitization with at least 8 bits (0–255) or preferably 10 bits (0–1023) is needed in spectral work. The intrinsic digitization of the Diode Array makes this solution a very attractive one, and it should certainly be investigated in detail.

The digitization also sets higher demands for the photometric stability of the device and some convenient means for accurate and fast calibration must be available. When used for identification only, the positional accuracy over the whole plate may not be so critical, but a determination of radial velocities calls for a very high accuracy that may only be accomplished over small plate areas.

The memory of the *on-line computer* should preferably be larger than the presently normal size (8–16 Kwords) for minicomputers because of the storage demand for the transmission readings and the program, but this is merely a cost problem.

The *software* for automatic classification is much more complicated than that for circular image processing. This means that, even when the necessary hardware becomes available, it will still take some time and certainly much experimentation before the automatic classification program will run satisfactorily.

The speeds that have already been attained range from about 7.5 h (GALAXY in search phase, 16 μm spot) to 6.5 min (LUCY, spot size 13 μm, spot speed 5μm μs^{-1}, line distance 5 μm) for the registration of 100 cm^2. Whereas the accuracy of the GALAXY measurements is high enough for astronomical purposes, this is not the case for LUCY.

TABLE I

Some numerical values for the automatic classification of objective prism spectra

Size of spectra	$200 \times 5000\ \mu$m^2
Distance between parallel scans	20 μm, i.e. 10 scans along a spectrum
Distance between transmission readings	2 μm
Computer cycling time	1 μs
Spot speed	0.2 μm μs^{-1}
Background registration, focussing, search for and registration of the spectrum	$\sim$ 0.1–0.2 s
Scanning time per spectrum	0.25 s
Classification time (criteria evaluation)	$\sim$ 0.6 s
Total time per spectrum	$\sim$ 1 s

Without prior knowledge of the amount and detailed substructure of the software and the scanning speed that will ultimately be adopted for the automatic classification of objective prism spectra, the numerical example in Table I gives only an indication of the obtainable speed.

The estimate of the classification time is based on a extrapolation of the figures in West (1971).

5. Some Implications of Automatic Spectral Classification

On the preceding pages we have discussed the possibility of complete automation of the classification of objective prism spectra. With its many advantages it must be foreseen as a future, very powerful tool in astronomical research. It is appropriate to add some words about the possible applications of an automatic classifier.

Clearly, the most obvious astronomical program that can be carried out with an automatic classifier is a *large-scale investigation of the galactic structure*. The automatic classifier would be an excellent mean for producing a large, very homogeneous material of spectral classes which would in turn be invaluable in the study of the spatial distribution of various stellar types. In this connection numerous other important problems could also be investigated; among these one might mention the determination of the luminosity functions of different population types, the peculiarity frequencies, spectral variability, discovery of spectroscopic binaries, etc. Of special importance is the possibility of direct calibration to basic stellar parameters that is offered by the automatic classification techniques.

It is here relevant to emphasize that the performance of an automatic classifier completely depends on the accumulated knowledge in the field of stellar classification and that an astronomically useful automatic classification is only possible in continued, close contact with the astronomers that are actively engaged in conventional spectral classification. In other words, the automatic classifier can best be regarded as an extremely efficient means of accomplishing what could otherwise only have been carried out in a tremendous effort of astronomers with normal equipment.

Let us briefly illustrate with an example the amount of information that could be processed with an automatic classifier. The 1 m ESO Schmidt telescope that will be in operation at Cerro La Silla in Chile from the beginning of 1972 will be equipped with an ultraviolet transparent prism, giving a dispersion of about 580 Å mm^{-1} at Hγ.

With a plate limiting magnitude of 14^m (a conservative estimate on IIa–O plates, 200 μm widening), about 11 000 stars may be expected on a 29×29 cm^2 (5.5×5.5 deg^2) plate at galactic latitude $0°$ (1300 stars at $90°$), cf. Seares *et al.* (1925). Neglecting overexposure and overlapping, the classification time would probably amount to several months (a few weeks) for a visual classifier, who classifies at a rate of 40 spectra h^{-1} and with time allowed for checking purposes. The corresponding time would be a few hours or less for an automatic classifier at 60 spectra min^{-1}, and there would be no need for checking. As a matter of fact, several plates of the same field could be processed on the same day in order to eliminate gross errors and to improve the accuracy.

A final problem that arises, concerns the storage of all the information that is obtained from the automatic classifier. Should lists be printed with the spectral classification and how shall the stars be identified? How can such a large material best be made available to the astronomical community? One might doubt whether there is any reason to publish extensive lists of spectral classes that are mainly of statistical value; it would most certainly be more economic and practical to make the original plates and the computer print-out available on request.

The Schmidt telescope has a very low ratio between observing and evaluation time. The spectral plates that were mentioned above will be obtained in 20–30 min. If such plates could be processed and prepared for astronomical interpretation of the spectral classes found during some hours, then a much more reasonable ratio would be obtained.

Experience in other fields of astronomy and in other sciences as well has since long proved that the introduction of automatic methods not only facilitates the current research; it also opens up entirely new possibilities. The classification of objective prism spectra will be no exception from this principle.

References

Abt, H. A., Meinel, A. B., Morgan, W. W., and Tabscott, J. W.: 1968, *An Atlas of Low-Dispersion Grating Stellar Spectra*, Kitt Peak National Observatory and Steward Observatory, Tucson.

Anders, H., Jacobs, D., Krishner, W., Lewis, A., Lingjaerde, T., Lord, D., Oropesa, J., Van Prag, A., Stumpe, B., Turek, L., Wiscott, D., and Zurbrechen, R.: 1970, in D. H. Lord and B. W. Powell (eds.), *Proc. of the International Conference on Data Handling Systems in High Energy Physics*, Cambridge, March 23–25, 1970, CERN Document 70–21, Geneva, p. 427.

Andrews, H. C.: 1970, *Computer Techniques in Image Processing*, Academic Press, New York and London.

Benot, M., Evershed. B. W., Messerli, R., and Powell, B. W.: 1968, *The HPD Mark 2 Flying-Spot Digitizer at CERN*, CERN Document 68–4, Geneva.

Bonsack, W. K.: 1970, *Bull. Am. Astron. Soc.* **2**, 298.

Brault, J. W. and White, O. R.: 1971, *Astron. Astrophys.* **13**, 169.

Cassatella, A., Maffei, P., and Viotti, R.: 1973, this volume, p. 127.

Clausen, J. V.: 1973, this volume, p. 134.

Fisher, C.: 1971, in *New Sci. and Sci. J.*, September 23, 1971, p. 676.

Furenlid, I.: 1971, *Astron. Astrophys.* **10**, 321.

Graselli, A. (ed.): 1969, in *Automatic Interpretation and Classification of Images*, Proceedings of the NATO Summer School, Pisa-Terrenia, Italy, August 26–September 6, 1968, Academic Press, New York and London.

Gratton, I., Martini, A., Martino, F., Natali, G., and Viotti, R.: 1971, *Publ. Roy. Obs. Edinburgh* **8**, 142.

Griffin, R. F.: 1967, *Astrophys. J.* **148**, 465.

Hambleton, J.: 1971, *Publ. Roy. Obs. Edinburgh* **8**, 157.

Høg, E.: 1971, personal communication.

Hutchings, J. B.: 1971, *Publ. Roy. Obs. Edinburgh* **8**, 185.

Hutchison, R. B.: 1971, *Astron. J.* **76**, 711.

Kibblewhite, E. J.: 1971, *Publ. Roy. Obs. Edinburgh* **8**, 122.

Landi Dessy, J., Jaschek, C., and Jaschek, M.: 1972, *A Spectral Atlas at 42 Å mm^{-1}*, Cordoba, Argentine, to be published.

Latham, D. W.: 1971, *Publ. Roy. Obs. Edinburgh* **8**, 183.

Leushin, W. W., Jacomo, A. A., and Rusak, N. P.: 1970, *Izv. Krym. Astrofiz. Observ.* **41–42**, 384.

Lord, D. H. and Powell, B. W. (eds.): 1970, *Proc. of the International Conference on Data Handling Systems in High Energy Physics*, Cambridge, March 23–25, 1970, CERN Document 70–21, Geneva.

Lutz, T. E. and Yoss, K. M.: 1969, *Astron. J.* **74**, 91.

McCarthy, M. F., Treanor, P. J., Bertiau, S. J., and Bertiau, F. C.: 1966, in K. Lodén, L. O. Lodén, and U. Sinnerstad (eds.), 'Spectral Classification and Multicolour Photometry', *IAU Symp.* **24**, 59.

Morgan, W. W., Keenan, P. C., and Kellman, E.: 1943, *An Atlas of Stellar Spectra with an Outline of Spectral Classification*, Univ. Chicago Press, Chicago.

Newcomb, J., La Bonte, A., and Luyten, W. J.: 1970, *The Automated Plate Scanner and Measuring Machine*, University of Minnesota, Minneapolis, Minnesota.

Peat, D. W. and Pemberton, A. C.: 1970, *Observatory* **90**, 141.

Rosenfeld, A.: 1969, *Picture Processing by Computer*, Academic Press, New York and London.

Rossen, J.: 1969, Thesis, Copenhagen University Observatory.

Rusconi, L. and Sedmark, G.: 1971, *Publ. Roy. Obs. Edinburgh* **8**, 188.

Samson, W. B.: 1969, *Publ. Roy. Obs. Edinburgh* **6**, No. 10.

Seares, F. H., Van Rhijn, P. J., Joyner, M. C., and Richmond, M. L.: 1925, Contributions Mt. Wilson Obs. No. 301.

Seitter, W. C.: 1970, *Atlas for Objective Prism Spectra – Bonner Spektral-Atlas I*, Ferd. Dümmlers Verlag Bonn.

Stock, J. and Osborn, W.: 1973, this volume, p. 290.

Strand, K. Aa.: 1971, *Publ. Roy. Obs. Edinburgh* **8**, 119.

Thompson, G. I.: 1967, *Publ. Roy. Obs. Edinburgh* **5**, No. 12.

Thompson, G. I.: 1971a, *Publ. Roy. Obs. Edinburgh* **7**, No. 2.

Thompson, G. I.: 1971b, *Publ. Roy. Obs. Edinburgh* **8**, 179.

Walker, G. S.: 1971, *Publ. Roy. Obs. Edinburgh* **8**, 103.

Watanabe, S. (Editor): 1969, in *Methodologies of Pattern Recognition*, Proceedings of the International Conference on Methodologies of Pattern Recognition, Honolulu, Hawaii, January 24–26, 1968, Academic Press, New York and London.

West, R. M.: 1970, *Bull. Abastumani Astrophys. Observ.* **39**, 29.

West, R. M.: 1971, *Bull. Abastumani Astrophys. Observ.* **42**, 109.

Yoss, K. M.: 1966, in K. Lodén, L. O. Lodén, and U. Sinnerstad (eds.), 'Spectral Classification and Multicolour Photometry', *IAU Symp.* **24**, 111.

Yoss, K. M.: 1973, this volume, p. 125.

Yoss, K. M. and Lutz, T. E.: 1968, *Astron. J.* **73**, S41.

MICROPHOTOMETRY-COMPUTER CLASSIFICATION OF OBJECTIVE-PRISM SPECTRA

K. M. YOSS

University of Illinois, Ill., U.S.A.

Abstract. Three-dimensional classification of late-type stars is obtained through computer-processed digitized microphotometer data. Plate transmissions are converted to relative intensity through spot-sensitometer calibration data. A pseudo-continuum consisting of several straight-line segments is formed by connecting high points in the spectrum. Absorption-line strengths and line ratios are then measured. The temperature-and-luminosity-sensitive ratios are relatively insensitive to seeing effects. For plates with a dispersion of 108 Å mm^{-1}, preliminary results indicate an accuracy in derived absolute magnitude comparable to, and possibly better than, that of MK classification.

The CN anomaly serves as the abundance index. The CN index was defined in a manner to give the largest figure of merit (the ratio of total range of the index to the mean error). The adopted method is the same as that of Yoss and Lutz (1971) and has a figure of merit over twice that of the index similar to that of Griffin and Redman (1960).

For field stars, plate X and Y positions are converted to equatorial coordinates and printed out in order of increasing right ascension, making quick and easy identification of the Henry Draper numbers.

Three-dimensional classification of late-type stars is obtained through computer-processed digitized microphotometer data. Plates taken with the University of Michigan Curtis Schmidt, with a dispersion of 108 Å mm^{-1} at Hγ, and while the instrument was still in Michigan, were used for this preliminary study.

Plate transmissions are converted to relative intensity through spot-sensitometer calibration data. Feature identification depends on locating maximum or minimum points relative to a manually-recorded position of a particular identified feature. A pseudo-continuum consisting of several straight-line segments is formed by connecting high points in the spectrum (at 4027, 4091, 4221, 4245, 4318, and 4502 Å). Absorption-line strengths and line ratios are then measured. The temperature- and luminosity-sensitive ratios are relatively insensitive to seeing and sky-fog effects. Half a dozen ratios (for example, 4325/4340 Å) are used to derive spectral type to within one sub-class.

The absolute-magnitude calibration depends on about 40 standard stars (generally two exposures per star) for which absolute magnitudes are known with mean errors $\leqslant \pm 0.5$ mag. Three ratios (4077/4071, 4176/4132, and 4215/4272 Å) are found useful for deriving absolute magnitude, with mean errors of ± 0.6 mag. (after the effects of the errors in the published values are removed). In spite of the fact that two of the ratios involve lines falling within the CN absorption band, there appears to be no dependence of the ratios on CN strength (or CN anomaly).

The CN anomaly serves as the abundance index. The CN index is defined in a manner to give the largest figure of merit (the ratio of total range of the index to the mean error of measurement). A two-segment CN continuum is fitted at 4091, 4163, and 4211 Å.

The fractional absorption between the two continua is then the measured value, and the CN anomaly is the departure from the average value as defined by a linear line extending over the entire $M_v - \mathrm{CN}$ range. The figure of merit is 17, compared to 8 for an index similar to that used by Griffin and Redman (1960) and 8 also for a visually estimated CN strength from similar plates (Yoss, 1961).

For aid in rapid identification of field stars, the plate X and Y positions are converted to equatorial coordinates and printed out in formats identical to the Henry Draper and BD catalogues. Generally prior identification of only one bright star in the field is sufficient to give coordinates accurate to 0.1 min in right ascension and to $2'$ in declination. Derived spectral types and photographic magnitudes are also printed, to aid in identification. These magnitudes, determined from the summed relative energy curves of the spectra, with the zero point determined from the value for the identified star, generally agree to within 0.1 mag. of the Henry Draper values. Several stars can be identified per minute in this manner. The microphotometry takes less than two minutes per star.

References

Griffin, R. F. and Redman, R. D.: 1960, *Monthly Notices Roy. Astron. Soc.* **120**, 287.
Yoss, K. M.: 1961, *Astrophys. J.* **134**, 809.
Yoss, K. M. and Lutz, T. E.: 1971, *Mem. Roy. Astron. Soc.* **75**, 21.

BLUE-INFRARED NARROW-BAND PHOTOMETRY
FROM OBJECTIVE PRISM SPECTRA

A. CASSATELLA, P. MAFFEI, and R. VIOTTI

Laboratorio di Astrofisica Spaziale – Frascati, Italy

(Read by L. Gratton)

Abstract. A technique for narrow-band photometry in the 3600–8400 Å spectral region from objective prism spectra is briefly described and applied to the Hyades.

For the purpose of making a quantitative study of a large number of low dispersion stellar spectra in young associations, a technique for the rapid reduction of spectra using the new digitized microphotometer MI.DI. and the IBM 1130 computer of the Laboratorio di Astrofisica Spaziale has been developed and applied, as a preliminary test, to the Hyades. A detailed description of the microphotometer and of its applications is given by Gratton *et al.* (1971).

The work is based on 8 IN Kodak plates obtained with the 65/90/210-cm Schmidt telescope of the Asiago Astrophysical Observatory using a 4° objective prism which gives a reciprocal dispersion of 608 Å mm^{-1} at Hγ. Half of the plates were hypersensitized with a flash lamp in order to reach the faintest stars – the average gain in the limiting magnitude being about 1^{m}5 or better which makes the use of hypersensitized plates rather convenient, provided the number of observations be doubled in order to have the same S/N ratio of the normal plates.

The photometric calibration was made with the Asiago spectrosensitometer on IN plates taken from the same box of plates used for the observations. The observational and calibration material was processed all together. Each calibration was read by the microphotometer at several different wavelengths and all the recordings were elaborated using a suitable program which gives the single characteristic curves and the averaged curve together with the internal accuracy of the calibration itself. It was found that the characteristic curve is not too sensitive to the wavelength and to the exposure time (in the range 10 min to 2 h), and does not vary between the plates of the same box, but it is very different between normal and hypersensitized plates (Cassatella and Viotti, 1971).

The correction for the selective instrumental sensitivity and the atmospheric extinction was made in the usual way by recording a number of standard Hyades stars among those whose monochromatic fluxes were measured with a photoelectric scanner by Oke and Conti (1965). The plate corrections $c(\lambda)$ were derived by comparing the observed photographic monochromatic magnitudes with the photoelectric ones. The procedure is described in detail in Figure 1 which shows the block diagram of the reduction program. Figures 2, 3 and 4 give the uncorrected monochromatic magnitudes for three stars of the Hyades.

Ch. Fehrenbach and B. E. Westerlund (eds.), Spectral Classification and Multicolour Photometry, 127–133. All Rights Reserved.

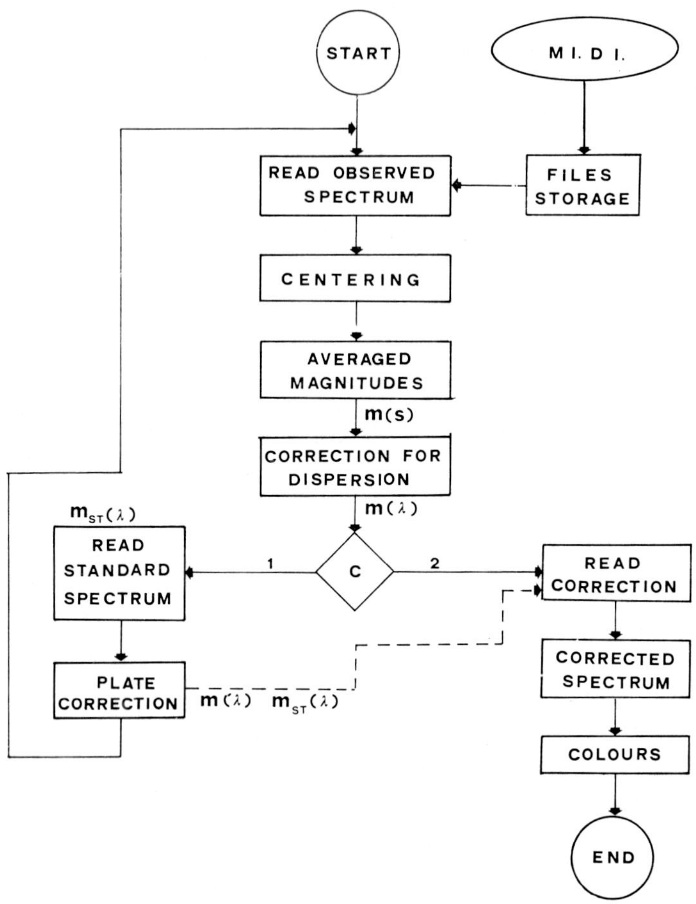

Fig. 1.   Block diagram of the reduction programs.

Since more than two standard stars are present in each plate, a photometric calibration of the plate could be derived directly by comparing the plate densities at a given wavelength with the corresponding monochromatic magnitudes of the standard stars. It was found that the slope of the characteristic curves in this manner are practically the same as those computed from the calibration plates. Since in the latter case we have a much larger number of points for drawing the characteristic curves, it was found more convenient to follow in this study the latter method.

The following Figure 5 gives the corrections $c(\lambda)$ derived for one IN plate. As soon as the $c(\lambda)$ of each plate were computed, the second part of the program in Figure 1 was used for the reduction of the spectra, and the corrected magnitudes were derived. Since one of the applications of this program is the determination of the infrared excess for stars in young associations, a blue-infrared color index $C\gamma$ was introduced defined as

$$c_\gamma = m_{4464} - \tfrac{1}{2}(m_{7100} + m_{8400}),$$

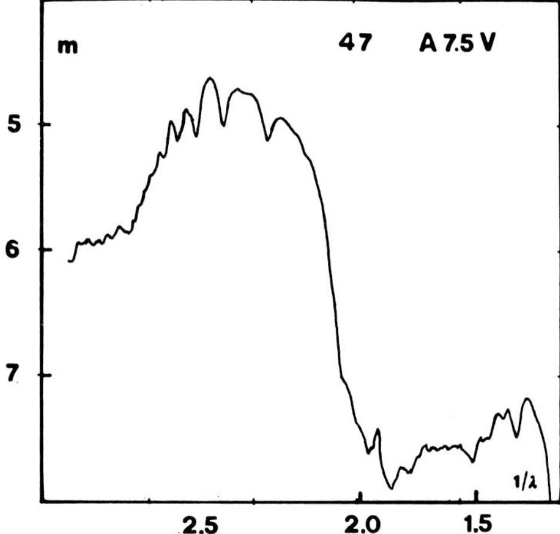

Fig. 2.   Tracing of star N 47 of the Hyades (A7.5V). In ordinates
monochromatic magnitudes corrected for the dispersion.

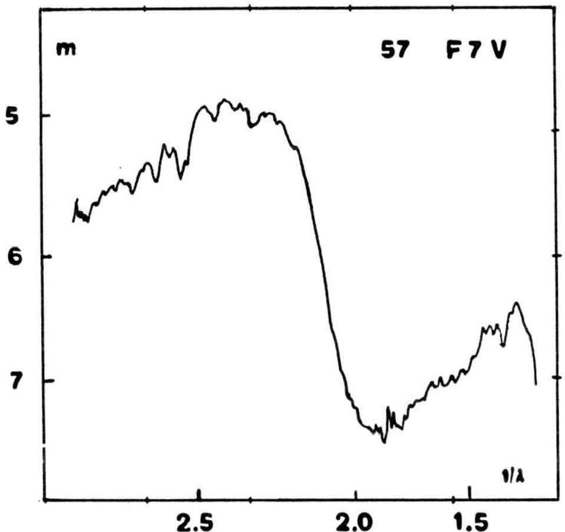

Fig. 3.   Tracing of star N 57 (F7V).

where m_λ are the mean magnitudes in a band centered on λ. The widths of the blue and of the two infrared bands are 50 Å and 100 Å respectively. The three bands were selected among those measured by Oke and Conti, in such a way to have a maximum plate density and to avoid as much as possible the stellar strong absorption bands and lines and the atmospheric bands.

74 color indices were computed for 30 Hyades stars, and the average C_y are plotted in Figures 6 and 7. The mean plate correction for the C_y colors was $2^m\!.33$ and the variation from plate to plate (hypersensitized or not) was no greater than $0^m\!.2$. The circles are main sequence A6–K8 stars; the triangle is a K0III star (γ Tau). Figure 6

                    A. CASSATELLA ET AL.

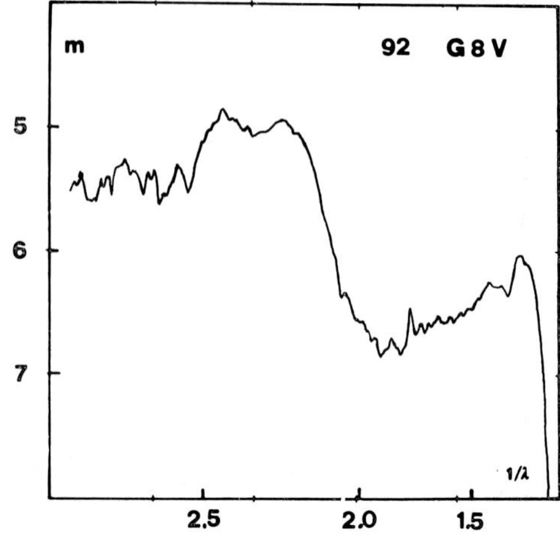

Fig. 4.   Tracing of star N 92 (G8V).

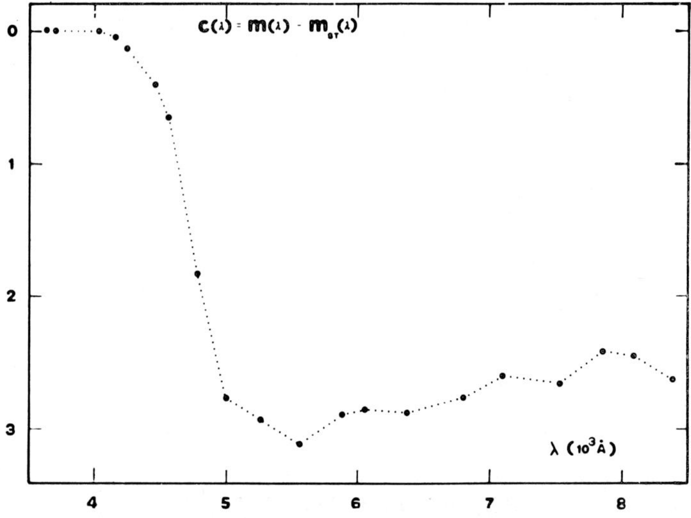

Fig. 5.   Plate corrections (in magnitudes) for a single IN plate.

gives the two color diagram C_γ, $B-V$ for the Hyades; the source for the $B-V$ is
Johnson and Knuckels (1955). Figure 7 shows the plot of C_γ as a function of the spectral type given by Morgan and Hiltner (1965) and by Ramberg (1941). The full lines
are the polynomial best fits of the data. The mean scatter of the points from the lines
is $0\overset{m}{.}05$ and $0\overset{m}{.}08$ respectively. The mean error for a single measurement for normal
plates is between $0\overset{m}{.}04$ and $0\overset{m}{.}08$ depending on the faintness of the star, and is about
30–50% larger for the hypersensitized plates. These errors are in a good agreement with
the expected accuracy for photographic photometry with objective prism spectra (see

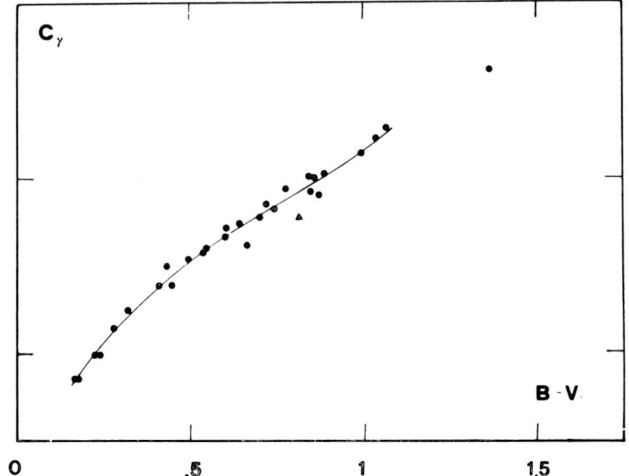

Fig. 6.   Two color diagram for the Hyades. (See text for the symbols.)

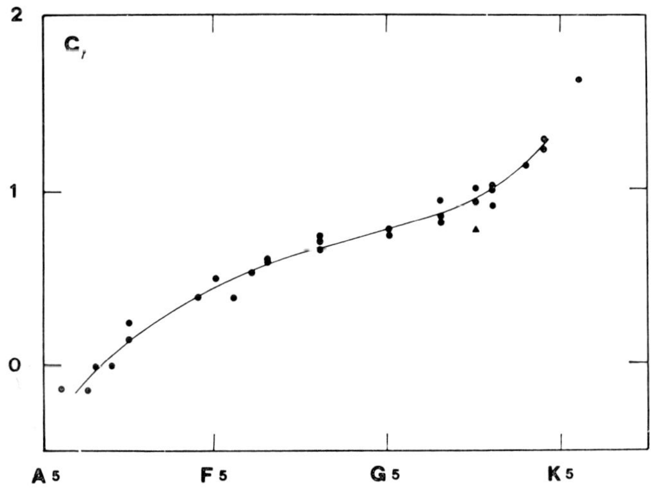

Fig. 7. $C\gamma$-spectral type diagram.

e.g. Gratton, 1939). The larger dispersion in Figure 3 may be accounted for errors in spectral classification. The color magnitude diagrams V, C_γ and $m(8400)$, C_γ are given in Figures 8 and 9.

The results in Figures 6–9 show that a satisfactory narrow band photometry can be made with objective prism plates, with the great advantage that one can make simultaneously the photometry of a very large number of stars in several different spectral regions and, in addition, a spectral classification can be provided from a visual inspection of the spectra or, automatically, with a suitable reduction program. For this

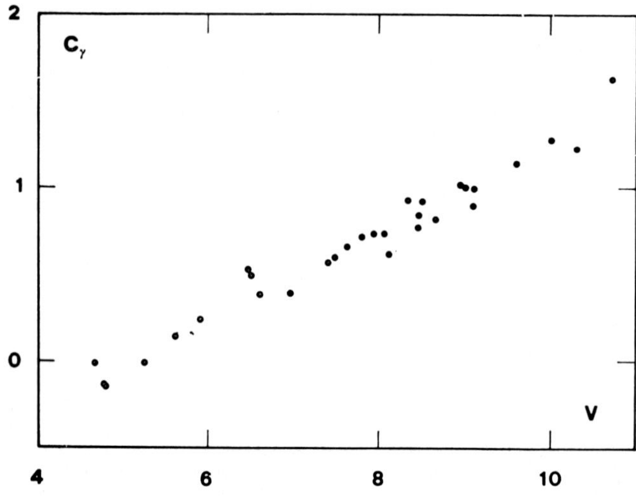

Fig. 8.   Color-magnitude diagram.

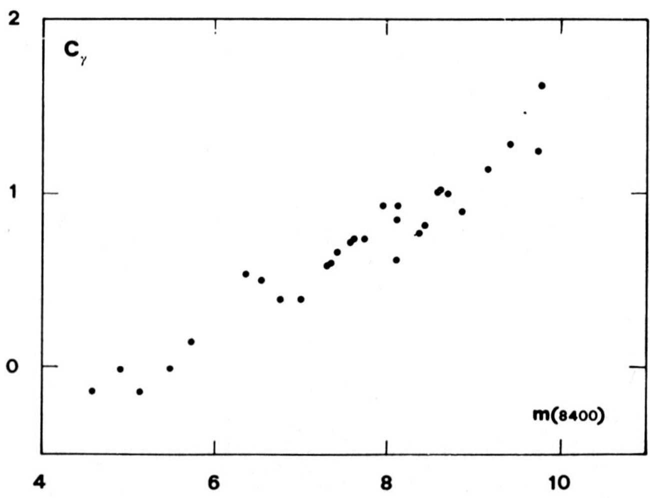

Fig. 9.   Color-infrared magnitude diagram.

reason the use of automatic measuring devices allows for a rapid reduction of the observational material. As for example, 100 spectra may be recorded with a digitized microphotometer in a few hours, including all the dead times. The time for the reduction of a single spectrum is 2 min for reading the punched tape, and 1 min for all the computations. The total time would be drastically reduced by using on-line connection between microphotometer and computer, so that the complete study of an association at the objective prism could be made in less than one day.

References

Cassatella, A. and Viotti, R.: 1971, Rapporto Interno No. 23, Laboratorio di Astrofisica Spaziale, Frascati.

Gratton, L.: 1939, *Stockholm Obs. Ann.* **13**, No. 3.

Gratton, L., Martini, A., Martino, E., Natali, G., and Viotti, R.: 1971, *Publ. Roy. Obs. Edinburgh* **8**, 142.

Johnson, H. L. and Knuckels, C. F.: 1955, *Astrophys. J.* **122**, 209.

Morgan, W. W. and Hiltner, W. A.: 1966, *Astrophys. J.* **141**, 177.

Oke, J. B. and Conti, P. S.: 1966, *Astrophys. J.* **143**, 134.

Ramberg, J. M.: 1941, *Stockholm Obs. Ann.* **13**, No. 9.

PRELIMINARY INVESTIGATION OF QUANTITATIVE SPECTRAL CLASSIFICATION BY MEANS OF OBJECTIVE-PRISM SPECTRA OBTAINED WITH THE BROERFELDE SCHMIDT TELESCOPE

J. V. CLAUSEN

Copenhagen University Observatory, Brœrfelde, Denmark

(Read by R. M. West)

Abstract. Different methods, used for measuring the strength of hydrogen lines in objective-prism spectra, are discussed and the parameters are compared with the photoelectric β-index. It is found that for A5V–G1V stars β can be given with a mean error of $\pm 0^{\mathrm{m}}020$ (one spectrum).

1. Introduction

The object of this paper is to present some results from an examination of objective-prism spectra obtained with the Brœrfelde Schmidt-telescope. The program was carried out in order to get an impression of to what extent these spectra can be used for quantitative spectral classification. Different methods, used for measuring the strength of hydrogen lines in spectra of A5V–G1V stars, are discussed and the parameters are compared with the photoelectric β-index.

2. The Telescope

The 50/77/150-cm Brœrfelde Schmidt-telescope was constructed with the principal purpose to obtain objective-prism spectra for accurate spectral classification. The instrument is equipped with a 15° objective-prism made of F2 glass, giving a dispersion of 102 Å mm^{-1} at Hγ, and the photographic magnitude-limit is about 9^{m}5. The diameter of the field covered by the telescope is 5°3.

The relatively short focal length ensures that, under normal conditions, i.e. seeing better than 3″, the resolution of the spectra is plate-limited.

Generally, spectra obtained so far have indeed turned out to be of good quality.

3. The Investigation

Spectra obtained on 21 Kodak IIa-D plates covering 4 overlapping fields in the Hyades have been examined. The plates were taken on three different nights. Exposure times of 150 (telescope diaphragmed to 25%-opening), 225 and 600 s were used in order to get well exposed spectra over the B magnitude-range 5^{m}–9^{m}. The spectra were widened to 0.30 mm.

159 spectra of 43 Hyades main-sequence stars covering the MK-type range A5V–G1V were selected for the investigation.

The spectra were recorded on the digitized microphotometer at the Copenhagen University Observatory, described by West (1972). The slit-width was 25 μ corresponding to 2.5 Å at Hγ and the distance between the transmission-readings punched on paper tape was 10 μ. The plate-speed was 200 μ s^{-1}. The dark current and fog transmission were read before and after each reading of a spectrum.

The photographic calibration was performed by using existing *UBV* photometry (Johnson and Knuckles, 1955), *uvby* photometry (Crawford and Perry, 1966) and spectrophotometry (Oke and Conti, 1966) for Hyades-stars. The calibration curves could be well represented by the Baker-formula log $I = B_1 \cdot \log(1/T - 1) + B_2$, where I is the intensity, T the transmission, B_1 and B_2 the constants to be determined. The variation of B_1 with wavelength was taken into account and calibration curves were established separately for each plate.

Even for moderately high-dispersion objective-prism spectra the only parameter normally used for measuring line-strength is the line-depth. The main reason for this is that the line-depth is easily derived from the spectral registrations, whereas it used to be very time-consuming to derive other parameters which might be more accurately reproducable, at least for the stronger lines. This limitation is easily removed to-day when the output from the microphotometer can be fed directly into an electronic computer as is done in the present case.

Three different parameters describing the strength of each of the lines Hβ, Hγ and Hδ were computed.

(1) 'line-depth' d in percentage of continuum intensity. Quasi-continua varying linearly with wavelength were used.

(2) 'equivalent width' w. Quasi-continua varying linearly with wavelength were used together with fixed line-limit wavelengths.

(3) $I(n, b)$-index defined by Furenlid (1971). n and b are the half-widths of the narrow and broad 'mathematical filters' centered on the line.

Several line-limit wavelengths were tried in computing w, as also different sets of n and b were tried in computing I. Lines with bottom-transmission outside the range covered by the photographic calibration were rejected.

The results from the different plates were compared. No systematic differences appeared to be present.

The mean value of the parameters are compared with the photoelectric β-index (Crawford and Perry, 1966). The relations between the parameters and β were generally found to be linear for A5V–G1V stars, see Figure 1; the relations between β and the I-indices for Hγ and Hδ are, however, slightly curved for $\beta \lesssim 2.660$ corresponding to F5V. The reason for this is certainly the greater influence on I of lines near to Hγ and Hδ.

From Table I it is seen that, for the Balmer lines (i.e. relatively strong lines), the I-index is superior to the two other parameters for measuring line-strength, especially when comparing with the line-depth.

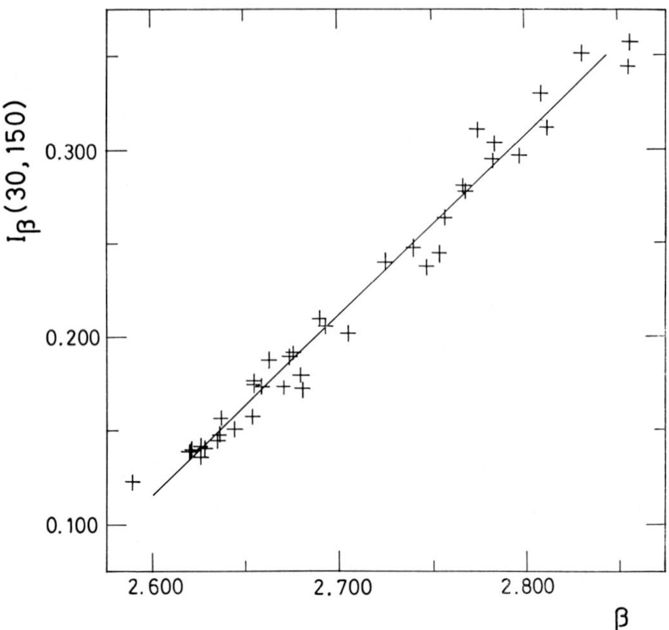

Fig. 1. β vs I_β (30, 150) relation for A5V–G1V Hyades stars.

It is seen that for A5V–G1V stars β can be given with a mean error of $\pm 0\overset{m}{.}020$ (one spectrum). For A5V–F5V stars, Hβ, Hγ and Hδ can be used with equal success. For F5V–G1V stars Hβ is preferable, the reason for this was already mentioned. The Hβ-region is, unfortunately, not well recorded by the widely used IIa-O emulsion.

TABLE I

Calculated mean errors in β (one spectrum)

Parameter	Internal m.e.	External m.e.
d_β	$\pm 0\overset{m}{.}03$	$\pm 0\overset{m}{.}05$
w_β	0.02	0.03
I_β (30, 150)	0.009	0.020
d_γ	0.04	0.06
w_γ	0.02	0.04
I_γ (30, 120)	0.010	0.020
		($\beta \geqslant 2.660$)
I_γ (10, 30)	0.015	0.033
		($\beta \geqslant 2.660$)
d_δ	0.04	0.06
w_δ	0.02	0.04
I_δ (30, 105)	0.012	0.019
		($\beta \geqslant 2.660$)
I_δ (10, 30)	0.015	0.028
		($\beta \geqslant 2.660$)

This preliminary investigation was restricted to A5V–G1V stars. We believe, however, that results of similar accuracy can be obtained for stars of all luminosity classes. Such an extension of the program is planned and will include also early-type objects. It should be recalled that for early-type stars, β measures the luminosity, and for later A stars and for F stars β is an indicator of spectral type.

The main purpose of our next program will be to establish schemes for quantitative 2- and 3-dimensional classification based on systematic examinations of the total line-information contained in objective-prism spectra of the present dispersion. The preliminary results discussed here indicate that the strength of the hydrogen lines measured by I-indices might be a valuable contribution to these schemes.

References

Crawford, D. L. and Perry, C. L.: 1966, *Astron. J.* **71**, No. 3.
Furenlid, I.: 1971, *Astron. Astrophys.* **10**, 321.
Johnson, H. L. and Knuckles, C. F.: 1955, *Astrophys. J.* **122**, 209.
Oke, J. B. and Conti, P. S.: 1966, *Astrophys. J.* **143**, 134.
West, R. M.: 1972, *Bull. Abastumani Astrophys. Obs.* **43**, 109.

GENERAL REMARKS ON QUANTITATIVE SPECTRAL CLASSIFICATION

B. STRÖMGREN

Copenhagen University Observatory, Denmark

At the *IAU Symp.* **24** held in Saltsjöbaden in 1964 I had an opportunity to discuss in a general way the procedures for stellar classification on the basis of narrow-band photometry. During the seven years that have elapsed there have been important contributions in this field by a number of astronomers. In this introductory talk I shall discuss, first, questions pertaining to classification systems and the various choices that define them. I also wish to deal, briefly, with problems referring to tests of classification systems and to the calibration of classification indices, as well as with some questions regarding the goals of quantitative spectral classification.

Consider the choice of indices that define a quantitative classification system, i.e. the choice of location and width of the wavelength bands to be measured. The choices have generally been made on the basis of experience previously gained through the development of methods for spectral classification through visual inspection, but they have been guided also by results of numerous tests of particular indices obtained through photometric investigations over several decades. The choice has to be made on the basis of a decision regarding the minimum band width that is practicable with a view to light economy; and it will depend on whether the goal is a classification system applicable for all, or nearly all, types of stars or to stars in a selected region of the Hertzsprung-Russell diagram. Finally the desired accuracy characterized by the degree of homogeneity of the star groups resulting from the classification is relevant.

When the bands that define the indices of a classification have been selected, the next choice pertains to the method of isolation of the bands for intensity measurements to be made, generally, by photoelectric techniques. Interference filters have been widely used for the purpose, mostly in such a way that the band intensities in question are measured consecutively. Use of a spectrograph for the isolation of the bands has obvious advantages, as demonstrated particularly by work in Cambridge, England. When this method has been chosen it is advantageous to avoid the use of an entrance slit that cuts into the stellar image because of the well-known adverse influence on photometric accuracy. This means that the angular dispersion of the spectrograph has to be relatively high.

As an example, let me refer to a grating spectrograph-photometer for photoelectric four-colour photometry in the *uvby* system that has been in use for some years at Kitt Peak National Observatory on a 36-inch reflector. It was designed for the purpose of obtaining simultaneous observations in the four colours. The angular dispersion is sufficient for excellent definition of the 200 Å bands in question without the use of an entrance slit, the proper adjustment of the spectrograph optics being checked every observing night through scanning across the Hδ line of a suitable star (e.g. a bright

A star) with a narrow slit that is interchangeable with the slot defining the v-band. The focal lengths of collimator and cameras (separate for the four bands) were chosen in such a way that the photomultipliers could be placed side by side in the instrument box.

A more compact spectrograph-photometer of this type has been constructed for use with the Danish 50-cm reflector on La Silla (ESO), Chile. Here the spectrograph serves only the purpose of separating the bands so that the four photomultipliers can again be placed side by side while the wavelength bands are defined through standard *ubvy* filters placed directly in front of the photomultipliers. Thus, to obtain compactness there was some sacrifice with regard to light economy.

Consider next the choice of photometric technique. Photographic photometry has been used successfully in quantitative spectral classification by Bertil Lindblad and his associates, and by Chalonge and his collaborators. However, the advantage of photoelectric photometry in terms of accuracy and time economy are such that this technique has been the preferred one in a number of classification systems developed for observations of one star at a time. For quantitative classification of faint stars in fields the technique of electronic-camera photometry holds great promise, and applications in the nearer future can be expected.

The specification of the sensitivity functions $\varphi(\lambda)$ for the bands that define a quantitative classification system is important. For narrow and intermediate-width bands $\varphi(\lambda)$ depends largely on filter transmission function or exit-slot location, although the transmission curves for the rest of the optics as well as the photomultiplier sensitivity function are of some influence. In actual practice the classification system is often defined through high-precision values of its indices for a sufficiently large number of standard stars, while the relevant sensitivity functions $\varphi(\lambda)$ are only known to some good approximation. Then, when it is desired to duplicate the system with another instrument the sensitivity functions are chosen so that they are close to those defining the original system, in which case accurate reduction to the system defined by the indices for the standard stars can be carried out using simple linear transformations (for an example cf. Crawford and Barnes, 1970).

A principal aim of quantitative spectral classification is subdivision of the stars into sub-groups, homogeneous within specified tolerances. The test for homogeneity may be made through the addition of further indices or, on a broader basis, through high-dispersion measures of several representative stars from each sub-group.

Experience obtained through the last decade has shown that two-dimensional classification is adequate for many purposes for B and A stars, provided that peculiar stars, and for certain applications rapid rotators, can be segregated either through parallel visual inspection of spectra or through the use of a suitable additional index. However, for stars of class F5 and later three-dimensional classification is necessary because the range of variation of the metal-content parameter is much wider. On the other hand the role of the rotation parameter is much reduced here, and peculiar stars are less frequent. In all classification schemes aiming at high accuracy the role of composite spectra must be considered, at least statistically.

In many applications of quantitative spectral classification the aim goes beyond division into homogeneous sub-groups in that it is desired to derive quantities such as effective temperature T_e, absolute visual magnitude M_v, and relative metal content directly from the indices of the classification systems. This means that the indices have to be calibrated.

I cannot here go into details with regard to the calibration of classification indices yielding effective temperature. Let me refer, however, to the progress that has been made possible through the determination of stellar angular diameters using intensity interferometry (cf. Hanbury Brown, 1968) and as a result of the feasibility of intensity measures in the rocket-ultraviolet. It should be mentioned in this connection that the number of stars with directly determined T_e-values of good accuracy is still so limited that accurate calibration taking into account variations of temperature-scale with atmospheric gravity g and relative metal content [Fe/H] can only be carried out using results of model-atmosphere calculations.

With regard to calibration of indices giving the absolute magnitude M_v the procedures are still based on the use of trigonometric-parallax stars as well as cluster- and association-stars. The necessity of including the metal-content parameter in these discussions has become clearer.

With regard to calibration of classification indices in terms of relative metal content there has been considerable progress during the last few years. As an example, let me refer to the calibration of the metal-line index m_1 of photoelectric $uvby$ photometry. The previous calibration (cf. Strömgren, 1966; also in this connection Wallerstein, 1962) has been considerably improved through recent investigations by Nissen (1970a, b, 1972). Values of [Fe/H] were determined by Nissen to high accuracy (± 0.06 me) for a number of main-sequence F and G stars through quantitative spectral analysis based on photoelectric measures of intensities in narrow well-defined bands containing only faint lines (the strength of which is not affected by atmospheric microturbulence). It is clear from Nissen's investigation that the metal-line index m_1 when calibrated yields [Fe/H]-values of very good accuracy for the great majority of main-sequence F1–G2 stars, with only a few per cent of the stars possibly deviating due to peculiarities. In particular, there is no doubt that stars of intermediate composition (metaldeficient by a factor of, say, 4 or more relative to the Hyades stars) can be segregated quite reliably with the help of the index m_1.

Calibration in terms of basic atmospheric parameters T_e, g and Z (or [Fe/H]) has proved possible using accurate model atmospheres (cf. Gingerich, 1969; Bell, 1971). In future investigations the inclusion of two parameters characterizing stellar rotation will presumably become more common.

The combination of model-atmosphere calculations and stellar-interior calculations of evolutionary sequences can yield a direct calibration of classification indices in terms of the basic stellar parameters mass age and chemical composition (for an example cf. Strömgren, 1970).

Let me finally refer to the application of quantitative spectral classification to the determination of very accurate colour excesses. Calibration based on nearby unreddened stars makes it possible to determine intrinsic colour indices from classification indices that are insensitive to reddening, and comparison with measured colour indices then gives the colour excess (cf. e.g. Crawford and Perry, 1966; Strömgren, 1972).

Procedures for quantitative spectral classification vary according to the principal goal in their application. Schemes for very accurate classification of brighter stars will differ from those aimed at classification of very faint stars. Certain schemes are tailored for use in galactic research whereas others are meant particularly for investigation of the physics of individual stars.

In this section of the Symposium there will be a number of papers giving results obtained with different systems (systems utilizing the Cambridge indices, developments of the Borgman-Walraven system, the $uvby + H\beta$ system, the Geneva system, the Brorfelde system, the DDO system, the Straižys system, and others) and some of the problems briefly referred to in these introductory remarks will be discussed in detail.

References

Bell, R. A.: 1971, *Monthly Notices Roy. Astron. Soc.* **154**, 343.
Brown, R. Hanbury: 1968, *Ann. Rev. Astron. Astrophys.* **6**, 13.
Crawford, D. L. and Barnes, J. V.: 1970, *Astron. J.* **75**, 978.
Crawford, D. L. and Perry, C. L.: 1966, *Astron. J.* **71**, 206.
Gingerich, O. (ed.): 1969, *Proc. Third Harvard-Smithsonian Conf. Stellar Atmospheres, Part III*, MIT Press, Cambridge, Mass. and London, England.
Nissen, P. E.: 1970a, *Astron. Astrophys.* **6**, 138.
Nissen, P. E.: 1970b, *Astron. Astrophys.* **8**, 476.
Nissen, P.E.: 1972, *Astron. Astrophys.* **19**, 261.
Strömgren, B.: 1966, *Ann. Rev. Astron. Astrophys.* **4**, 433.
Strömgren, B.: 1970, *Mitt. Astron. Ges.*, Nr. 27, 15.
Strömgren, B.: 1972, *Quart. J. Roy. Astron. Soc.* **13**, 153.
Wallerstein, G.: 1962, *Astrophys. J. Suppl.* **6**, 407.

APPLICATIONS OF THE $UB_1B_2V_1G$ PHOTOMETRIC SYSTEM

M. GOLAY

Observatoire de Genève, Genève, Switzerland

Abstract. This photometric system which was described for the first time in 1963 by Golay has been in use at the Geneva Observatory since 1959. We are involved in a long term observational programme having as objective the study of cluster stars, peculiar stars and interstellar extinction. This research aims at the determination of physical parameters such as effective temperature, gravity, abundance, or those of spectral type, absolute magnitude, [Fe/H]. To permit such an analysis the photometric effects of rotation, of multiplicity and of interstellar extinction were given particular attention. Attempting to attain the above mentioned purposes, we have had to give great care to the precision of the measurements during their acquisition as also during their reduction. Finally, in view of allowing a valid comparison between observations and the theory of stellar atmospheres, particular care had to be given to the maintenance of the pass-bands and to the knowledge of their form.

We intend here to give the main bibliographical references relative to the articles which present the general properties of the system, its applications and the catalogues.

1. Pass-Bands and Catalogue

The pass-bands adopted in 1959 have been analysed by Rufener and Maeder (1971a) by means of an original method based on the observations in the 7 colours $UBVB_1B_2$ V_1G of stars whose energy distribution as a function of wavelength is particularly well known. The filters U, B and V are similar to those of the usual U, B, V system. On the other hand, the filters B_1, B_2, V_1, G have pass-bands width approximately equal to half those of filters B and V. The mean wavelengths of the system are

	U	B	V	B_1	B_2	V_1	G
λ_0 (Å)	3456	4245	5500	4024	4480	5405	5805

The response curves are given with a step of 50 Å.

About 1500 stars are published in the last edition of the catalogue prepared by Rufener (1971), and more than 2500 will be published in the next edition which is now being prepared. This catalogue will also contain 250 stars of the southern sky. The latest edition contains a table giving the standard deviations as a function of the weight published for each series of magnitude measurements made in the 7 colours. The most frequent weight is $P = 3$, which gives the following standard deviations (in thousandths of a magnitude):

σU	σB	σV	σB_1	σB_2	σV_1	σG
6.1	3.2	3.9	3.2	3.3	3.7	4.4

2. System of Apparent $[V]$ Magnitudes

The next edition of the photometric catalogue will also contain the apparent $[V]$ magnitudes. We use square brackets for the $[U]$, $[B]$ and $[V]$ magnitudes or $[U-B]$, $[B-V]$ indices obtained with the pass-bands described above so as to distinguish them from Johnson's U, B, V, $(U-B)$, $(B-V)$ system. Actually, the $[V]$ magnitudes are very similar to the V magnitudes. Let us point out that the two systems V and $[V]$ are independent and that our $[V]$ measurements do not result through the application of a transformation formula. The method used consists in treating differential magnitude measurements by the method of least squares; the resulting sequence is then reduced to the V standard scale by means of one point only ($[V]=V=5.450$ for the star HD 23288). The method and its application to the Praesepe and Pleiades clusters have been published by Rufener and Maeder (1971b). The standard deviation $\sigma_{[V]}$ for these two clusters reaches 0.006.

3. General Properties of the $UB_1B_2V_1G$ System

Here, we only mention a few of the publications devoted to the examination of our photometric system's properties. Let us begin by pointing out that the published catalogues give the 7 colours relative to colour B. In the applications, we use various indices and linear combinations of indices. These are:

$$B_2 - V_1$$
$$[d] = (U - B_1) - 1{,}430\,(B_1 - B_2) + \varepsilon d$$
$$[\varDelta] = (U - B_2) - 0{,}832\,(B_2 - G) + \varepsilon\varDelta$$
$$[g] = (B_1 - B_2) - 1{,}357\,(V_1 - G) + \varepsilon g$$
$$[m_2] = (B_1 - B_2) - 0{,}69\,(B_2 - V_1) + \varepsilon m.$$

The first 4, $[B_2-V_1]$, $[d]$, $[\varDelta]$, $[g]$ are studied particularly by Golay (1972). The coefficients are chosen to obtain parameters which are weakly dependant on interstellar extinction. The corrective terms ε are there to eliminate the residual effects of interstellar extinction by relying on an approximate knowledge of the star's spectral type, of its luminosity class and of the interstellar extinction law. The parameter $[m_2]$ defined by Hauck (1968), together with $[d]$ and $[B_2 - V_1]$ allows a three dimensional representation of A0 to G5 stars which gives account of effective temperature, luminosity and blocking by metallic lines. The parameters $[m_2]$ and $[g]$ measure blocking by lines situated in the interval 3800–4800 Å. The parameters $[d]$ and $[\varDelta]$ measure the Balmer discontinuity, $[d]$ in a manner weakly dependant on blocking in the above mentioned interval, $[\varDelta]$, on the other hand, is sensitive to the importance of this blocking. The relation between $[d]$ and $[\varDelta]$ thus allows us to obtain a stellar classification in spectral type and luminosity class as well as some indications concerning certain spectral peculiarities. When we add the intrinsic colour indice $[B_2-V_1]_0$ (which is practically independant of blocking between A0 and G5) to this $[d]$ vs $[\varDelta]$

diagram, a three dimensional representation of stars becomes possible. Figure 1 illustrates the properties of the $[d]$ vs $[\Delta]$ diagram. A great number of other diagrams can be constructed, for example $[d]$, $[g]$, $[\Delta]$, $[m_2]$ vs $[B_2 - V_1]$. The calibrations obtained for these diagrams in terms of effective temperature, [Fe/H], Mv, rotational velocity, have been published by Golay (1969). The photometric aspects of Ap, Am, AVI peculiar stars have been studied by Golay (1972) and by Hauck (1971a). Hauck (1971b) has shown that the positions of δ Scuti stars in a photometric diagram agree better with those of luminosity class IV rather than with those of Am stars. Figure 1 shows that supergiants hotter than A2 occupy a long and narrow band in the $[d]/[\Delta]$ diagram. The position of a star along this band depends on the importance of the Balmer discontinuity. On the other hand ,the $[d]$, $[\Delta]$ diagram is ambiguous for supergiants cooler than A3, wheras the $[d]$, $[B_2 - V_1]_0$ diagram allows a good separation of supergiants. Moreover, again according to Hauck and Nicollier (Goy, 1971b), the separation of hot dwarf stars with emission lines from supergiants appears clearly in $[d]$ vs $[\Delta]$.

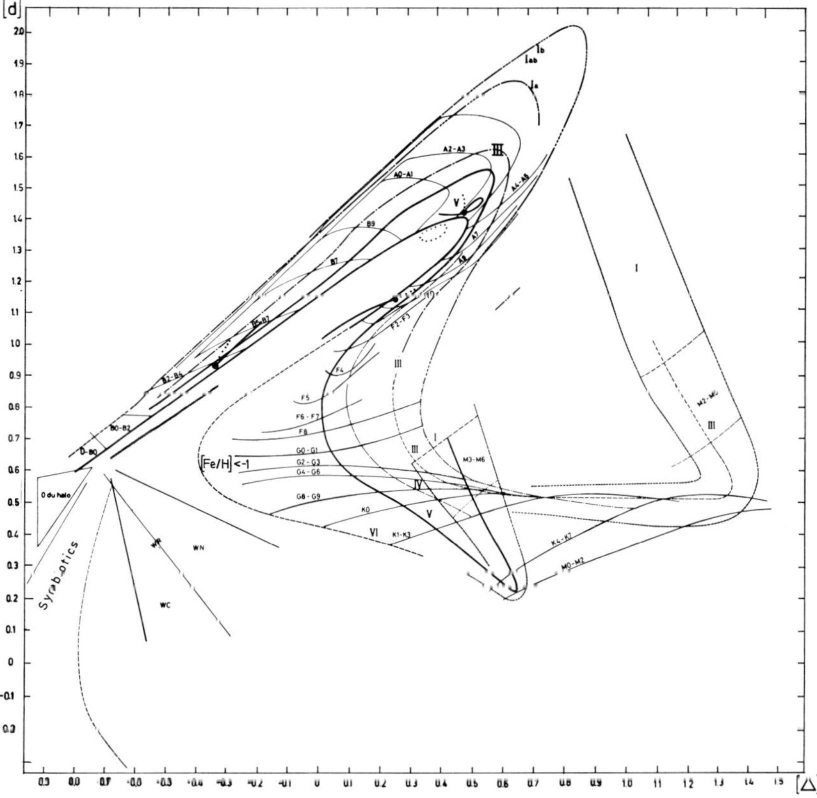

Fig. 1. $[d]$ vs $[\Delta]$ diagram with σd, $\sigma\Delta = 0$. The wavy line is the locus of rapidly rotating B0–B4 stars. The dotted loop illustrates the effect of binarity for various mass ratios. The loci of stars of equal mass rotating with velocities between 0 and breakup velocity are indicated in three cases. In each case the continuous line concerns the inclination $i = 90°$ and the dotted line the inclination $i = 0°$.

The $[d]$ vs $[\Delta]$ diagram illustrates the positions occupied by O, Wolf-Rayet, symbiotic stars. The analysis of the colours of reddened O stars and the determination of the individual interstellar extinction laws which lead to the observed values of reddening is carried out by Goy (1971a, b). The study of strongly reddened galactic clusters suggests a relationship between the density of interstellar matter and the form of the extinction law.

4. Stellar Models

In view of determining effective temperatures, gravities and abundances for F and G stars, mainly by means of information obtained in our 7 colours, Peytremann (1970) has developed stellar atmosphere models which include the opacities of metallic lines. He has computed a grid of model atmospheres with effective temperatures between 5000 K and 8500 K and gravities between $\log g = 2$ and 4.5. Various metal abundances relative to the Sun have been considered, namely $\sim 0 - 0.1 - 1$. The distribution of abundances is that of the Sun. The velocity of microturbulence was chosen equal to 2 km s^{-1}. Except for the treatment of the lines, the models are classic. The atmospheres are in radiative and convective equilibrium; the convective flux is computed with a ratio of mixing length to scale height of $l/H = 1$. These models are frequently used to analyse the various diagrams of the $UBVB_1B_2V_1G$ photometry. Other models have been used by Maeder and Peytremann to study the photometric effects of rotation. The models used are based on those of Kurucz (Harvard College Observatory) and take into account the hydrogen lines of the Balmer and Lyman series. The surface of the star is divided into several zones having each an effective temperature and a gravity: the models were established for three masses ($5M_\odot$, $2M_\odot$, $1.4\ M_\odot$) and 4 values of rotation (expressed in fractions of breakup velocity) $\omega = 0$, 0.5, 0.8, 0.9, 0.99. Six values of inclination were considered for each velocity. These models were compared with photometric observations of the effects of rotation (Golay, 1968) and were found to be in agreement with these for stars hotter than A7–F0 (except for the interval B0–B3). Beyond F0, it seems necessary to introduce non-uniformly rotating models. The photometric effects of rotation are illustrated in the $[d]$ vs $[\Delta]$ diagram of Figure 1. Maeder (1971) has particularly examined the effect of rotation on the age determination of 7 galactic clusters measured in the $UBVB_1B_2V_1G$ system. He has shown that neglecting rotational effects can lead to overestimations of age reaching 70%. In the same article, Maeder shows that the correcting of the colour indices $B_2 - V_1$ for effects due to rotation depends of the product $v \sin i$.

5. Stellar Classification

The $[d]$ vs $[\Delta]$ diagram of Figure 1 contains the loci of equal MK spectral type and of equal luminosity class. By considering the intrinsic colour index $(B_2 - V_1)_0$, we find that it is possible to extend this representation to cool stars (according to Grenon, 1972). Stars of MK class are distributed on a nondevelopable surface in the 3-dimensional representation $[d]$, $[\Delta]$, $(B_2 - V_1)_0$. According to Golay (1972), peculiar stars lie outside this surface, either above, or below (or within the knee between B9–A9). Only those of much attenuated peculiarity are close to the surface. A fourth dimension with the indice $(V_1 - G)_0$ is of great interest for the classification of cool stars. In the spectral interval O–G8, two stars having the same position in the $[d]$, $[\Delta]$, $(B_2 - V_1)$ space have also the same 'pseudo-continuum'. We call 'pseudo-continuum' the representation obtained by plotting the monochromatic magnitudes derived from the U, B, V, B_1, B_2, V_1, G, heterochromatic magnitudes as a function of the corresponding effective wavelengths. The spectral energy distribution of a star is then given by a polygonal line (the process and its application are described by Golay (1971, 1972)). The comparison of the pseudo-continua of stars reddened by interstellar matter leads to the study of the difference between extinction laws. The comparison of pseudo-continua of non reddened stars helps to locate the pass-band (or bands) coinciding with the phenomenon responsible for differing positions in the $[d]$, $[\Delta]$, $(B_2 - V_1)$ diagram. Finally, the comparison of pseudo-continua of stars of same spectral type but having different rotational velocities allows the study of photometric effects due to rotation (according to Golay (1968, 1972)). Thanks to the great number of stars measured accurately in the U, B, V, B_1, B_2, V_1, G, system, it is possible to seek out the stars for which the 6 colours relative to G do not differ by more than an arbitrary value δ. This work was first done by Golay et al. (1969) and is resumed for each new extension of the catalogue. The computer scans the whole catalogue and separates the stars into natural groups. This grouping has been carried out for 3 values of δ: $\delta = 0.01$ mag., $\delta = 0.02$ and $\delta = 0.05$. It is interesting to point out for $\delta = 0.01$, the stars of the same group (2 to 3 stars per group and over 100 possible groups) always have the same spectral type and the same luminosity class. The few discrepancies encountered disappeared when we got a better classification from the spectroscopists. The same applies to peculiarity, Am characteristics or peculiar lines. For Am stars, as an example, if we have access to a classification which distinguishes marginal Am stars from typically Am stars (Cowley et al., 1969), stars of a given group are either all marginal or all Am. This still applies when of Cowley et al. (1969) add the indices s; all the stars of the group also share this caracteristic. When we consider $\delta = 0.02$, it is clear that the groups contain a greater number of members and that the identity of spectroscopic characteristics is no longer as sharply defined as for $\delta = 0.01$. When present, the Am characteristic is the one most frequently common to all stars of the group. It sometimes coexists with an Ap type. By searching in literature, it has almost always been possible to find a peculiarity for the stars of a group which contains at least one star recognized as an Ap. It then appears that the different characteristics Am, Ap, can be distinguished from stars of MK class by characteristic deviations of the pseudo-continua of at least 0.02 mag. in one, or in several colours.

Let us go back to Figure 1. We notice that the main sequence stars between B9 and A5 are widely dispersed. According to the calibration of the $[d]$ vs $[\Delta]$ diagram as described in Golay (1972) and reproduced in Golay (1971) (calibration based on models by Mihalas with blocking by hydrogen lines), this dispersion is equivalent to a dispersion in $\log g$ of 4.2 to 3.5. Photometric effects due to rotation can also be a cause of dispersion in this rather complicated region of the $[d]$ vs $[\varDelta]$ diagram. Such effects would nevertheless explain but half of the observed dispersion. Photometric effects of binarity of a few stars would not suffice either to explain this dispersion. A-type stars recognized as being magnetic show a dispersion of the same order (but are less uniformly distributed and seem to be more concentrated on the line of highest gravities). It still remains to be proved that the observed dispersion (which also appears in Chalonge's D, λ_1 spectrophotometry) is only an effect due to gravity which is not detectable in the MK classification.

Figure 2 illustrates the position of Am and Am: (marginal Am) stars classified by Cowley *et al.* (1969). We now see, as was not the case in the earlier spectral classifications of Am stars, that marginal metallic stars are clearly separable from typically metallic stars.

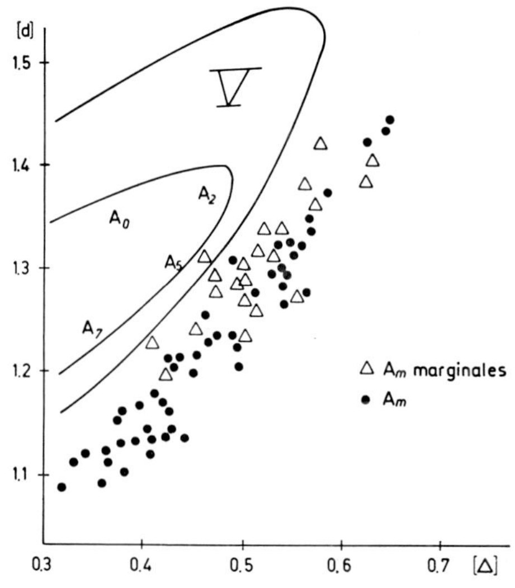

Fig. 2.   [d] vs [$\varDelta$] diagram of Am stars.

6. Variability

In 10 yr of continued observations, numerous stars used as standards have been measured a great number of times. We have shown in (Golay, 1972) that the parameter $[g]$ is an indicator of absolute magnitude, for hot stars, comparable to β of the u, v, b, y, β photometry, and that $[d]$ is an indicator of temperature. The $[g]$ vs $[d]$ diagram thus appears as an Hertzsprung-Russell diagram for hot stars. Maeder and Rufener (1972) have shown that σ, the indicator of quality used by Rufener (1971) to characterize a series of measurements in 7 colours, is often unusually large for a whole group of stellar types which are situated in the hatched region of the diagram in Figure 3. This region coincides with that where the phenomenon of mass-loss is to be expected. A similar diagram can be made for cooler stars. The parameter $[d]$ is then an indicator

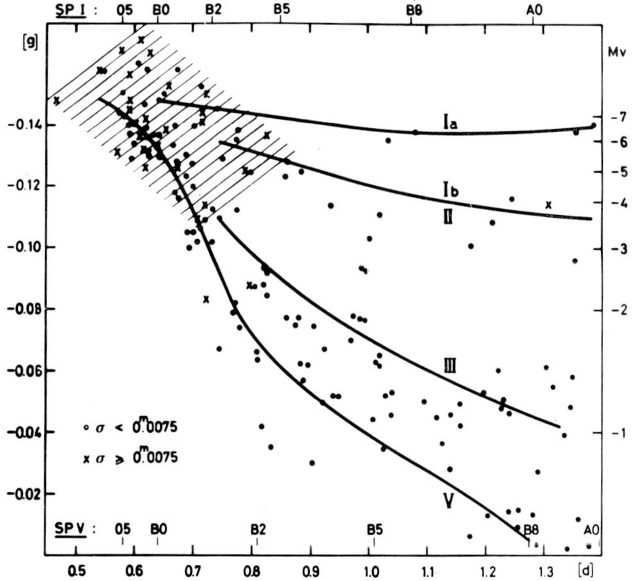

Fig. 3. [g] vs [d] diagram of hot stars. σ designates the indicator of quality of the 7 colour measurements. Figure taken from Meader and Rufener (1972).

of absolute magnitude and $(B_2 - V_1)_0$ an indicator of temperature. There also, the indicator of quality σ is distinctly more important for certain supergiants and suggests the existence of microvariability.

References

Cowley, A., Cowley, C., Jascheck, M., and Jascheck, C.: 1969, *Astron. J.* **74**, 375.
Golay, M.: 1963, *Publ. Obs. Genève* **A64**, 419.
Golay, M.: 1968, *Publ. Obs. Genève* **75**, 105.
Golay, M.: 1969, in Owen Gingerich (ed.), *Seven Color Photometry in Theory and Observation of Normal Stellar Atmospheres*, MIT Press.
Golay, M.: 1971, in L. N. Mavridis (ed.), *Structure and Evolution of the Galaxy*, D. Reidel Publishing Company, p. 34.
Golay, M.: 1972, *Vistas in Astronomy* **14**, in press.
Golay, M., Peytremann, E., and Maeder, A.: 1969, *Publ. Obs. Genève* **76**.
Goy, G.: 1971, *Publ. Obs. Genève*, **A78**.
Goy, G.: 1972, in press.
Goy, G. and Maeder, A.: 1969, *Publ. Obs. Genève* **78**.
Grenon, M.: 1972, *Obs. Genève*, in preparation.
Hauck, B.: 1968, *Publ. Obs. Genève* **75**.
Hauck, B.: 1971a, *Conferenze Dell'Osservatorio Astronomico di Milano-Merate*, Series 1, No. 12.
Hauck, B.: 1971b, *Astron. Astrophys.* **11**, 79.
Maeder, A.: 1971, *Astron. Astrophys.* **10**, 354.
Maeder, A. and Rufener, F.: 1972, *Obs. Genève*, in preparation.
Peytremann, E.: 1970, Thesis, Université de Genève.
Rufener, F.: 1971, *Astron. Astrophys. Suppl.* **3**, 181.
Rufener, F. and Maeder, A.: 1971a, *Astron. Astrophys. Suppl.* **4**, 43.
Rufener, F. and Maeder, A.: 1971b, *Publ. Obs. Genève*, **A78**.

CLASSIFICATION OF EARLY TYPE STARS ON THE DOMINION ASTROPHYSICAL OBSERVATORY PHOTOMETRIC SYSTEM

S. C. MORRIS and G. HILL

Dominion Astrophysical Observatory, Victoria, B.C., Canada

and

G. A. H. WALKER and H. I. B. THOMPSON

Department of Geophysics, University of British Columbia, Vancouver 8, B.C., Canada

Abstract. A classification method has been developed for early–type stars observed on the Dominion Astrophysical Observatory photometric system. Two reddening–independent parameters, $Q(35)$ and $Q(38)$, are used. $Q(35)$ is a measure of the Balmer discontinuity, while $Q(38)$ is a measure of the strength of the upper members of the Balmer series. A preliminary calibration of $Q(35)$ and $Q(38)$ in terms of spectral types and luminosity classes is given, and applications to several groups of stars are shown.

A four-channel photometer was put into use by the Dominion Astrophysical Observatory (DAO) in 1968 and has been used in a variety of observational programs since then. The instrument (described in detail by Walker *et al.*, 1970) has two passbands which reproduce B and V of the UBV system (Johnson *et al.*, 1966) and two ultraviolet bands, one 200 wide placed beyond the Balmer discontinuity, the other 150 wide, which includes the upper members of the Balmer series (H9–H16). We designate our passbands in terms of their mean wavelengths, quoted to the nearest hundred angstroms. Thus, 35 refers to the $\lambda\,3400$–$\lambda\,3600$ passband; $c(35$–$44)$ is the colour derived from $\lambda\,3400$–$\lambda\,3600$ and $\lambda\,3900$–$\lambda\,4900$ passbands; and similarly for $E(35$–$44)$, etc.

The instrument was designed in the hope that with one instrument we would be able to derive luminosities and temperatures for early-type stars whose light suffers interstellar reddening. In practice we can observe effectively down to 11^{m} with a 16-in. telescope and un-refrigerated photomultipliers. The multichannel nature of the instrument has important advantages in observational efficiency and in relative immunity from the effects of variable grey extinction due to clouds.

In developing a new photometric system our first task was to obtain colours and magnitudes for a set of standard stars. These have recently been published (Hill *et al.*, 1971). In this paper we present a preliminary calibration of the photometric data in terms of spectral types for early-type stars. Ultimately this calibration will be made in terms of M_v and T_e. In what follows we shall describe how the calibration was derived and show how several groups of stars compare with the basic calibration.

In Figure 1 we show a colour-colour plot $[c(35$–$44)$ vs $c(44$–$54)]$ for: (a) relatively unreddened standard stars, (b) 'Petrie-Lee' field stars (Petrie and Lee, 1966) of various degrees of reddening and (c) stars in Cepheus IV, an apparently young cluster of large and variable reddening. The ratio $E(35$–$44)/E(44$–$54)$ is derived from the slope of the

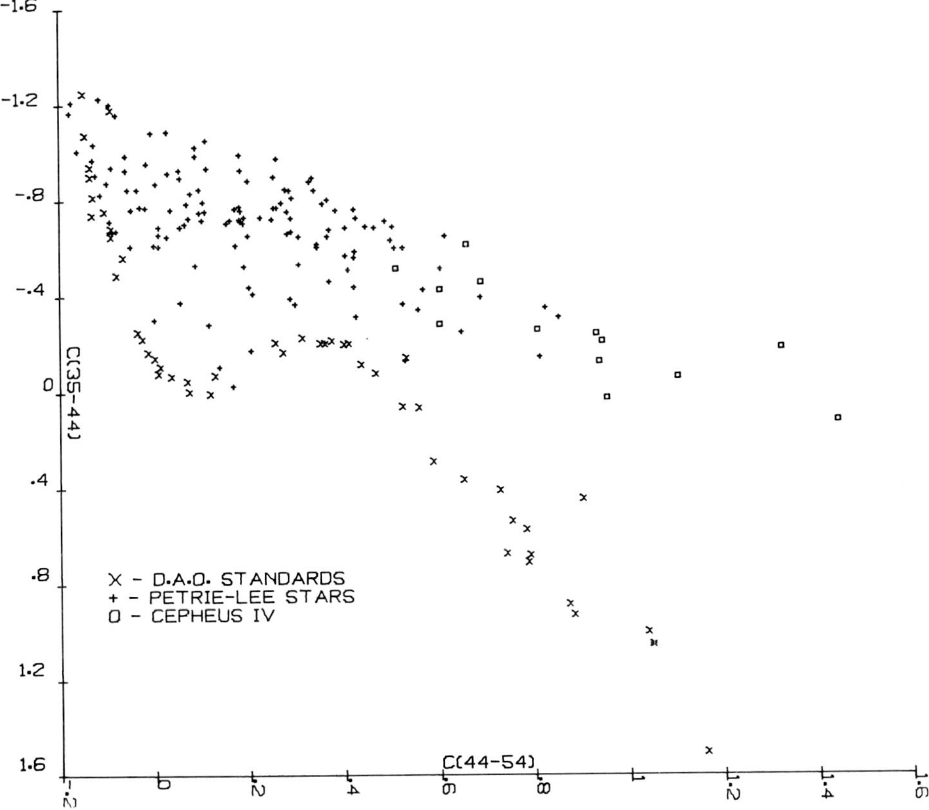

Fig. 1.   Colour — colour plot, showing the effects of interstellar reddening.

upper envelope of this plot, and has a value of 0.83; the stars in Cepheus IV were not used in this derivation. The ratio $E(38\text{–}44)/E(44\text{–}54)$ was derived in similar fashion; it is equal to 0.62.

We define two reddening-free parameters $Q(35)$ and $Q(38)$ by the relations

$$Q(35) = c(35\text{–}44) - \frac{E(35\text{–}44)}{E(44\text{–}54)} \cdot c(44\text{–}54)$$

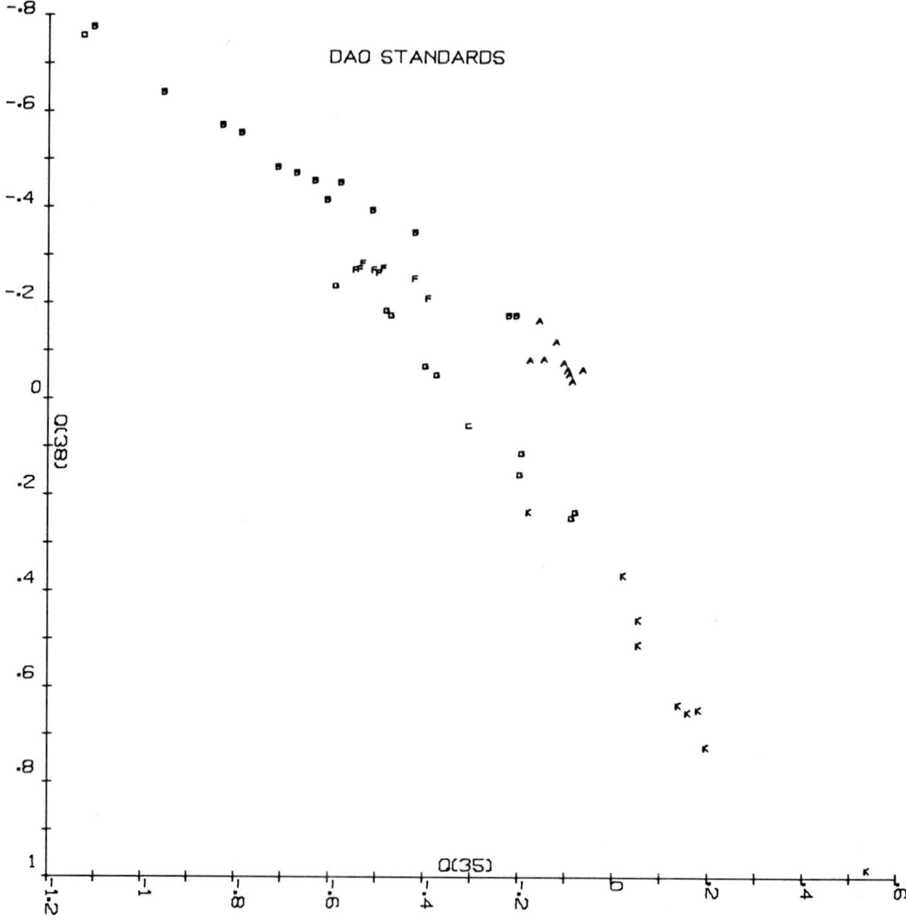

Fig. 2. $Q - Q$ plot for DAO standards. The symbol plotted indicates spectral type; a composite spectrum is denoted by c.

and

$$Q(38) = c(38\text{-}44) - \frac{E(38\text{-}44)}{E(44\text{-}54)} \cdot c(44\text{-}54)$$

We assume that the ratios of the colour-excesses are constant around the sky. In Figure 2 we show a '$Q - Q$' plot [$Q(35)$ vs $Q(38)$] for the DAO standards, for stars ranging in spectral type from late O to late K. It is clear that this classification method

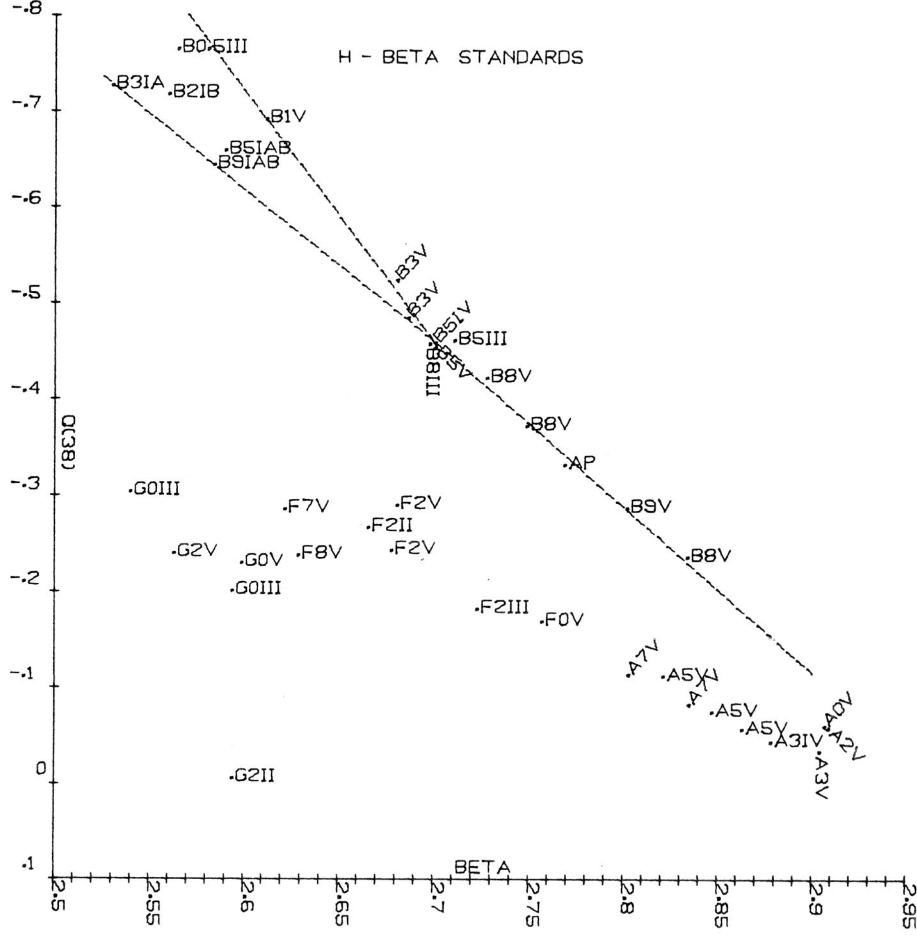

Fig. 3.   The relation between $Q(38)$ and Hβ index.

will probably not be suitable for types A and F. The relation between what we expect to be the luminosity sensitive index $Q(38)$ and Crawford's β-index (Crawford and Mander, 1966) is seen in Figure 3. For luminosity class V stars there is a linear relation from B0–B5 and another linear relation from B5–B8. The change of slope is probably caused by the inclusion of a number of He I lines, principally λ 3819, in the 38 band. This line tends to strengthen as the H lines weaken in the earlier main-sequence stars. The supergiants follow a different line. In these stars, λ 3819 varies much less

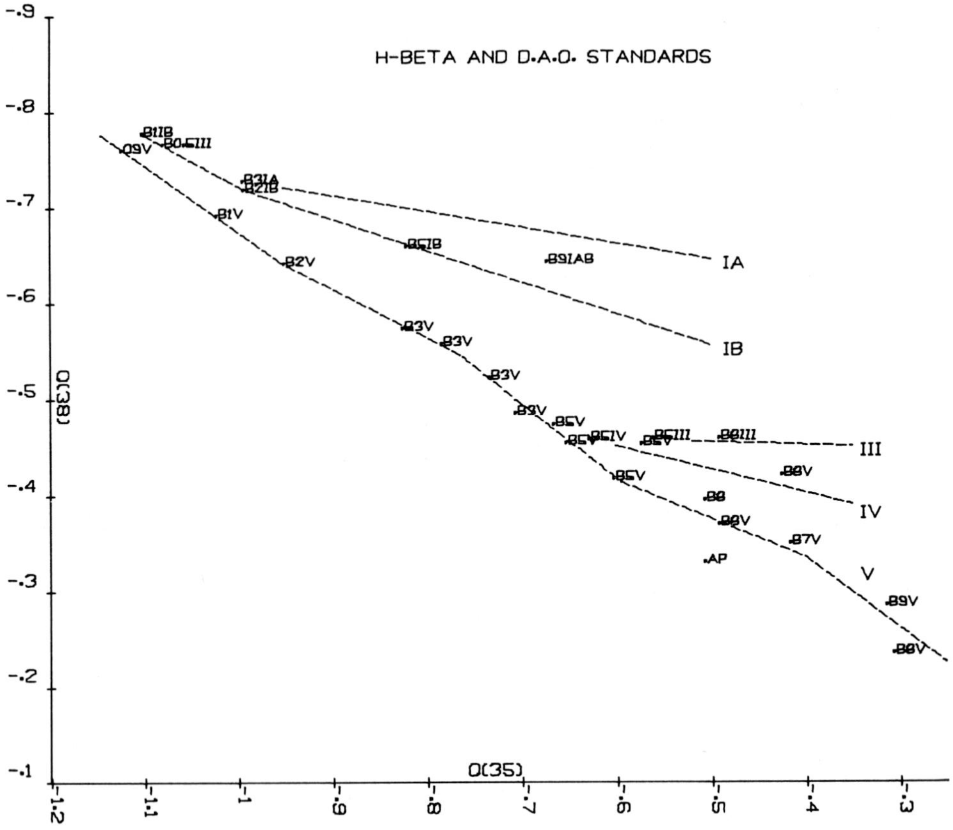

Fig. 4. $Q - Q$ plot showing spectral types and luminosity classes.

rapidly with temperature and disappears from the spectrum somewhere between O5 and O8.5. If one plots measures of Be stars on this diagram, most of them fall on a line parallel to, but about 0.2 below, the extension of the class V line. Figure 4 is a $Q - Q$ plot for the D.A.O. standards and our observations of Crawford's Hβ standards, both of which groups have been exceptionally accurately observed. The main sequence line is very well defined although a little irregular; the loci of the other classes is indicated, but is based on as yet inadequate data. Nevertheless, it seems clear that a

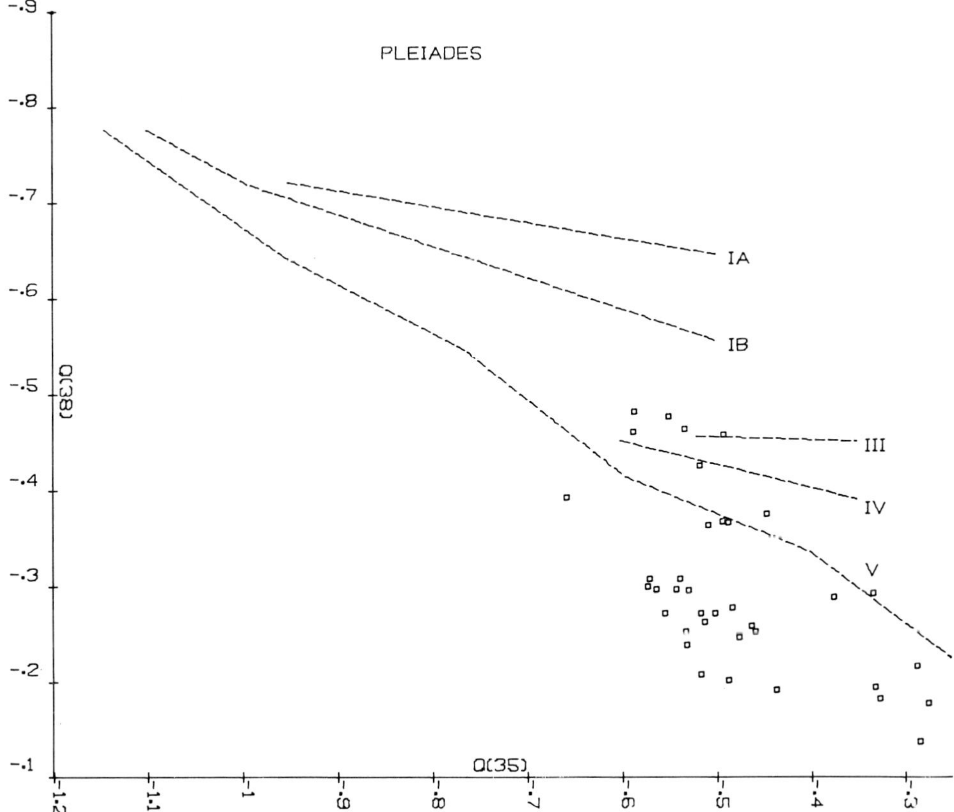

Fig. 5. $Q - Q$ plot for the Pleiades. In this and succeeding figures the luminosity class lines are those derived in Figure 4.

good calibration will be possible with the additional data that we have in hand. The wide separation of the luminosity classes in the $Q - Q$ plane is noteworthy and indicates that this method should yield good classification accuracy.

We now compare observations of various clusters with the preliminary calibration. The cluster observations are displayed on a background derived from Figure 4. Figures 5–7 show results for the Pleiades, NGC 7243 and NGC 2169. The similarity found here between the Pleiades and NGC 7243 agrees with the results of Hill and

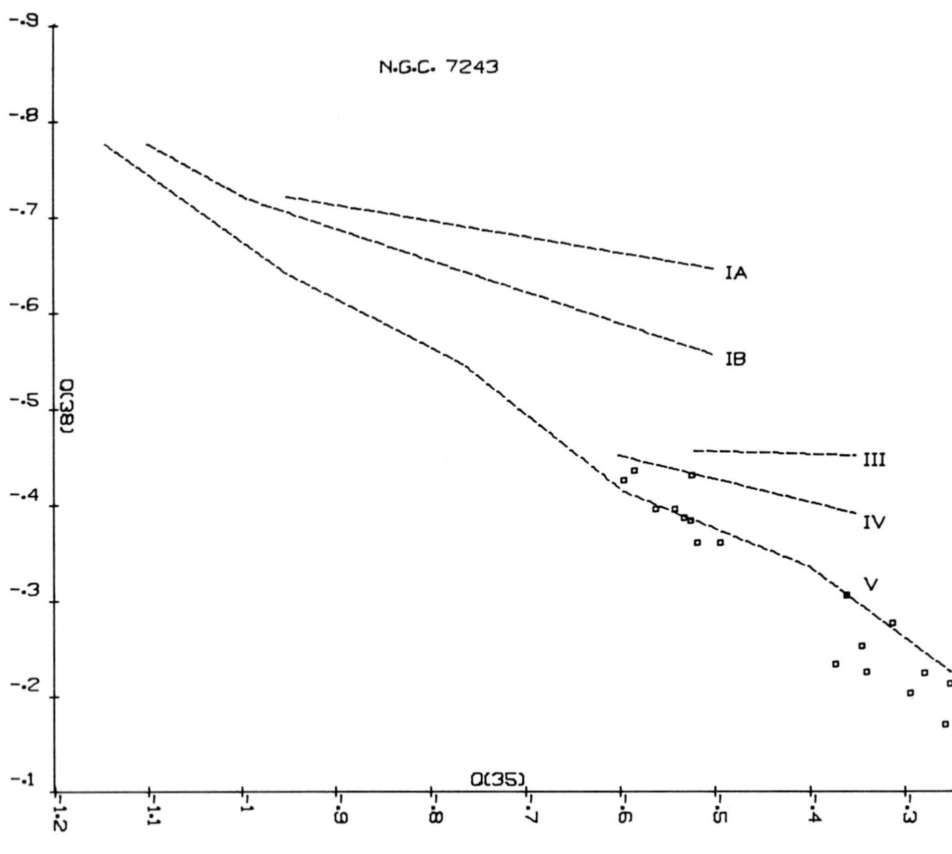

Fig. 6. $Q - Q$ plot for NGC 7243.

Barnes (1971); in particular, very similar ages are indicated by the positions of the turnoff points. NGC 2169 is seen to be a young, relatively un-evolved cluster. In h and χ Persei (Figure 8) the transition from class V to class Ia–Ib is notable. A number of deviant points is apparent. Those for which we have spectral types are emission-line objects, or, in one case, an A1Ia star. Cepheus IV is interesting (Figure 9) because of an undoubted systematic deviation from the class V line. Such a difference could be caused either by anomalous values for the ratios of the colour excesses in the direction

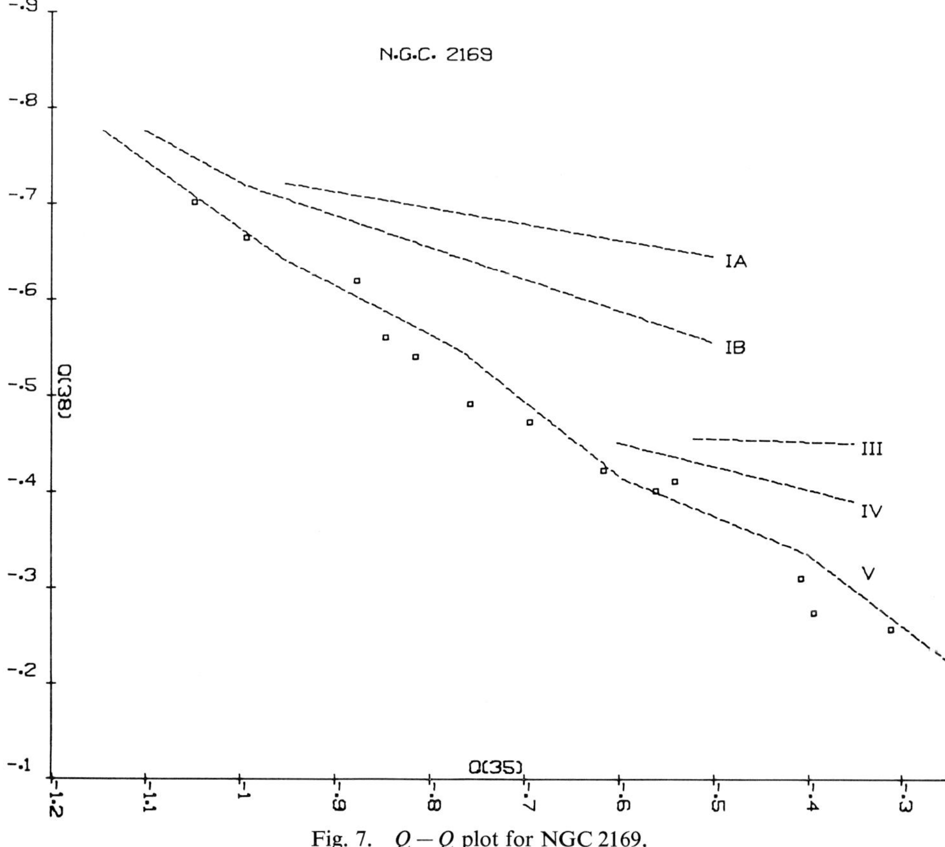

Fig. 7. $Q - Q$ plot for NGC 2169.

of Cepheus IV, or by the stars in the association really being fainter than class V. Measures of these stars plotted in Figure 1 appear to define an extension of the normal reddening line (as is also the case for the $c(38-44)$ vs $c(44-54)$ plot); this is evidence against anomalous reddening in the direction of the cluster. Walker (1965) examined the luminosity sensitive He II line, λ 4686, in the spectra of members of Cepheus IV and found evidence that the stars are subdwarfs, thus supporting the second possible explanation. From its involvement with extensive nebulosity, this

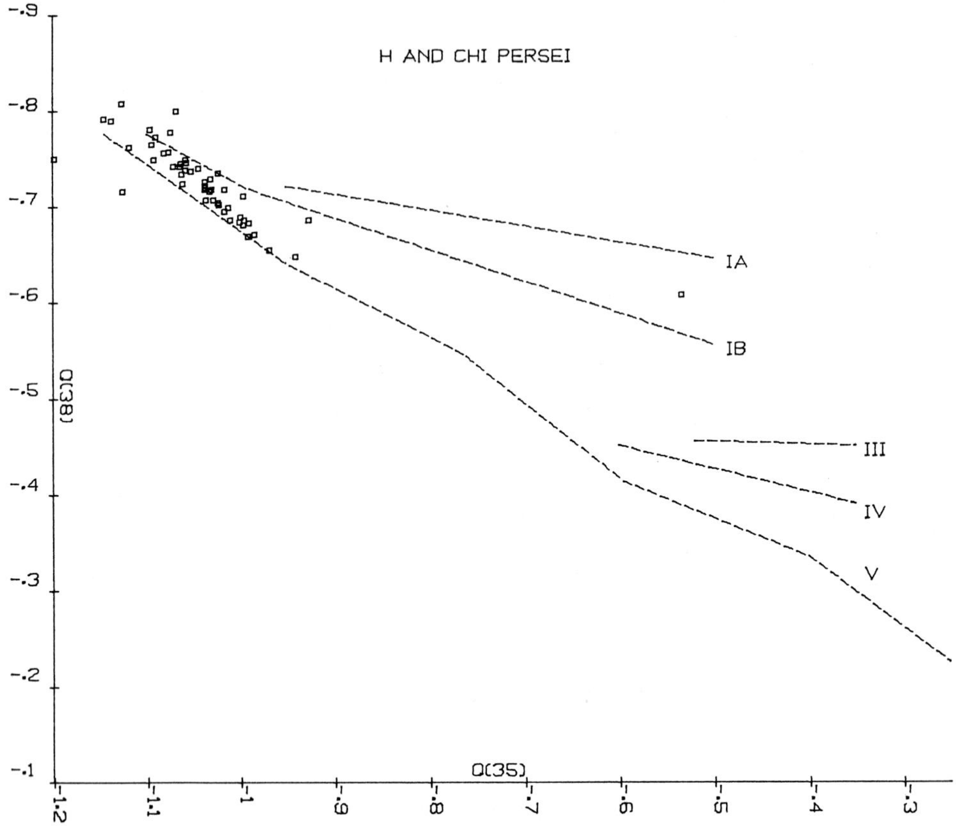

Fig. 8. $Q - Q$ plot for h and χ Persei.

group of stars appears to be very young. During its hydrogen-burning lifetime, a B star will brighten by about one magnitude. Thus the class V line in the plots should have a breadth of about one magnitude, and the irregularity of the class V line in our plot for standard stars is a result of defining it by stars with a variety of ages. The deviation of measures of stars in Cepheus IV from the line defined by measures of standards is thus probably an age effect.

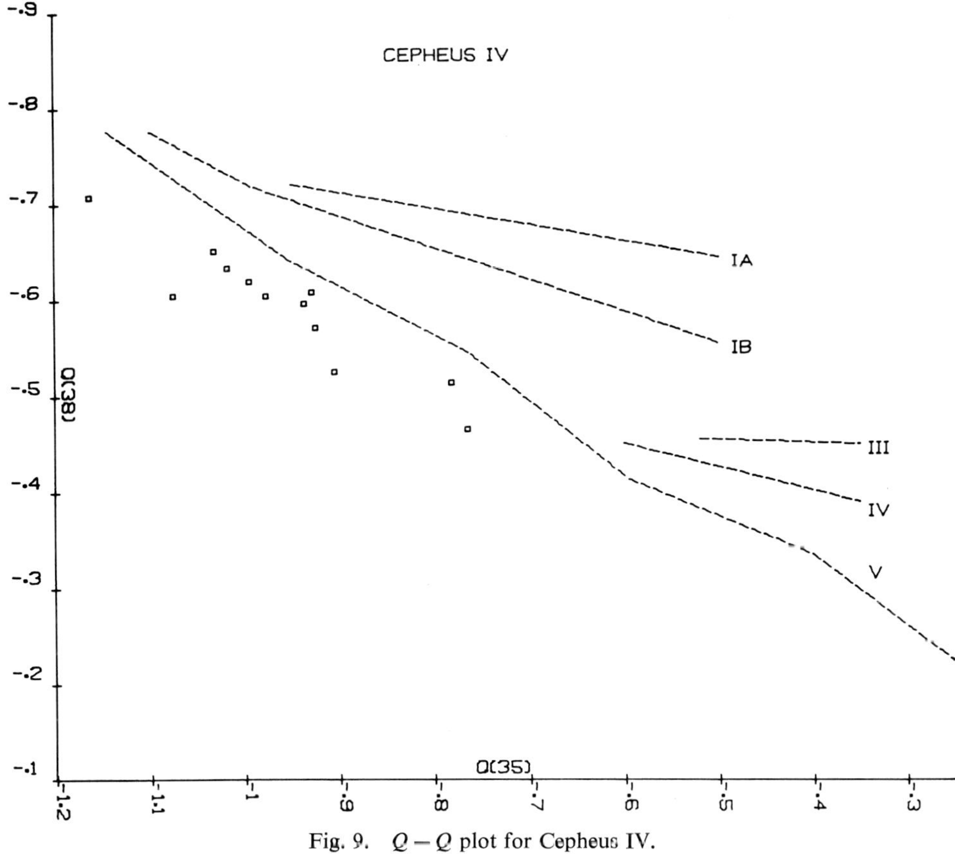

Fig. 9. $Q - Q$ plot for Cepheus IV.

References

Crawford, D. L. and Mander, J.: 1966, *Astron. J.* **71**, 114.

Hill, G. and Barnes, J. V.: 1971, *Astron. J.* **76**, 110.

Hill, G., Morris, S. C., and Walker, G. A. H.: 1971, *Astron. J.* **76**, 246.

Johnson, H. L., Mitchell, R. I., Iriarte, B., and Wisniewski, W. Z.: 1966, *Commun. Lunar Planetary Lab* **4**, 99.

Walker, G. A. H.: 1965, *Astrophys. J.* **141**, 660.

Walker, G. A. H., Andrews, D. H., Hill, G., Morris, S. C., Smyth, W. G., and White, J. R.: 1970, *Publ. Dominion Astrophys. Obs.* **13**, 415.

$C(35-38)$ is a measure of the Balmer discontinuity.

The original set of observations on the system were done by McClure and van den Bergh (1968) using the 188-cm reflector of the David Dunlap Observatory. Altogether, 210 bright stars of a wide range of spectral type, luminosity class and metal abundance, distributed throughout the northern sky were observed. A new set of standards is now being set up in the equatorial region using the facilities of the Kitt Peak National Observatory, to make the system more convenient to the southern hemisphere observer.

3. The Determination of Reddening

Young B stars are the most reliable reddening indicators. Often, however, such stars are not available near the object of interest. This is particularly the case in old stellar populations such are found at high galactic latitudes and in old open clusters. In these cases, it is much more convenient if reddening can be determined using red giant stars which are usually the brightest stars in the population.

The DDO system, combined with the $B-V$ colour index of the UBV system has been used successfully to determine reddening of red giant stars in a number of fields. This is possible because the G band break index $C(42-45)$ has about equal sensitivity to effective temperature as the wide band $B-V$ index whereas the separation in wavelength of the filter bandpasses, and therefore the effect of interstellar reddening, is only about one quarter that of $B-V$. McClure and Racine (1969) have developed a method for obtaining reddening, from DDO photometry of red giants, resulting in the following equation:

$$E(B-V) = 2.175(B-V) - 2.380C(45-48) - 1.420C(42-45) + 1.841.$$

The values of reddening computed from this equation are applicable for G and K giants in the colour range $0^m\!.80 < (B-V)_0 < 1^m\!.55$, luminosity class (LC) in the range $II < LC < IV$, and for stars of a moderate range of metal abundance. The random error of an individual reddening determination has been found to be a few hundredths of a magnitude, and therefore, with only a few stars an accurate mean value of reddening is obtained. Values of reddening determined from DDO photometry to date are given in Table II.

TABLE II

Reddening values from DDO photometry

Field	$E(B-V)$	Reference
M3	0.00	McClure and Racine (1969)
M13	0.00	McClure and Racine (1969)
M31	0.11	McClure and Racine (1969)
M33	0.03	McClure and Racine (1969)
N. Gal. Pole	0.00	McClure and Crawford (1971)
M67	0.06	Janes (1971)
NGC 2360	0.09	McClure (1971a)
NGC 3680	0.09	McClure (1971a)
NGC 2477	0.31	Hartwick, Hesser and McClure (1971)

Reddening corrections can be made to the DDO colour indices by applying the following colour excess ratios:

$$\frac{EC(41\text{–}42)}{E(B-V)} = 0.066; \qquad \frac{EC(42\text{–}45)}{E(B-V)} = 0.234; \qquad \frac{EC(45\text{–}48)}{E(B-V)} = 0.354.$$

4. Spectral Classification from DDO Indices

The $C(45\text{–}48)$ index, when plotted vs the G band break index $C(42\text{–}45)$, has been found empirically to show good surface gravity dependence for late-type stars. This is illustrated in Figure 1 where luminosity class Ib, III and V field stars are plotted. The spectral classification of a late-type star observed on the DDO system can be determined from its position in Figure 2 where the indices have been calibrated in terms of spectral type and luminosity class. Since the blanketing lines in this figure are almost parallel to the intrinsic sequences, the classification is insensitive to differences in metal abundance.

Calibration of the position in Figure 1 in terms of effective temperature and surface gravity is more difficult due to the uncertainties in model atmosphere and binary star

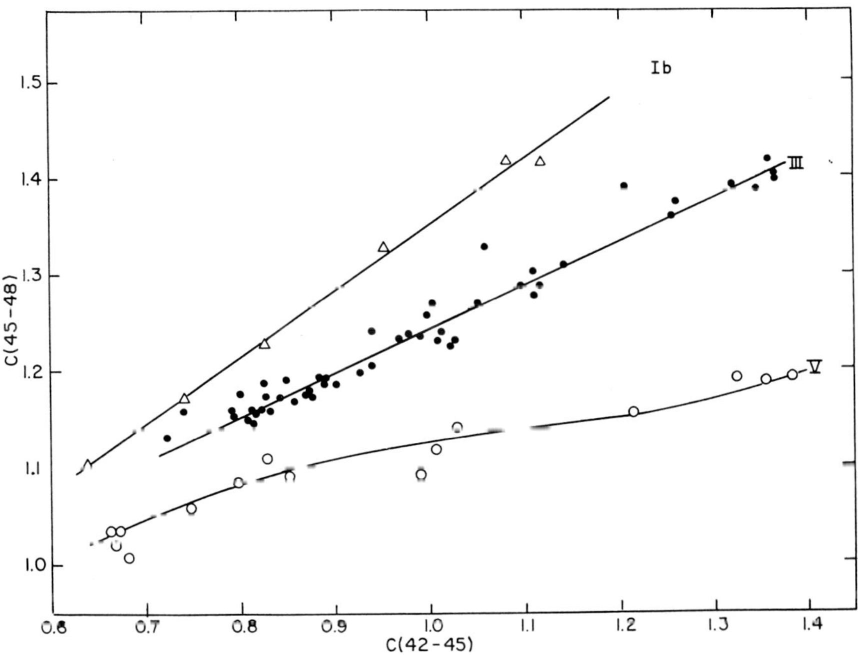

Fig. 1. A plot of the surface gravity dependent colour index $C(45\text{–}48)$ vs the temperature dependent colour index $C(42\text{–}45)$ for a random sample of late-type luminosity class III stars (dots), class V stars (open circles), and class Ib stars (triangles). The class III stars are all brighter than $V = 4\overset{m}{.}0$, and therefore, should have no reddening. Reddening corrections have been applied to the supergiants, however.

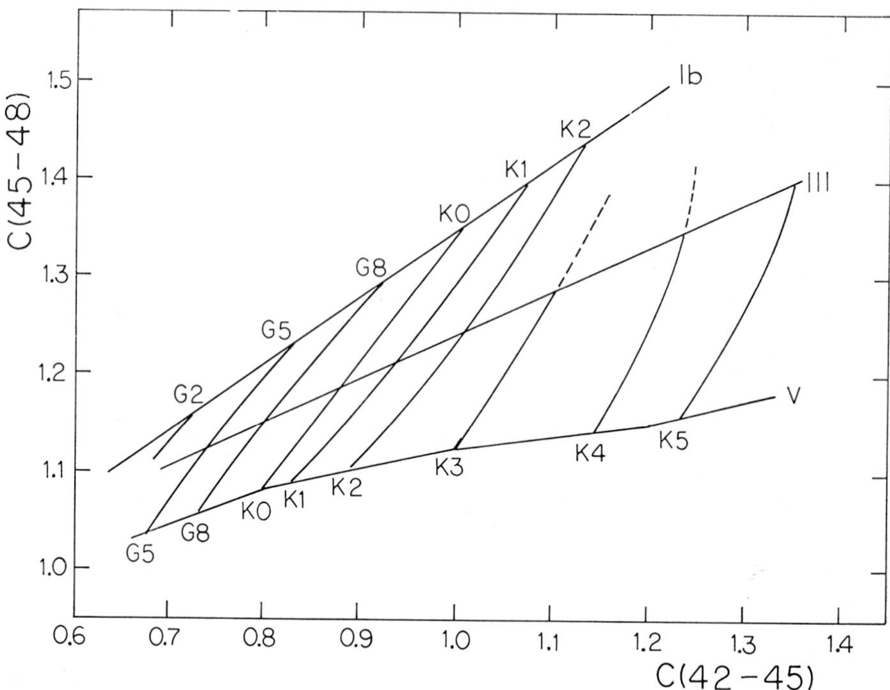

Fig. 2. The calibration of the indices of Figure 1 in terms of spectral type and luminosity class. The blanketing lines are almost parallel to the intrinsic sequences in this diagram.

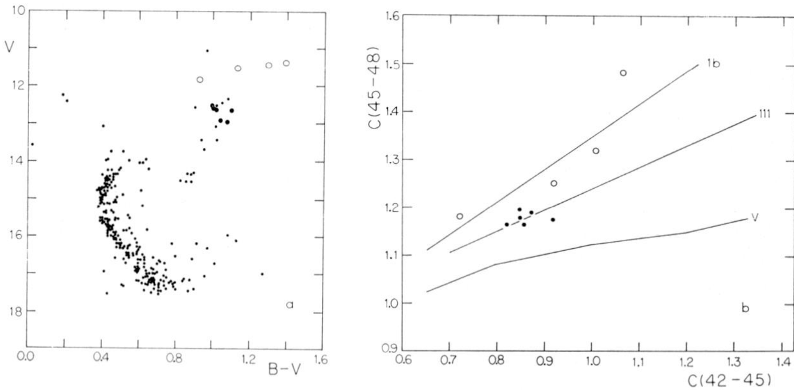

Fig. 3. (a) The colour-magnitude diagram of the old open cluster NGC 2420 from preliminary photographic photometry of Yale 40-in. plates by McClure (1971b). The large dots are stars on the giant branch that were observed on the DDO system, and the open circles are stars that appear above the giant branch that also have DDO photometry. (b) The C (45–48) vs C (42–45) diagram for the same NGC 2420 stars. The four stars that lie above the giant branch in (a) show very low surface gravities in (b).

results for late-type stars. In addition, blanketing corrections must be applied for this calibration. This will be discussed in a subsequent paper by Osborn. The value of the DDO indices in studying surface gravities of giant stars is shown in Figure 3 where giants in the old open cluster NGC 2420 are shown. Figure 3a shows the colour-magnitude diagram of the cluster prepared from preliminary photographic photometry with the Yale 40-in. telescope. The stars plotted as large dots and open circles also have DDO photometry, and are shown in the DDO surface gravity diagram in Figure 3b. The DDO indices indicate that the very luminous giants (open circles) in the colour-magnitude diagram have particularly low surface gravities. This is expected if they have masses equal to or lower than those of the fainter giant branch stars.

5. The Cyanogen Band Strength

The CN band shortward of λ 4216 is very sensitive to both surface gravity and chemical composition in late-type stars, and therefore, to be useful as a classification parameter, the effects of both these variables must be considered together. For example, the CN band cannot be used as a surface gravity index without a prior knowledge of the metal abundance of the star and vice versa. In the DDO system, a separate surface

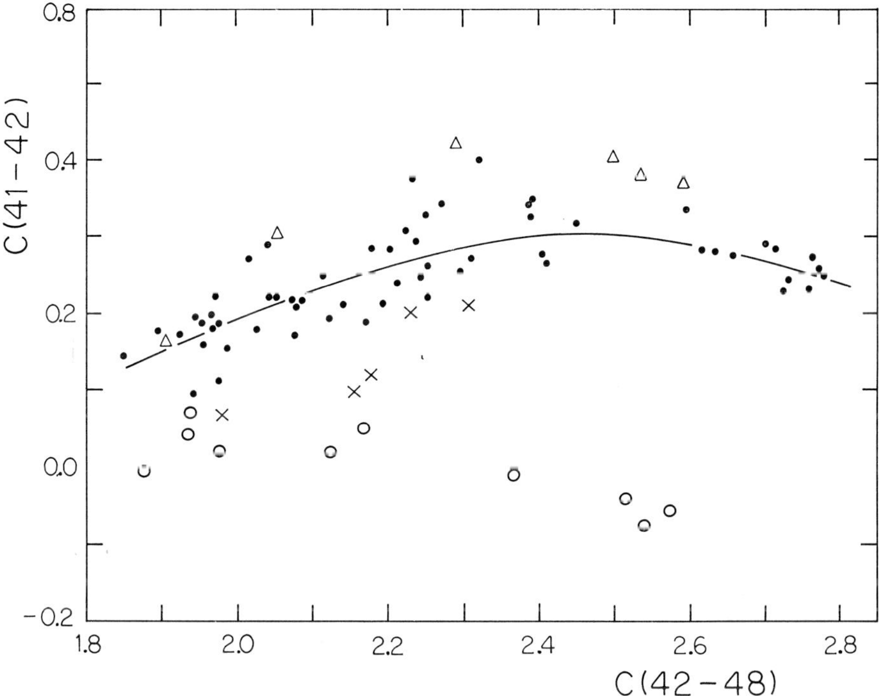

Fig. 4. The CN index C (41–42) versus the colour index C (42–48) for the random sample of bright luminosity class III stars (dots), class V stars (open circles), and class Ib stars (triangles). Also shown are a few luminosity class III stars that have space velocities greater than 100 km s^{-1} (crosses).

gravity index already exists, as discussed in the last section. Therefore, advantage is taken of the excellent sensitivity of the CN band to abundance of the elements in the star, and the band is used as a metallicity parameter after correction for the effects of surface gravity.

Figure 4 shows the CN index C (41–42) vs the temperature index C (42–48) for field stars near the Sun. Also included are a few high velocity giants that have DDO photometry. Two features to note in this diagram are (1) there is a strong dependence on metal abundance shown by the position of the high velocity giants, and (2) there is a strong dependence on surface gravity. In fact, the latter dependence resembles closely the dependence shown in Figure 1. Because of this resemblance, a simple surface gravity correction can be made to the CN index of Figure 4 by using the residual from

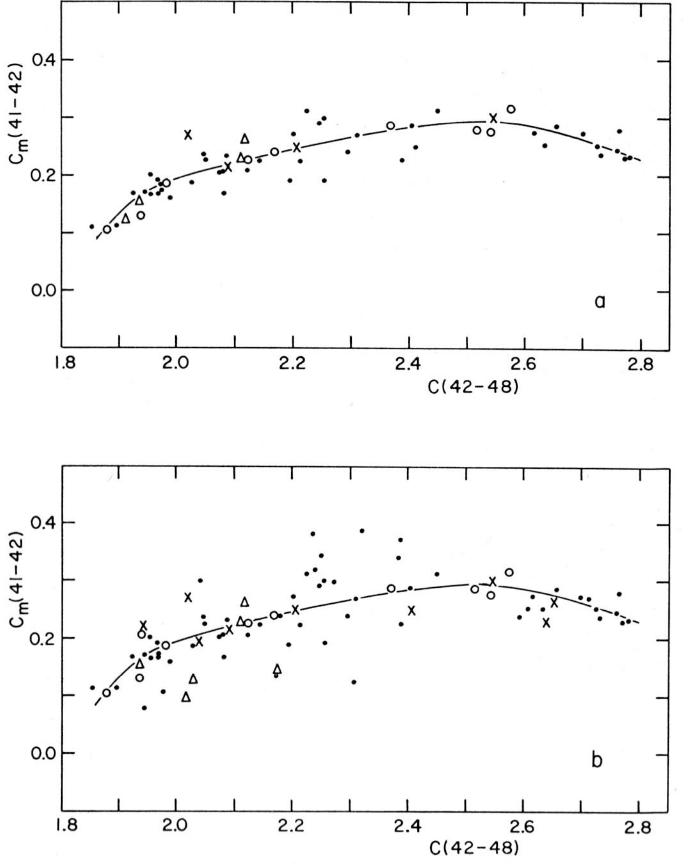

Fig. 5.   (a) A plot of the surface gravity free CN index C_m(41–42) vs the colour index C (42–48). Included are bright luminosity class III stars (dots), class II–III and II stars (crosses), class III–IV and IV stars (triangles), and class V stars (open circles). The sample is limited to stars with ultraviolet excesses in the range $-0^m.05 \leqslant \delta(U-B) \leqslant 0^m.05$. (b) The same plot as (a), but with no limitation as to ultraviolet excess. As in Figure 5, there appears to be no separation of luminosity classes, but the increased scatter in this diagram indicates good sensitivity of the index to metallicity.

the mean class III line in Figure 1. The CN index is thereby normalized to represent the CN strength of a luminosity class III star of the same metallicity. McClure (1970) has formed a CN index in this way, that is insensitive to variations of surface gravity. It can be expressed by the equation

$$C_m(41-42) = C(41-42) - 1.66[C(45-48) - 0.45C(42-45) - 0.792]$$

and is plotted vs $C(42-48)$ in Figure 5. Figure 5a is limited to stars with $|\delta(U-B)|$ $\leqslant 0^{m}.05$, whereas Figure 5b includes all bright stars observed regardless of ultraviolet excess. There appears to be no dependence on surface gravity but the increased scatter in Figure 5b is an indication that the index is indeed sensitive to metallicity. This is further illustrated in Figure 6, where a sample of strong CN stars that have been selected spectroscopically by Schmitt (1971) as exhibiting strong CN bands, plus a number of high velocity stars are plotted.

The residual above or below the mean line drawn in Figure 5 is a measure of metallicity and can be correlated with spectroscopically determined [Fe/H] values. This correlation is shown in Figure 7. The scatter is well within the errors expected from the accuracy of the parameters plotted. As illustrated by this figure, the CN strength is very useful for studying metallicities of population I stars. It is not so useful for

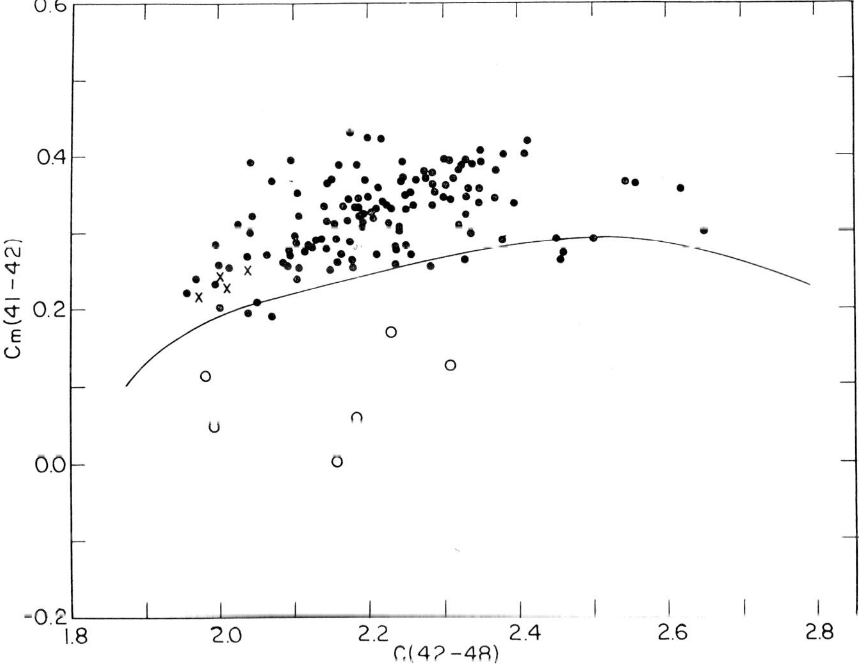

Fig. 6.   A plot of $C_m(41-42)$ vs $C(42-48)$ for a sample of spectroscopically selected strong CN giants (dots), a sample of high-velocity giants (open circles), and the four Hyades giants (crosses).

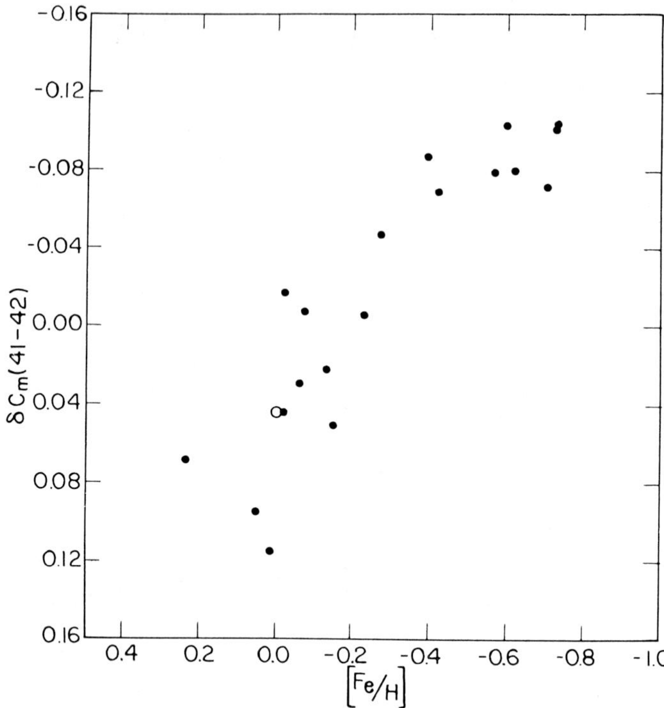

Fig. 7.   A correlation between the CN residual δC_m and logarithmic iron-to-hydrogen ratios for giant stars from the list published by Wallerstein and Helfer (1966). The [Fe/H] values are with respect to the Hyades giants, and the open circle in this figure represents a mean of the four Hyades giants.

halo populations, on the other hand, because the CN band is very weak in such stars.

6. Application to Population Studies of Field Stars

When dealing with field giants one quite often has no prior knowledge of the distance of the stars, and the Population I mass luminosity law does not necessarily hold. The DDO system is very useful for studying field stars, therefore, because it gives a direct indication of surface gravity, temperature, and metallicity without any prior knowledge of the above two parameters.

DDO photometry has been applied recently by McClure and Crawford (1971) to the study of G and K giant stars at the north galactic pole. The DDO surface gravity diagram for all late-type stars down to 13th apparent magnitude in a region within a few degrees of the pole is shown in Figure 8. One would expect that in a high latitude field the stellar population would be predominately old. Indeed, Figure 8 resembles the HR diagram of an old cluster. There are pronounced giant and dwarf sequences with subgiants between. In other words, a type of H-R diagram has been constructed for field stars without any knowledge as to their distances. One feature to note here is that there are very few stars of luminosity class brighter than class III. Osborn (1971)

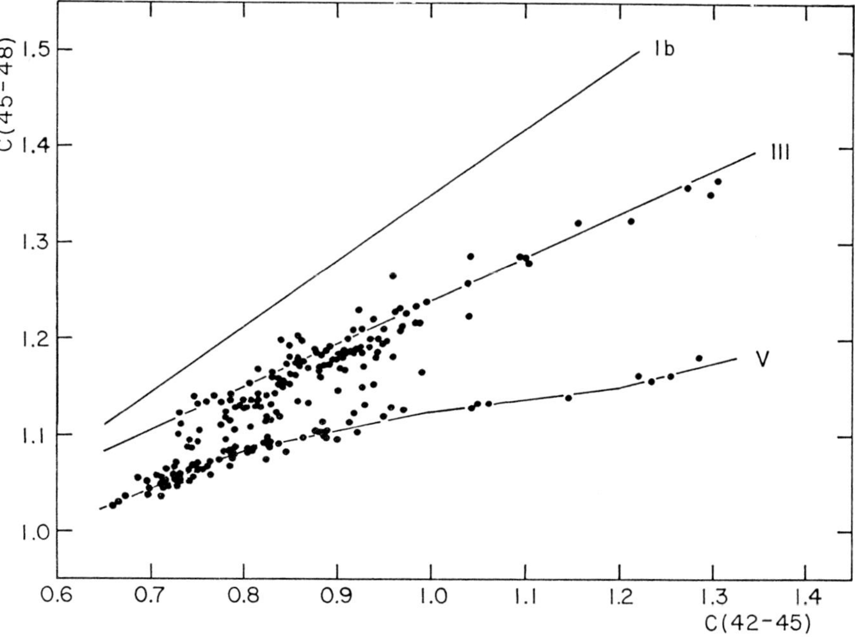

Fig. 8.   A plot of the colour index $C\,(45\text{–}48)$ vs the colour index $C\,(42\text{–}45)$ for late-type field star.
near the north galactic pole. A main sequence and giant branch extend to late spectral types.
Numerous subgiants are found earlier than K1 $[C\,(42\text{–}45) < 0\overset{m}{.}95]$. The figure resembles the colour-
magnitude diagram of an old open cluster, and this 'HR' diagram was obtained without any prior
knowledge of the distances to these stars.

has found that halo-type stars, as expected due to their low masses, exhibit very low
surface gravities in this diagram. Therefore, the sample of stars shown here appears
to be predominantly disc population resembling an old open cluster, even though the
stars extend to well over a kiloparsec above the galactic plane. The absence of halo
stars appears to be due to the sharp fall-off of star numbers along the giant branches of
globular clusters near the blue limit of the present study. There is still some question,
as pointed out by Helfer and Sturch (1971), as to the calibration at the blue end of
Figure 8 in terms of position in a UBV colour-magnitude diagram. In particular, it
appears that stars on the horizontal part of the subgiant branch in an old cluster such
as M67 (subgiants with $C\,(42\text{–}45) < 0\overset{m}{.}8$) may fall close to the dwarf sequence.

Figure 9 shows the cyanogen residual δC_m plotted for the NGP stars vs Z distance
determined from the spectral classification of Figure 2. The metallicity appears to
weaken with distance above the galactic plane as expected, but there is large scatter,
and one or two anomalous strong CN stars were found at large distances above the
plane.

A sample of bright field stars near the Sun has also been studied by Janes and Mc-
Clure (1971) to see whether strong CN stars have different kinematics than normal
solar neighbourhood giants. The kinematics for the two groups are shown in Figure 10.

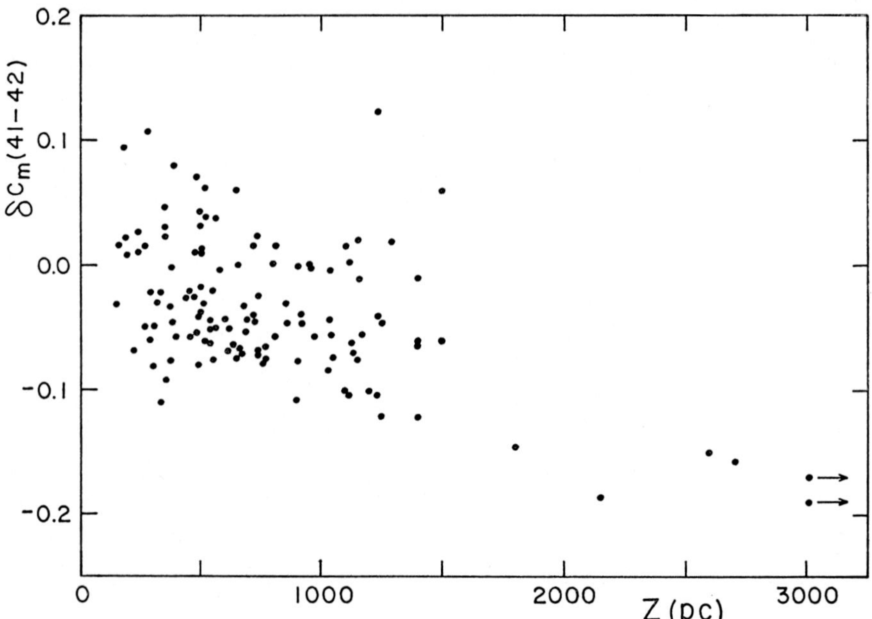

Fig. 9.   The CN residual δC_m for giant stars in the north galactic pole sample versus distance
above the galactic plane.

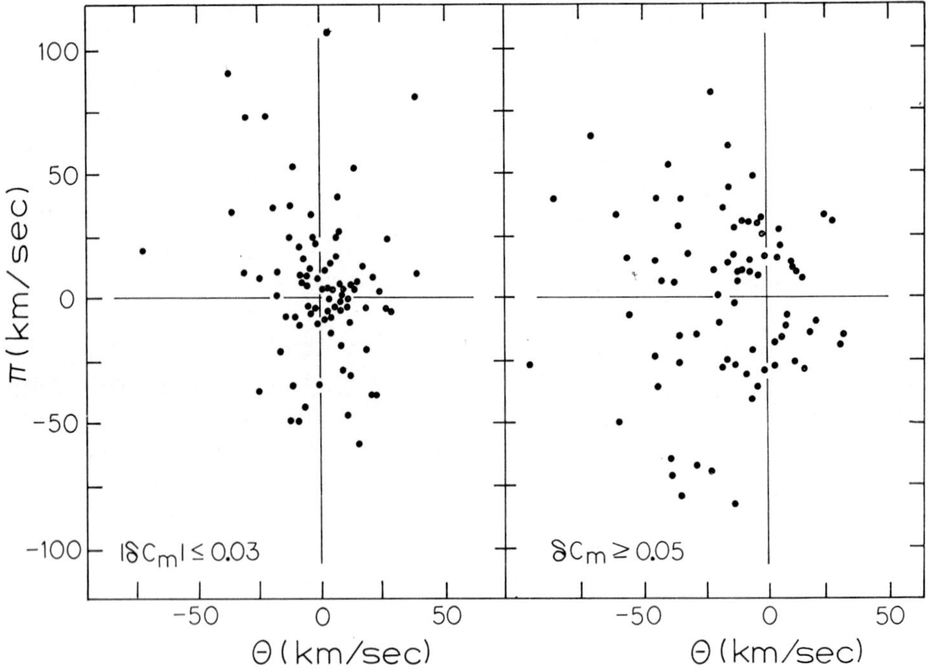

Fig. 10.   The distribution of space velocities of field G and K giants in the (π, θ) plane. 'Normal stars,
are on the left (a), and strong CN stars are on the right (b).

The normal giants in Figure 10a ($|\delta C_m| \leqslant 0\overset{m}{.}03$) show a concentration of velocities, as expected, near the local standard of rest. The strong CN stars in Figure 10b, on the other hand ($\delta C_m \geqslant 0\overset{m}{.}05$) show no such concentration. In this diagram, there appears to be a clumpy distribution forming somewhat of a ring around the local standard of rest, and an underlying widely scattered distribution including some relatively high velocity stars. The clumpy distribution resembles a relatively young population, pre-sumably giants that have evolved from main sequence A-type stars such as Hyades or younger-type giants. In fact, NGC 2477, a southern cluster similar in age to the Hyades has been found from DDO photometry by Hartwick, Hesser and McClure (1971) to have CN strong stars. The underlying scattered distribution of stars has a mean θ velocity that is negative, and resembles an old stellar population. There may be a second explanation for these stars, however. They could be the tail end of a population of giants with less energetic orbits, in the mean, than the Sun's. From observations of McClure (1969), Spinrad, Gunn, Taylor, McClure, and Young (1971) of galaxy nuclei, and from Searle's (1971) observations of H II regions in galaxies, it is probable that the metallicity of the Galaxy increases towards the center. Therefore, it is conceivable that K giant stars whose orbits bring them closer to the center of the Galaxy than the Sun's orbit would have stronger CN bands than normal.

7. Application to the Study of Open Clusters

Red giant stars are the brightest and therefore the most easily observed stars in old open clusters. The giants in several old to intermediate age clusters have now been observed on the DDO system. An example of the DDO surface gravity diagram for the old metal poor cluster NGC 2420 has been shown in Section 4. Observations of a more normal old cluster M67 are shown in Figure 11. Here the coolest giants and

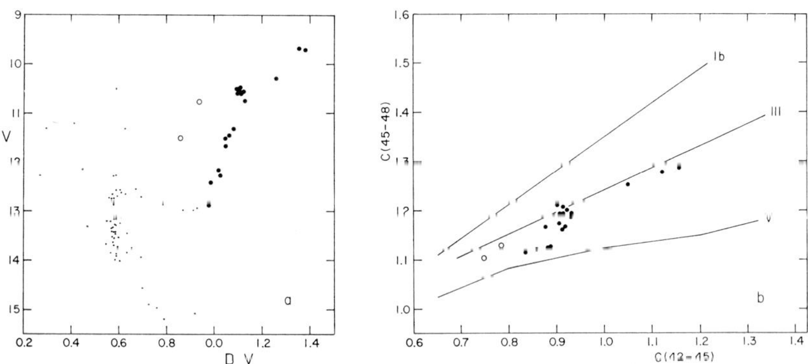

Fig. 11.   (a) The colour-magnitude diagram for M67. The large dots are giant stars that have been observed on the DDO system. The open circles are stars lying above the subgiant branch that also have DDO photometry. (b) Preliminary values from Janes (1971) for the C (45–48) versus C (42–45) diagram of giants in M67. The lower luminosity giants show high surface gravities in this diagram, and the two stars that lie above the subgiant branch in the colour-magnitude diagram have lower surface gravities than subgiants.

clump stars resemble field luminosity class III stars whereas fainter stars on the sloping part of the giant branch in the colour magnitude diagram fall well below the class III relation.

Table III shows the mean CN residuals δC_m for the giant stars in old and intermediate age clusters observed on the DDO system to date. The clusters are arranged in order of age.

TABLE III

CN strengths for open clusters

Cluster	l^{II}	b^{II}	r(pc)	δC_m	Reference
NGC 5822	322°	+ 4°	700	0ᵐ01	McClure (1971b)
NGC 2477	254	− 6	1300	0.08	Hartwick *et al.*
Hyades	180	−22	40	0.04	McClure and van den Bergh (1968)
NGC 2360	230	− 1	1100	−0.01	McClure (1971a)
NGC 2420	198	+20	2600	−0.07	McClure (1971b)
NGC 3680	287	+17	800	0.02	McClure (1971a)
NGC 6819	74	+ 8	2000	0.01	McClure (1971b)
M67	216	+32	800	0.04	Janes (1971)

Of particular interest are NGC 2477 and NGC 2420. NGC 2477 is a cluster of about Hyades age (Hartwick *et al.*, 1971) and is situated very close to the galactic plane. It is the only cluster found so far to have giant stars with stronger CN residuals than the Hyades giants. NGC 2420 is a cluster that appears from its colour-magnitude diagram to be slightly younger than M67, but lies more than 1 kpc above the galactic plane. As well as the giants having weak CN bands, the main sequence stars in Arp's photoelectric sequence for this cluster have a mean ultraviolet excess of $\delta(U-B) = 0ᵐ09$. All other clusters studied to date, including M67, appear to have giants with DDO CN indices in the range between the Hyades value and the mean of the solar neighbourhood giants.

Acknowledgements

I would like to acknowledge the support and initial direction of Dr S. van den Bergh who was instrumental in choosing the filter bandpasses for the DDO system and overseeing the development of the standard system of stars. Since the initial observing time allotted at the David Dunlap Observatory for setting up the standard stars, most of the observations have been made at the Kitt Peak National Observatory and Cerro Tololo Inter-American Observatory where the resident astronomers have been very generous in allotment of telescope time and overall support of the project. Observations made with the Yale 40-in. telescope by Mr K. A. Janes have also been very helpful in the preparation of this paper.

References

Griffin, R. F. and Redman, R. O.: 1960, *Monthly Notices Roy. Astron. Soc.* **120**, 287.

Gyldenkerne, K.: 1958, *Ann. Astrophys.* **21**, 77.
Hartwick, F. D. A., Hesser, J. E., and McClure, R. D.: 1971, in press.
Helfer, H. L. and Sturch, C.: 1971, private communication.
Janes, K. A.: 1971, in preparation.
Janes, K. A. and McClure, R. D.: 1971, *Astrophys. J.* **165**, 561
Keenan, P. C. and Keller, G.: 1953, *Astrophys. J.* **117**, 241.
Lindblad, B. and Stenquist, E. 1934, *Stockholm Obs. Ann.* **11**, No. 12.
McClure, R. D.: 1969, *Astron. J.* **74**, 50.
McClure, R. D.: 1970, *Astron. J.* **75**, 41.
McClure, R. D.: 1971a, in press.
McClure, R. D.: 1971b, unpublished.
McClure, R. D. and Bergh, S. van den: 1968, *Astron. J.* **73**, 313.
McClure, R. D. and Crawford, D. L.: 1971, *Astron.J.* **76**, 31.
McClure, R. D. and Racine, R.: 1969, *Astron.J.* **74**, 1000.
Osborn, W. H.: 1971, unpublished Ph. D. Thesis, Yale University.
Schmitt, J. L.: 1971, *Astrophys. J.* **163**, 75.
Searle, L. T.: 1971, *Carnegie Inst. Washington, Yearb.* **69**, 75.
Spinrad, H., Gunn, J. E., Taylor, B. J., McClure, R. D., and Young, J. W.: 1971, *Astrophys.J.* **164**, 11.
Wallerstein, G. and Helfer, H. L.: 1966, *Astron. J.* **71**, 350.

RESULTS FROM THREE DIMENSIONAL SPECTRAL CLASSIFICATION OF POPULATION II STARS USING THE DDO PHOTOMETRIC SYSTEM

W. OSBORN*

Inst. Venezolano de Astronomia and Univ. de Los Andes, Mérida, Venezuela

Abstract. The David Dunlap Observatory (DDO) intermediate-band photometric system described in the previous paper has been used to observe a number of red giant branch, horizontal branch, and asymptotic branch members in each of the five globular clusters M3, M5, M10, M13, and M92. A calibration of the DDO system is described by which it was possible to determine the effective temperatures, surface gravities, [Fe/H] values, and masses of the observed stars. The mean [Fe/H] values for the clusters were found to be -1.01 for M3, -0.68 for M5, -1.44 for M10, -1.69 for M13, and -1.96 for M92. Evidence was found that the masses of the horizontal branch and asymptotic branch stars are systematically smaller than those of red giant branch members. Two stars were discovered to have CN bands that are anomalously strong for Population II objects. The observational results have been compared with the theoretical predictions of two detailed Population II evolutionary tracks and in general the agreement is good.

In the preceeding paper of this symposium McClure has described the David Dunlap Observatory intermediate-band photometric system (usually referred to as the 'DDO system') and has shown how the system may be used for three dimensional classification of G and K stars. Several applications of DDO photometry for studies of Population I objects, along with the results to date, were discussed. In this paper it is shown that the DDO system can also be applied successfully to problems involving Population II stars.

Observations with the DDO system have been used to determine effective temperatures, surface gravities, metal abundances, and mass estimates for a number of late-type globular cluster stars. The principle aim of this program was to obtain data suitable for comparison with the predictions of the detailed Population II evolutionary tracks that are being computed by several research groups, for example, Iben and Rood (1970) and Demarque and Mengel (1971a, b). The amount of observational data presently available for comparison with such calculations is small.

Selected for observation in the program were stars in each of the globular clusters M3, M5, M10, M13, and M92. These clusters are all bright, well-studied objects that were accessible to the available telescope and cover a range of metallicities from extremely metal poor to moderately metal rich. Within each cluster the stars were selected on the basis of four criteria: (1) the star is located within ten minutes of arc of the cluster center, (2) the star has a position in the broad-band cluster HR diagram appropriate for a red giant branch, red horizontal branch, or asympotic branch member, (3) the star is brighter than $V = 16$, and (4) on deep plates of the cluster the star has at least 20″ in separation from any other star. The first two of these criteria gave the

---

* Visiting student, Kitt Peak National Observatory, which is operated by the Association of Universities for Research in Astronomy, under contract with the National Science Foundation.

star a high probability of being a late-type physical member of the cluster while the last two criteria insured that DDO colors could be obtained that would be of sufficient accuracy to yield reliable data and free from contamination from background stars.

DDO photometry was obtained for a total of thirty-three stars in the five clusters. From their positions in the cluster HR diagrams, sixteen of the stars are members of the red giant branches, four are red horizontal branch members, four are asymptotic branch members, and six stars were classified as probable upper asymptotic branch members but could possibly be instead red giants. Three of the stars were shown by the DDO photometry to be foreground dwarfs. The observations were obtained with the 84-in. telescope (213-cm aperture) of the Kitt Peak National Observatory equipt with an integrating photometer and utilizing a twelve seconds of arc diaphragm. As well as the four basic filters of the DDO system – Filters 48, 45, 42, and 41, the characteristics and uses of which were given in the paper by McClure – the two supplementary ultraviolet filters of the system – Filters 38 and 35, which have peak wavelengths of 3813 and 3458 Å and half-widths of 331 and 370 Å respectively – were also observed. The latter two filters are used to form the color indices $C(38–42)$, a measure of the metallic line-blanketing, and $C(35–38)$, a measure of the Balmer discontinuity. These two additional indices are necessary for unambigous three dimensional classification of Population II stars because of the loss of sensitivity of the $C(41–42)$ index when the CN bands become weak.

To obtain the desired physical parameters of the globular cluster stars from their observed DDO colors, use was made of previously derived calibrations of the system

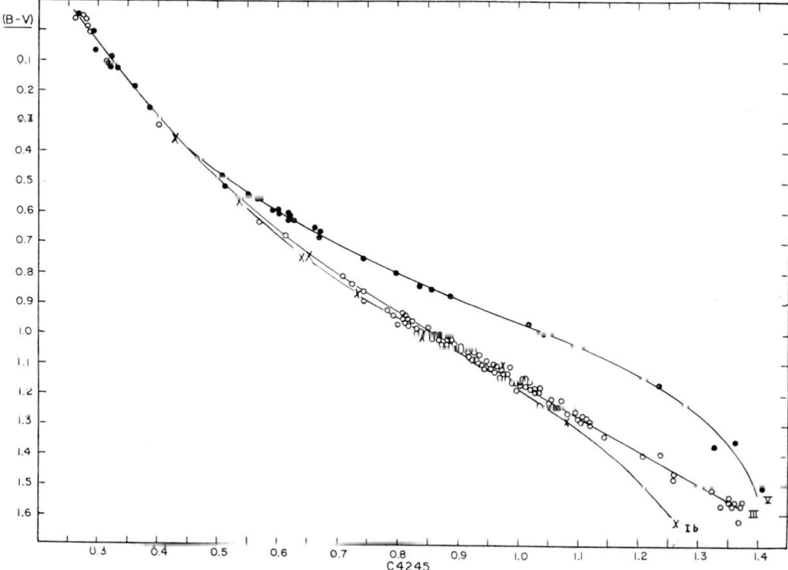

Fig. 1. A plot of $B-V$ vs the DDO temperature index $C(42–45)$. Only stars known to be of MK luminosity class Ib, III, or V and with normal UBV colors (*i.e.* no ultraviolet excess or deficiency) are plotted.

with which for the spectral range F5–K5 the DDO colors of a star (either Population I or Population II) can be used to determine its effective temperature, its surface gravity, and its metal abundance. The details of these calibrations and their derivations have been presented elsewhere (Osborn, 1971) but a brief summary will be given.

The effective temperature and surface gravity calibrations were based on previous calibrations of these quantities with $(B-V)$ and MK spectral type. The correlations adopted are given in Table I. Thus, for each case the determination of the relationship between either $(B-V)$ or MK spectral class and the appropriate DDO color index, which could be easily found from the over 500 stars for which all three types of information are available, led directly to the calibration. This process is illustrated in Figures 1 and 2. The first figure shows the relation between $(B-V)$ and the DDO temperature indicator $C(42–45)$ and the second figure the temperature calibration that results using Table I. Figures 3–5 show in a similar manner the derivation of the surface gravity calibration.

The metal abundance calibrations were obtained directly without recourse to previous calibrations. An exhaustive search of the literature was made for F5–K5 stars with published iron to hydrogen abundance ratios. The search yielded a list of 303 determinations for 231 stars, the data from which were placed on a self-consistent

TABLE I

Relation adopted between MK spectral type, $(B-V)$, effective temperature, and surface gravity

Sp.	$(B-V)$			Eff. temperature			Log gravity		
	Ib	III	V	Ib	III	V	Ib	III	V
F5	0.35	0.43	0.44	6630	6510	6550	1.7		4.4
F8	0.57	0.54	0.52	6020	6010	6120	1.7		4.4
G0	0.75	0.65	0.58	5670	5760	5880	1.6	3.4	4.4
G2	0.87	0.75	0.62	5540	5430	5720	1.5	3.3	4.4
G5	1.01	0.86	0.68	5000	5180	5590	1.3	3.1	4.5
G8	1.11	0.93	0.74	4830	4940	5470	1.2	2.8	4.5
K0	1.19	1.02	0.81	4580	4720	5240	1.1	2.5	4.5
K1	1.27	1.09	0.86	4400	4530		1.0	2.3	4.5
K2	1.34	1.17	0.91	4260	4350	4850	0.9	2.1	4.5
K3	1.46	1.26	0.96	4140	4050	4660	0.8	1.9	4.5
K4	1.54	1.40	1.01	4010	3840		0.7	1.7	4.5
K5	1.61	1.50	1.14	3690	3610	4140	0.6	1.5	4.5

system of [Fe/H] values and mean values derived for the stars with more than one determination. Then, using the stars on the list for which DDO photometry was available, the calibrations were obtained by plotting the [Fe/H] of the star versus its *color anomaly*, this being defined as the difference between the observed value of the abundance sensitive color index and the average value of the index for stars of similar spectral type and of nearly solar abundance. Note that for Population II stars the DDO system has two abundance indicators: $C(41–42)$ and $C(38–42)$. The two associated

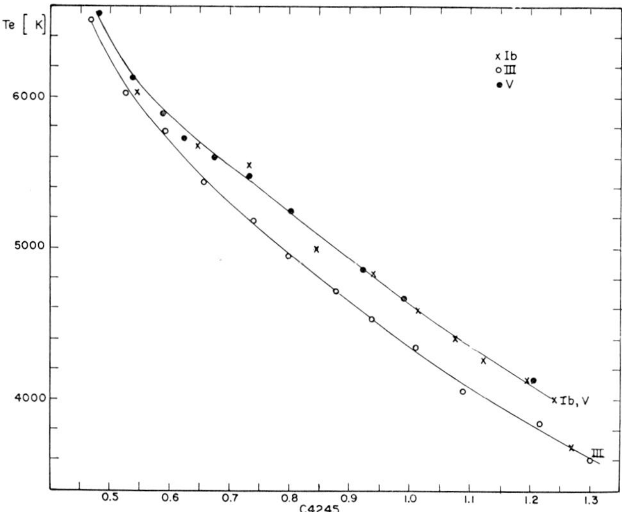

Fig. 2.   The relation between $C(42\text{–}45)$ and effective temperature that results from the previous figure using the data of Table I.

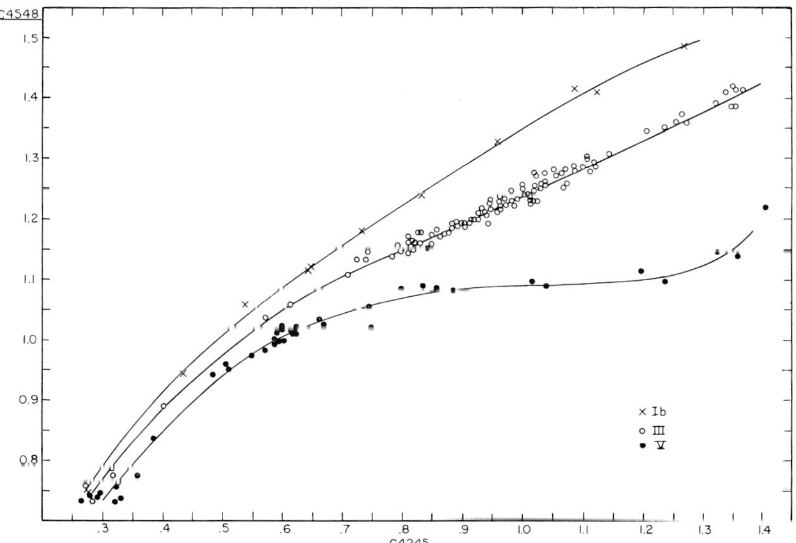

Fig. 3.   The $C(45\text{–}48)$ vs $C(42\text{–}45)$ diagram for luminosity class Ib, III, and V stars.

color anomalies are denoted δC_m (from C_m, the gravity corrected from of the $C(41\text{–}42)$ index) and $\delta 3842$. The abundance calibration derived from the $C(38\text{–}42)$ index is shown in Figure 6.

The derivation of the physical quantities of the program stars proceeded by first adopting apparent visual (AV) distance moduli and values of interstellar reddening

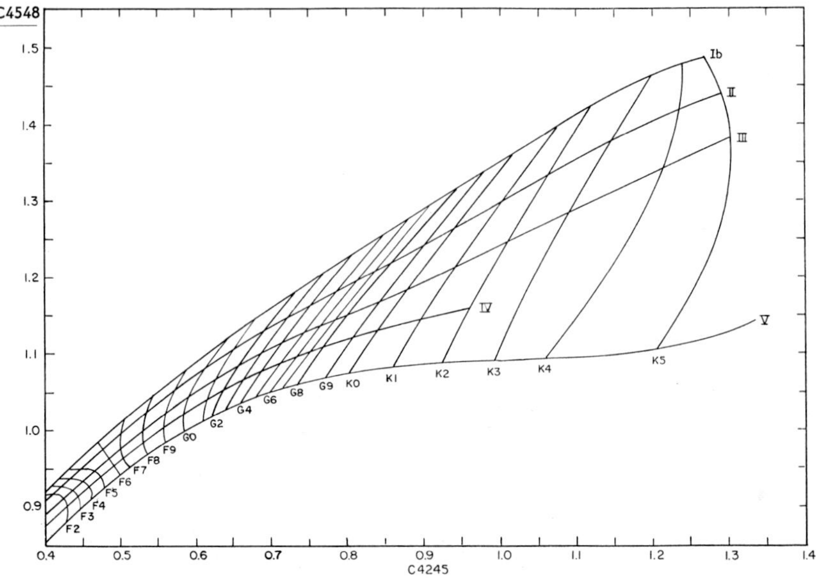

Fig. 4. The relation between MK spectral type and position in the $C(45–48)–C(42–45)$ diagram that was derived using the MK spectral classes of the stars plotted in Figure 3.

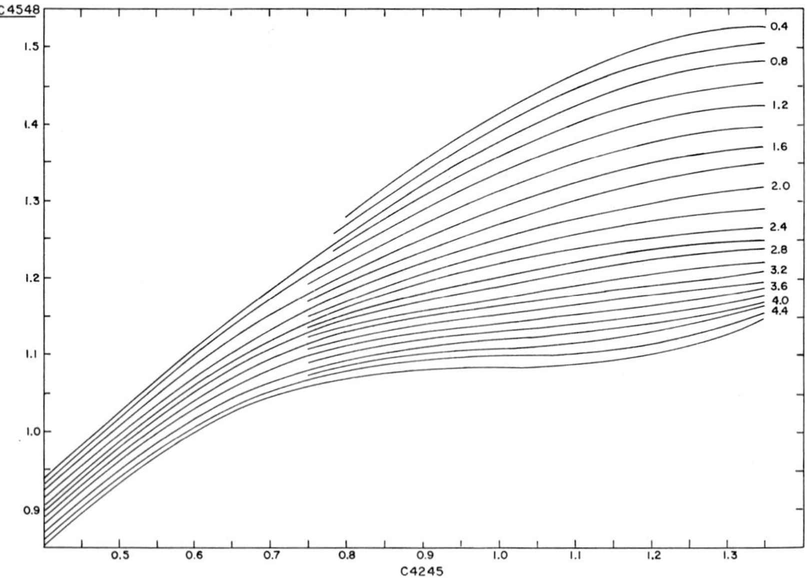

Fig. 5. The lines of constant $\log g$, from $\log g = 0.4$ to 4.6, in the $C(45–48)–C(42–45)$ diagram that were obtained using the surface gravities of Table I and Figure 4.

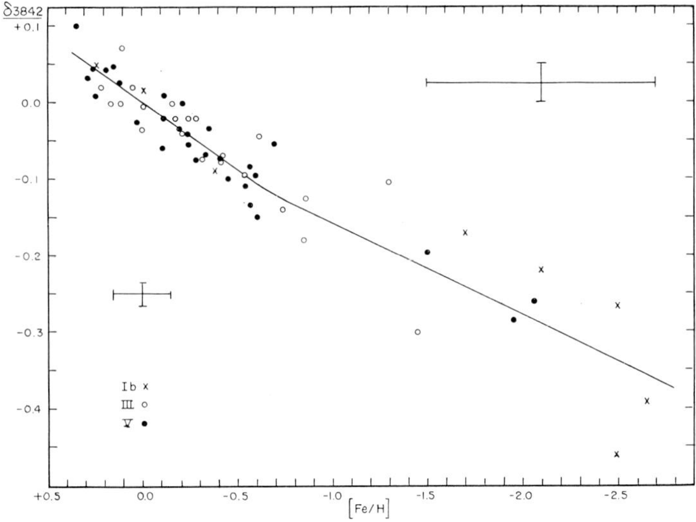

Fig. 6.   The correlation between $\delta3842$ and [Fe/H]. Typical error bars of the points for stars of nearly solar abundance and for extremely metal deficient stars are shown.

for the five clusters (Table II) with which absolute magnitudes were calculated for the stars and their observed DDO colors corrected for reddening. Then, application of the calibrations to the un-reddened colors gave an effective temperature, a surface gravity, and an [Fe/H] value for each star. By combining the temperature and gravity results with the absolute magnitude, a mass for the star could also be calculated.

TABLE II

Adopted distance moduli and reddening values

	M3	M5	M10	M13	M92
$(m - M)_{AV}$	15.3	14.8	14.4	14.4	14.5
$E(B - V)$	0.005	0.035	0.26	0.015	0.02

The principle results from the data are the following. First, mean [Fe/H] values have been found for the five clusters, the values being -1.01 for M3, -0.68 for M5, -1.44 for M10, -1.69 for M13, and -1.96 for M92. The uncertainity of each value is of the order of ±0.3, which is approximately a factor of two better than previous results. It is noted that M13 was found to be somewhat more metal poor than usually quoted, in particular significantly more metal poor than M3. The relative ordering of these two clusters has been in doubt for some time.

The second result is that the calculated masses for the stars in the post helium flash stages of evolution, that is the horizontal branch and asymptotic branch stars, were found to be systematically smaller than the masses derived for the pre-helium flash red giants. This systematic trend was found for all five clusters studied and a t-test for significance on the combined data gave a reliability at the 98% level. Nevertheless, in view of the small sample sizes and the uncertainities in the individual mass values, this finding stands in need of confirmation. If confirmed it would have important implications in regard to the interpretation of globular cluster HR diagrams using stellar evolution theory.

Third, two stars – one each in M5 and M10 – were discovered to have anomalously strong CN bands, as measured by the DDO $C(41$–$42)$ index, in relation to other globular cluster stars. In all but these two cases the globular cluster stars were found to have $C(41$–$42)$ indices similar to those exhibited by luminosity class IV stars, reflecting the well known fact that their CN bands are much weaker than what would be expected from their low surface gravities, which are comparable to those of class Ib or class II stars. For the two anomalous stars, however, $C(41$–$42)$ indicated CN bands considerably stronger, equivalent to luminosity class III bands. The surface gravities derived for the two stars were similar to those of other globular cluster stars of their temperatures hence the CN band enhancement does not appear to be a gravity effect. Other stars in globular clusters with enhanced carbon molecule features in their spectra have previously been reported in M13 (Popper, 1947), M92 (Strom and Strom, 1971), and ω Centauri (Harding, 1962; Stock and Wing, 1971).

Finally, the derived data have been compared with the detailed Population II evolutionary tracks of Demarque and Mengel (1971a, b). The agreement in general is good. The position in the physical HR diagram of the horizontal branch, the slope of the Hayashi line, and the effective temperature separation of the red giant and asymptotic branches are all as predicted. The red giants are found to be somewhat cooler than predicted but this could easily be caused by uncertainities in some of the physics used in the theoretical calculations, notably in the constants employed in the mixing length theory.

The work described in this paper was performed as part of a Ph. D. dissertation in the Department of Astronomy, Yale University. The observational material was obtained at the Kitt Peak National Observatory in the period January to November 1969.

References

Demarque, P. and Mengel, J. G.: 1971a, *Astrophys. J.* **164**, 316.
Demarque, P. and Mengel, J. G.: 1971b, *Astrophys. J.* **164**, 469.
Harding, E. A.: 1962, *Observatory* **82**, 205.
Iben, I. and Rood, R. T.: 1970, *Astrophys. J.* **159**, 605.
Osborn, W. H.: 1971, *Positions of Globular Cluster Stars in the Physical HR Diagram*, Ph. D. Thesis, Yale University
Popper, D. M.: 1947, *Astrophys. J.* **105**, 204.
Stock, J. and Wing, R.: 1972. *Bull. Am. Astron. Soc.* **4**, 324.
Strom, S. E. and Strom, K. M.: 1971, *Astron. Astrophys.* **14**, 111.

PHOTOMETRIC CLASSIFICATION OF STARS
IN THE VILNIUS OBSERVATORY

V. STRAIŽYS

Vilnius Astronomical Observatory, Lithuania, U.S.S.R.

(Read by A. Ardeberg)

Abstract. The programs of three-dimensional classification of stars using eight-color intermediate band photometric system are described. They include (1) investigation of interstellar reddening and absorption in the direction of objects significant from point of view of stellar evolution, (2) investigation of spatial distribution of stars of different spectral classes, luminosities and chemical composition, (3) the detection of unique objects.

The multicolor intermediate band photometric system developed in Vilnius observatory can be used for purely photometric determination of stellar spectral classes, absolute magnitudes, chemical composition parameter and interstellar reddening. The system is described in a series of papers which appeared in Vilnius Observatory Bulletins. The most recent general description of the system is given by Straižys (1970) and Straižys, Sviderskienė (1971). The response curves of the system are given by Straižys and Zdanavičius (1970), the calibration of reddening-free diagrams Q, Q and two-color diagrams in spectral classes and absolute magnitudes – by Sviderskienė and Straižys (1971), the method of classification of stars in chemical composition – by Bartkevičius and Straižys (1970a, b, c, d). The general scheme of three dimensional classification of any collection of stars, including the samples of different temperatures, luminosities, populations, interstellar reddenings and some kinds of peculiarities, is given by Straižys and Sviderskienė (1971). Now we generally use eight filters marked by U, P, X, Y, Z, V, T, S with mean wavelengths at 345, 374, 405, 466, 516, 544, 625 and 655 nm, however for different tasks less quantity of the filters is sufficient.

In 1967–1970 the photoelectric observations of about 900 stars for calibration purposes were obtained (Zdanavičius *et al.*, 1969; Bartkevičius and Metik, 1969; Sūdžius *et al.*, 1970; Straižys *et al.*, 1970; the remaining unpublished). Since 1971 we have started a research programs described below.

TABLE I

Numbers of BS stars having no published MK spectra

V	$\delta > 0$	$\delta < 0$
< 5.0	50	130
$5.0 - 5.5$	240	340
$5.5 - 6.0$	630	730
> 6.0	1040	1460
Total	1960	2260

One of our first programs accessible to small telescopes is two and three dimensional classification of BS stars without published MK spectra (Table I). Up to now photoelectric observations of about 200 stars are obtained.

The Vilnius observatory system gives the possibility to investigate the structure of the Galaxy in crowded regions of the Milky Way and to reach stars too faint for MK classification from the spectra. Now we are measuring the stars up to 13^m with photomultiplier EMI 9502 working as a photoncounter with a 70 cm reflector. The replacement of glass filters by the interference ones with maximum transmission about 70–80% will add at least 1^m. Thus in the near future we shall be able to obtain three dimensional classification of stars up to 15–$15^m.5$ with a 160-cm reflector.

Such faint stars accessible permit to solve the following problems.

(1) To investigate interstellar reddening and absorption in the direction of objects significant from point of view of stellar evolution. Such objects are globular clusters, pulsars, galaxies, quasars, novae and supernovae, some unique stars (both constant and variable), etc. In the most cases interstellar absorption up to these objects is known only approximately and it is impossible to obtain exact intrinsic energy distribution in their spectra. This problem can be solved by the use of small number (20–30) of faint stars around the object under investigation and distributed up to distance of the object, if it is inside our Galaxy, or up to Galaxy boundary, if it is extragalactic object.

(2) To investigate spatial distribution of stars of different spectral classes, luminosities and chemical composition into different directions from the Sun, including the variation of spatial structure of the Galaxy with z-coordinate, approaching the spiral arms or the center of the Galaxy. These investigations are, however, much more laborious because observations of hundreds of stars are needed. On the other hand, they will give much more exact results than earlier methods because the individual distance moduli of stars will be used instead of statistical ones.

(3) To reveal unique objects. During the investigation of interstellar absorption or spatial structure of the Galaxy the stars will be found which will be useless for the investigation due to their pecularity but very valuable from the astrophysical point of view. They can be extremely metal-deficient stars, CH-stars, barium stars, double and multiple stars of different types, variable stars of different types, emission-line stars, Wolf-Rayet stars, P Cygni stars, symbiotic stars etc. After photometric detection they can be investigated in detail by spectrophotometry and other methods.

The works in Vilnius observatory which are already started or are being planned are concentrated on the solution of the listed problems. In the near future we are planning to investigate the interstellar absorption in the direction of some infrared stars, globular clusters lying near the Milky Way and the galaxies of the Local group. The observations up to 13^m are already made in the region of infrared star NML Cyg and another region near the Great Cygnus Rift including open cluster IC 4996 and the star P Cygni.

In collaboration with Radioastrophysical Observatory of Latvian Academy of Sciences in Baldone and the Main Astronomical Observatory of Ukrainian Academy of Sciences near Kiev we are investigating the possibility to realize the Vilnius observatory system photographically. The application of electronography seems very promissing also.

Note added in proof. The results of two-dimensional classification of 530 stars are published by Zdanavičius *et al.*: 1972, *Bull. Vilnius Obs.*, No. 34, 3.

References

Bartkevičius, A. and Metik, L. P.: 1969, *Bull. Vilnius Obs.*, No. 26, 13.
Bartkevičius, A. and Straižys, V.: 1970a, *Bull. Vilnius Obs.*, No. 28, 33.
Bartkevičius, A. and Straižys, V.: 1970b, *Bull. Vilnius Obs.*, No. 30, 3.
Bartkevičius, A. and Straižys, V.: 1970c, *Bull. Vilnius Obs.*, No. 30, 16.
Bartkevičius, A and Straižys, V.: 1970d, *Bull. Vilnius Obs.*, No. 30, 33.
Straižys, V.: 1970, *Bull. Vilnius Obs.*, No. 28, 6.
Straižys, V., Drazdys, R., and Gurklytė, A.: 1970, *Bull. Vilnius Obs.*, No. 29, 10.
Straižys, V. and Sviderskienė, Z.: 1972, *Astron. Astrophys.* **17,** 312.
Straižys, V. and Zdanavičius, K.: 1970, *Bull. Vilnius Obs.*, No. 29, 15.
Sūdžius, J., Zdanavičius, K., Sviderskienė, Z., Straižys, V., Bartkevičius, A., Zitkevičius, V., Kavaliauskaité, G., and Kakaras, G.: 1970, *Bull. Vilnius Obs.*, No. 29, 3.
Sviderskienė, Z. and Straižys, V.: 1971, *Bull. Vilnius Obs.*, No. 31, 3
Zdanavičius, K., Sūdžius, J., Sviderskienė, Z., Straižys, V., Burnasov, V., Drazdys, R., Bartkevičius, A., Kakaras, G., Kavaliauskaité, G., and Jasevičius, V.: 1969, *Bull. Vilnius Obs.*, No. 26, 3.

PHOTOMETRIC CLASSIFICATION OF B-TYPE STARS

D. L. CRAWFORD

Kitt Peak National Observatory, Tucson, Ariz., U.S.A.

Abstract. Several photometric techniques for classification of B-type stars exist. This note describes another one, now nearly ready for publication.

We have observed most of the stars with spectral types B5 and earlier, and brighter than $m_v = 6.5$ with the *uvby* and Hβ systems. In addition, we have data for the following clusters or associations: h and χ Per, NGC 6231, α Per, Pleiades, IC 2602, IC 2391, III Cep, and Sco-Cen. Good MK types exist for most of these stars, and *UBV* data is available for many of them.

Figures 1 and 2 show the relations between $(b-y)$ and $(B-V)$ and between $(u-b)$ and $(U-B)$ for the B, A, and F-type stars we use as standards. Table I gives the mean values of several MK spectral types. The unreddened values were determined as des-

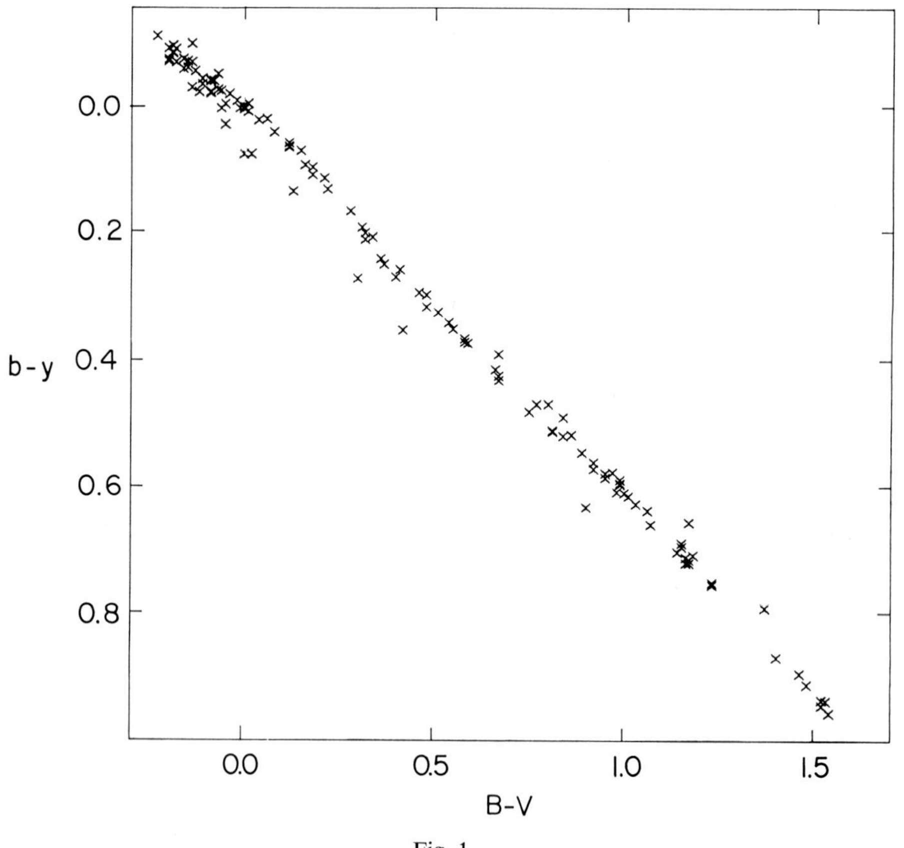

Fig. 1.

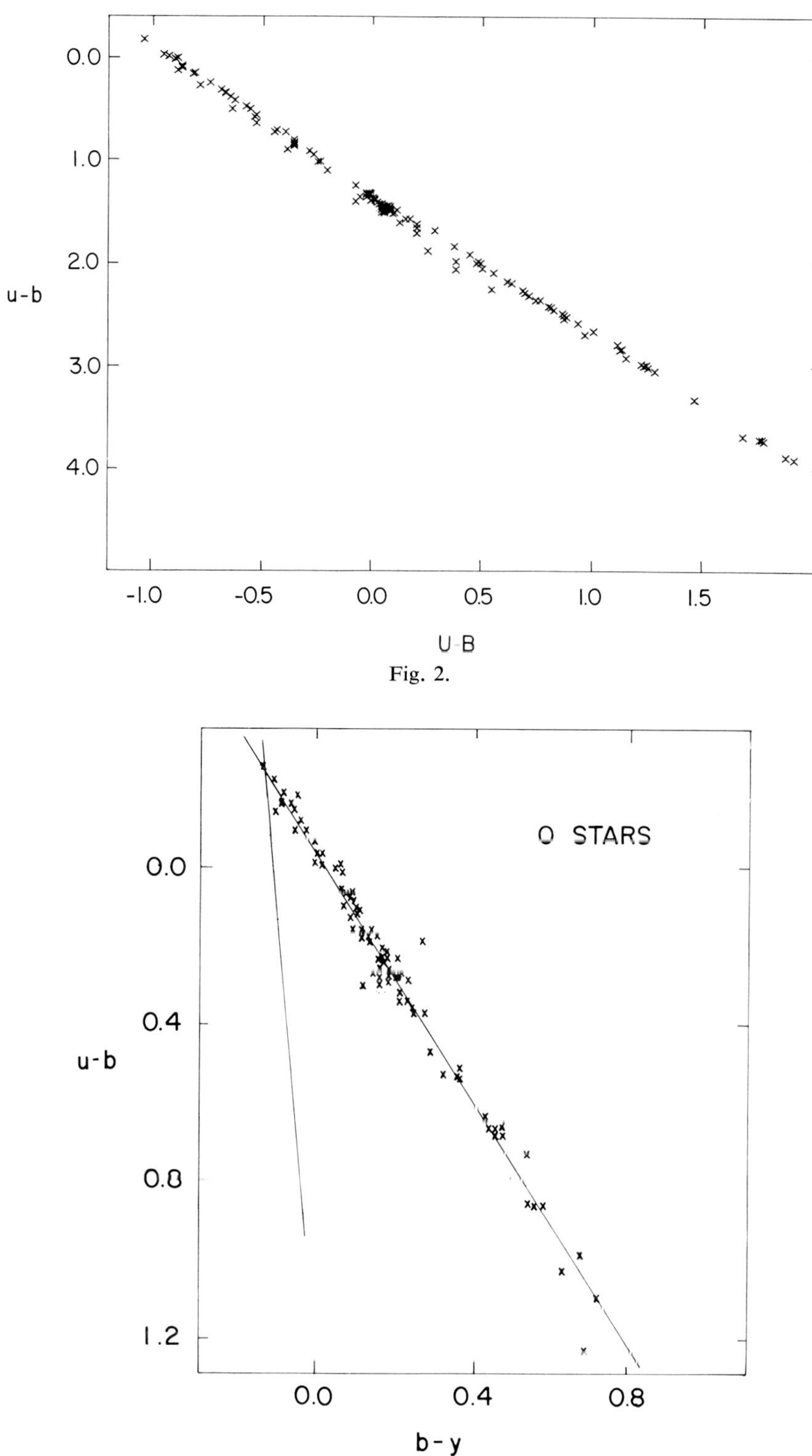

Fig. 2.

Fig. 3.

D. L. CRAWFORD

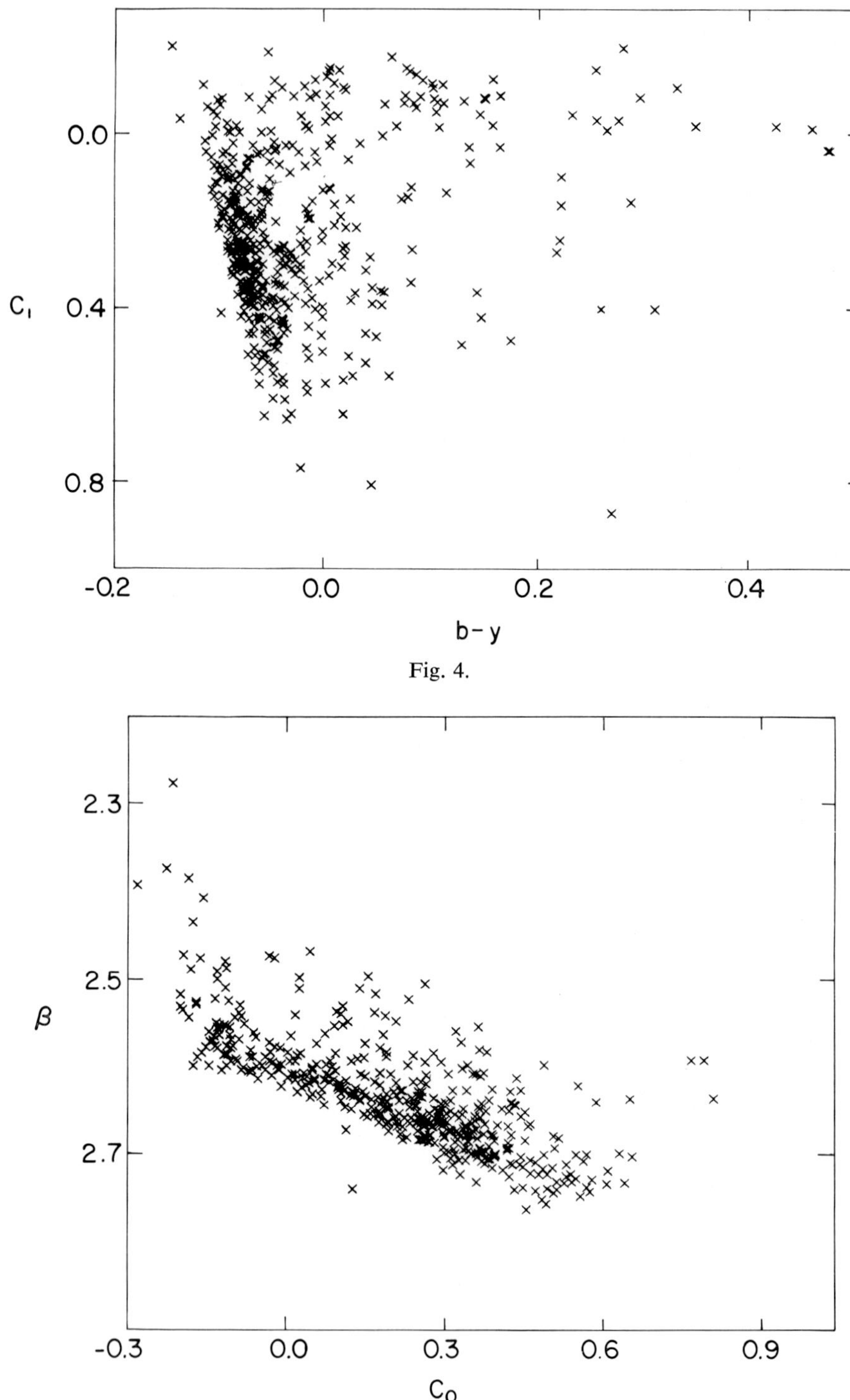

Fig. 4.

Fig. 5.

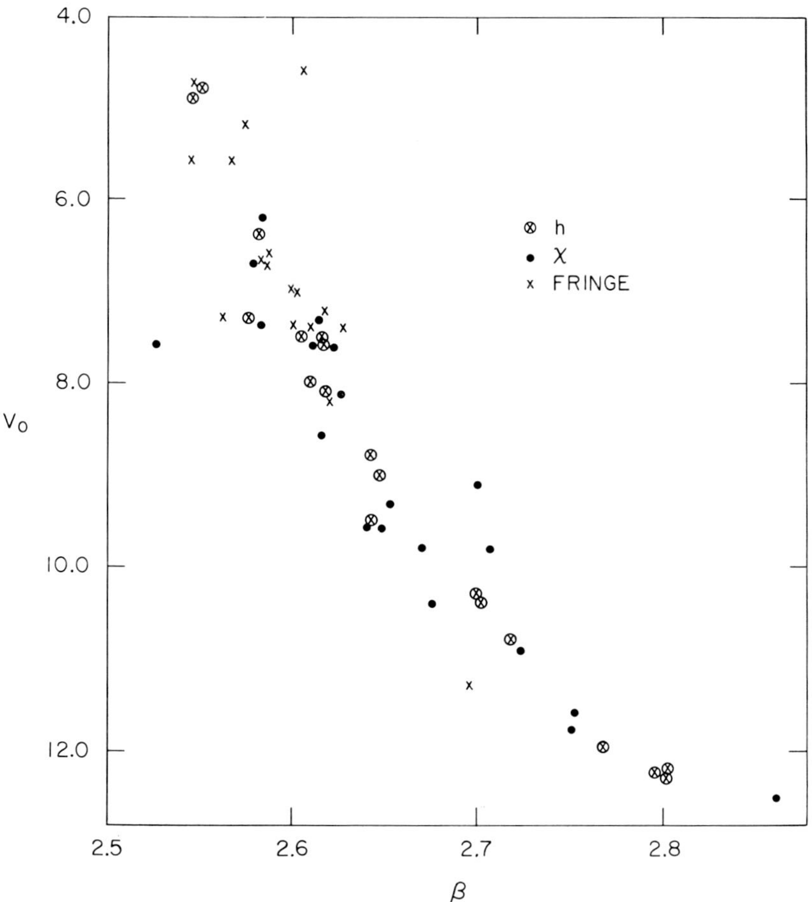

Fig. 6.

TABLE I

MK type	C_0	$(b-y)_0$	β (ZAMS)	M_v	$(U-B)_0$
$\leqslant$ O8	-0.15		2.575:	-5.5	-1.15
O9	-0.12	-0.13	2.590	-4.6	-1.10
B0	-0.07	-0.12	2.608	-3.9	-1.05
B1	0.02	0.11	2.629	2.9	0.96
B2	0.15	-0.10	2.658	-1.9	-0.84
B3	0.33	-0.09	2.701	-1.0	-0.67:
B4	0.37	-0.08	2.709	-0.8	-0.63·
B5	0.42	-0.07	2.720	-0.6	-0.59

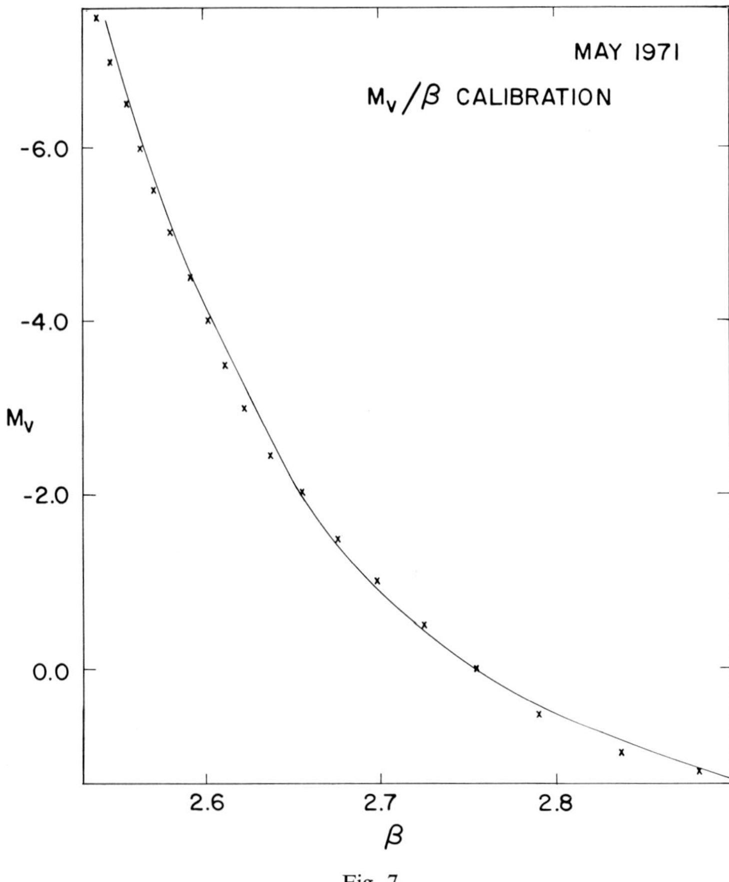

Fig. 7.

cribed by Crawford *et al.* (1970). Figure 3 shows the relations between $(u-b)$ and $(b-y)$ for a number of O-type stars.

The relation between the reddening lines for $(u-b)/(b-y)$ and the other indices are

$$E(c_1) = 0.2E(b-y)$$
$$E(m_1) = -0.3E(b-y)$$
$$E(b-y) = 0.7E(B-V)$$
$$E(u-b) = 1.7E(b-y).$$

Figure 4 shows the relation between c_1 and $b-y$ for the bright B-type stars. The rather sharp left envelope can be taken as the preliminary intrinsic color relation: $(b-y)_0$ in terms of c_1. Figure 5 shows the relation between $β$ and c_1 for the bright B-type stars. The $β$ parameter is primarily a measure of luminosity and the c_1 parameter a measure of temperature. The lower envelope is the zero age line; data for stars of the youngest clusters lie nearly along this envelope. Comparison of the location of the data points in this diagram with the star's MK type indicates a very good relation between the two systems, though there are certainly some deviating stars. Effects due to emission of $H\beta$ or $V \sin i$ effects apparently do not cause serious problems.

Time, nor the conference subject, do not permit a discussion of calibration problems here, but Figure 6 shows the relations between β and V_0 for the stars of the h and χ Per group. Similar relations exist for other groups, and a fitting of such individual relations leads to a calibration of M_v in terms of β (see Figure 7) – the second dimension for the photometric classification: the c_0 parameter being the first dimension.

Details of the classification and the calibrations will be published shortly in the *Astronomical Journal*.

Reference

Crawford, D. L., Glaspey, J. W., and Perry, C. L.: 1970, *Astron. J.* **75**, 822.

MULTICOLOR PHOTOMETRY OF THE GALACTIC CLUSTER
NGC 2362

CH. L. PERRY

Louisiana State University Observatory, Baton Rouge, La., U.S.A.

Abstract. U, B, V colors of CPD stars brighter than $12^{m}0$ in the central region of NGC 2362. Color excess. Discussion of the distance modulus of the cluster.

1. Introduction

The galactic cluster NGC 2362 is a sparsely populated but tightly knit aggregate of early-type stars dominated by the fourth magnitude O-type star τ CMa. The cluster has been of considerable astronomical interest. Because it is only a few million years old, NGC 2362 has played a vital rôle in the extension of the UBV zero-age main sequence toward the blue and bright end of the color-magnitude diagram; see, for example, an excellent discussion by Blaauw (1963), where references are given to earlier papers also involved with the development of the main-sequence fitting procedure. Likewise, the cluster will be an important addition in the development of the $uvby$ and Hβ photometric systems. Recent observational investigations of the cluster include a photometric study by Johnson (1950) and a combined UBV and MK classification study by Johnson and Morgan (1953). The present photometric investigation was undertaken because of both the intrinsic importance of the cluster and the variance in previous determinations of its distance.

2. Observations

All CPD stars brighter than $12^{m}0$ within an area of the sky between $7^{h}13^{m}00^{s}$ and $7^{h}14^{m}20^{s}$ (1875) of right ascension and between $-24°35'$ and $-24°50'$ of declination were chosen as the major part of the observing program. The four non-CPD stars brighter than $12^{m}0$, as observed by Johnson and Morgan, completed the program. Additional UBV observations were made to supplement the existing data. $uvby$ and Hβ photometry was obtained for the majority of the stars in the immediate cluster region.

Details of the instrumentation, observing techniques, and reduction procedures have been described by Perry and Hill (1969). Table I lists the internal mean errors of a single observation. Henceforth, mean errors are quoted throughout the paper. Table II gives, in succession, the CPD numbers, the UBV, $uvby$, and Hβ values obtained in this investigation, along with the number of nights on which observations of a given star were made; each star was usually observed twice during a given night. The V magnitudes on the UBV system, listed in the sixth column, were derived from the y deflections of the *uvby* photometry. *UBV* observations of the star $-24°2251$ were made but its photometric values are not included in Table II because its colors were outside the color range of the standard stars. It is a late-type object.

TABLE I
Internal mean errors of a single observation

V	$B-V$	$U-B$	V	$b-y$	m_1	c_1	β
$0^{m}\!019$	$0^{m}\!013$	$0^{m}\!017$	$0^{m}\!019$	$0^{m}\!013$	$0^{m}\!017$	$0^{m}\!019$	$0^{m}\!012$

The *UBV* photometries of Johnson and Morgan and Table II were combined into a homogeneous system. The *V* magnitudes derived from the *uvby* photometry were also included. Table III lists the CPD numbers, the Johnson numbers, the binary star numbers from the *Index Catalogue of Visual Double Stars* by Jeffers and van den Bos (1963), the final *UBV* values, along with the number of observations. The MK spectral types, listed in the last column, are due to Johnson and Morgan or Schild (1970). The photometry for the four stars indicated in the fourth column of the table refers to the combined light of the binary components.

3. Results

A. COLOR EXCESS OF NGC 2362

The intrinsic colors of the program stars were determined by four methods, namely, (1) the *UBV* Q-method, (2) Strömgren's (1966) $[u-b]$ calibration, (3) Crawford's (1970) c_0 calibration, and (4) MK spectral types. These methods will be discussed in some detail in turn.

The *UBV* color indices listed in Table III were employed to compute the individual intrinsic color indices via the equations $(B-V)_0 = 0.335(U-B) - 0.241(B-V)$ and $(U-B)_0 = 1.241(U-B) - 0.893(B-V)$. The coefficients in the above relations are appropriate for a reddening slope in the *UBV* color-color diagram of $E(U-B)/E(B-V) = 0^{m}\!72$. A discussion of the *UBV* Q-method is given, for example, by Morgan and Harris (1956). Individual color excesses in both colors were next calculated via the equations $E(B-V) = (B-V) - (B-V)_0$ and $E(U-B) = (U-B) - (U-B)_0$. The twenty-nine program stars, selected as clusters members, yielded mean cluster color excesses of $E(B-V) = 0^{m}\!100 \pm 0^{m}\!020$ and $E(U-B) = 0^{m}\!070 \pm 0^{m}\!014$. The questions of cluster membership is discussed in the next section.

Figure 1 illustrates the *UBV* color-color diagram for NGC 2362 plotted with the data in Table III. The solid line traces the standard relation derived by Johnson and Morgan, shifted by $0^{m}\!10$ in $(B-V)$ and $0^{m}\!07$ in $(U-B)$. The small scatter about the standard relation indicates that the reddening is nearly uniform over the cluster region.

An ultraviolet color index for the *uvby* photometric system may be derived via the

CH. L. PERRY

TABLE II

Multicolor photometry of NGC 2362

CPD	V	$B-V$	$U-B$	n	V	$b-y$	m_1	c_1	n	β	Notes
$-24°2182$	$9^{m}31$	$-0^{m}11$	$-0.^{m}76$	5							(1)
	11. 80	0. 48	0. 02	4							
$-24°2185$	8. 91	-0.13	-0.84	5							
$-24°2188$	10. 05	-0.08	-0.66	6							
$-24°2195$					$11^{m}40$	$0^{m}027$	$0^{m}092$	$0^{m}721$	2	$2^{m}724$	2
$-24°2196$					10.76	0.006	0.086	0.086	2	2.699	2
$-24°2197$										2.796	2
$-24°2198$					11.17	-0.001	0.123	0.534	2	2.738	2
$-24°2199$	11.46	-0.02	-0.32	4	11.46	0.036	0.091	0.584	2	2.768	2
$-24°2200$	11.93	-0.01	-0.22	4	11.93	0.036	0.107	0.745	2		
$-24°2201$	10.50	0.10	0.08	4						2.851	2
$-24°2203$					9.80	-0.030	0.101	0.262	4	2.676	2
$-24°2204$					9.98	-0.016	0.084	0.306	3	2.705	2
$-24°2205$					9.48	-0.038	0.084	0.152	5	2.656	2
$-24°2206$					11.75	0.031	0.128	0.817	2		
$-24°2207$					9.51	-0.044	0.086	0.105	5	2.650	2
$-24°2208$										2.798	2
	11.63	0.24	0.10	4							(2)
$-24°2211$	11.90	0.00	-0.20	3	11.90	0.032	0.128	0.733	2		
$-24°2212$					10.43	-0.008	0.088	0.373	2	2.674	2
$-24°2213$					8.76	-0.053	0.080	-0.021	5	2.620	2
					9.48	-0.047	0.055	0.105	5	2.658	2
											(3)
	11.91	0.01	-0.19	4	11.91	0.043	0.097	0.796	2		(4)
$-24°2216$	4.39	-0.15	-1.00	std	4.39	-0.045	0.065	-0.141	std	2.564	std
$-24°2217$	10.98	-0.04	-0.43	4	10.98	0.018	0.091	0.477	2	2.714	2
$-24°2218$	10.77	0.00	-0.42	7	10.77	0.028	0.108	0.500	2	2.741	2
$-24°2220$										2.736	2
$-24°2221$	10.60	0.15	0.09	4							
$-24°2222$	11.91	0.06	-0.19	4	11.91	0.080	0.104	0.717	2		

Table II (continued)

CPD	V	$B-V$	$U-B$	n	V	$b-y$	m_1	c_1	n	β	Notes
$-24°2223$					10.38	-0.026	0 117	0.363	2	2.685	2
$-24°2225$					10.15	-0.045	0 107	0.228	3	2.692	2
$-24°2226$	10.76	0.28	0.22	5						2.800	2
$-24°2228$					$8^{\mathrm{m}}18$	$-0^{\mathrm{m}}063$	$0\overset{\mathrm{m}}{.}070$	$-0^{\mathrm{m}}044$	5	$2^{\mathrm{m}}614$	2
$-24°2230$	$11^{\mathrm{m}}33$	$-0^{\mathrm{m}}04$	$-0^{\mathrm{m}}33$	6	11.33	0.023	0 089	0.644	2	2.746	2
$-24°2231$					9.30	-0.044	0 083	0.090	5	2.636	2
$-24°2232$	11.06	-0.05	-0.52	4							
$-24°2235$	10.91	0.16	0.12	3	10.91	0.130	0 143	1.088	2		
$-24°2236$	10.50	-0.03	-0.49	3	10.50	0.025	0 094	0.364	2	2.704	2
$-24°2240$										2.689	2
$-24°2244$	11.34	-0.01	-0.38	7	11.34	0.035	0.085	0.577	2	2.739	2
$-24°2246$	10.39	0.44	-0.02	5							
$-24°2248$	6.80	-0.15	-0.72	std	6.80	-0.073	0.089	0.219	std	2.634	std
$-24°2250$	9.79	-0.05	-0.55	5							
$-24°2253$	10.75	0.27	0.13	5							
$-24°2254$	10.70	0.01	-0.10	4							

Notes to Table II
(1) Johnson No. 64.
(2) Johnson No. 65
(3) Johnson No. 39
(4) Johnson No. 22.

CH. L. PERRY

TABLE III
Combined *UBV* photometry of NGC 2362

CPD	Johnson		V	n	B − V	n	U − B	n	Spt
−24°2182			9ᵐ31	5	−0ᵐ11	5	−0ᵐ76	5	
	64		11.80	6	0.48	6	0.02	4	
−24°2185			8.91	5	−0.13	5	−0.84	5	B8
−24°2188			10.05	6	−0.08	6	−0.66	6	
−24°2195	1		11.39	5	0.00	3	−0.28	3	
−24°2196	5		10.77	5	−0.06	3	−0.52	3	B5V:
−24°2197	3		11.05	3	0.09	3	−0.15	3	
−24°2198	2		11.18	5	−0.02	3	−0.38.	3	
−24°2199	66		11.46	8	−0.02	6	−0.32	6	
−24°2200	11		11.93	8	0.00	6	−0.22	4	
−24°2201	13		10.49	6	0.10	6	0.08	6	
−24°2203	9		9.82	7	−0.08	3	−0.61	3	B3V
−24°2204	12		10.00	6	−0.07	3	−0.57	3	B3:nn
−24°2205	48		9.50	8	−0.11	3	−0.69	3	B3V
−24°2206	15		11.76	5	0.03	3	−0.14	3	B9V:
−24°2207	14		9.54	8	−0.12	3	−0.73	3	B2V
−24°2208	16		10.57	3	0.07	3	−0.20	3	B9V:
	65		11.64	6	0.24	6	0.10	4	
−24°2211			11.90	5	0.00	3	−0.20	3	
−24°2212	21		10.44	5	−0.07	3	−0.51	3	B3Vn
−24°2213	20		8.77	9	−0.15	4	−0.87	4	B1.5V
	39		9.48	5	−0.09	3	−0.67	3	B2V
	22		11.92	8	0.01	6	−0.19	4	
−24°2216	23	250 − 07(AB)/10(Ap)	4.39	std	−0.14	std	−1.00	std	09III
−24°2217	24		10.99	8	−0.04	6	−0.43	6	B7nn
−24°2218	25		10.77	10	0.00	8	−0.42	8	
−24°2220	50		10.20	3	0.00	3	−0.36	3	B6V
−24°2221		250 − 06(AB)	10.60	4	0.15	4	0.09	4	
−24°2222	52		11.92	8	0.05	6	−0.19	4	
−24°2223	26		10.41	5	−0.07	3	−0.52	3	B5V
−24°2225	30		10.16	6	−0.07	3	−0.58	3	B3V
−24°2226	32		10.76	6	0.28	6	0.22	5	
−24°2228	30	250 − 09(D)	8.20	10	−0.17	5	−0.91	5	B1V
−24°2230	29		11.33	9	−0.04	7	−0.33	7	
−24°2231	31		9.31	8	−0.12	3	−0.76	3	B2V
−24°2232			11.06	4	−0.05	4	−0.52	4	
−24°2235			10.91	5	0.16	3	0.12	3	
−24°2236	34		10.50	7	−0.03	5	−0.49	5	B5V
−24°2240	36		10.76	3	0.02	3	−0.48	3	B7:nn
−24°2244	42		11.34	11	−0.01	9	−0.38	7	
−24°2246			10.39	5	0.44	5	−0.02	5	
−24°2248	46	250 − 50(AB)	6.80	std	−0.16	std	−0.72	std	B2IV
−24°2250		50 − 49(AB)	9.79	5	−0.05	5	−0.55	5	
−24°2253			10.75	5	0.27	5	0.13	5	
−24°2254			10.70	4	0.01	4	−0.10	4	

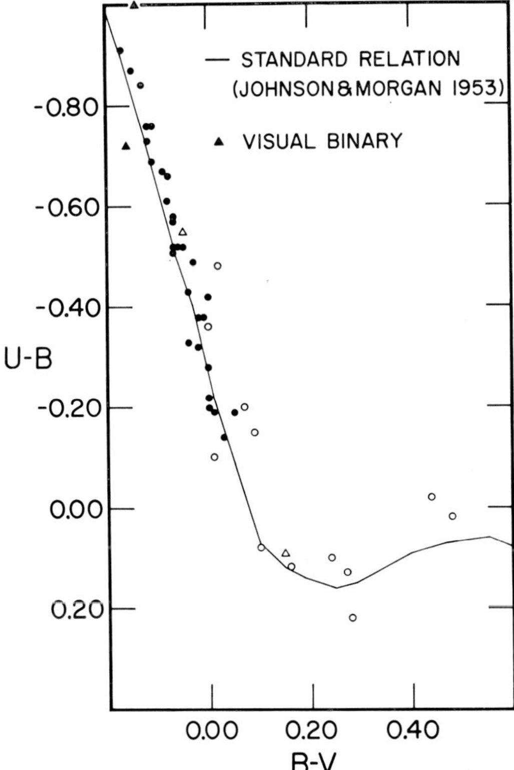

Fig. 1.   The *UBV* color-color diagram for NGC 2362. The solid line traces the standard relation of Johnson and Morgan (1953). Visual binaries are indicated.

equation $(u-b)=c_1+?m_1+?(b-y)$. Strömgren (1966) has formed an index $[u-b]$ which is independent of interstellar reddening; the index is given by the equation $[u-b]=(u-b)-1.84(b-y)$. On the basis of *uvby* photometry of nearby and hence unreddened B stars, Strömgren found that $(b-y)_0$ is a function of $[u-b]$ only. After values of $(u-b)$ and $[u-b]$ were computed for those stars in Table II with *uvby* photometry, the intrinsic $(b-y)_0$ colors were interpolated from Strömgren's calibration. Individual color excesses were next calculated via the equation $E(b-y)-(b-y)-(b-y)_0$. The nineteen program stars, selected as cluster members, yielded a mean cluster color excess of $E(b-y)=0\overset{m}{.}090\pm0\overset{m}{.}022$. Crawford (1964) has shown that the color excesses in the *UBV* and *uvby* photometric systems are related by the equation $E(B-V)=1.43E(b-y)$; the above value of $E(b-y)$ is therefore equivalent to $E(B-V)=0\overset{m}{.}129\pm0\overset{m}{.}031$.

On the basis of *uvby* photometry of bright and hence unreddened B stars, Crawford (1970) has devised an iterative procedure to determine the intrinsic $(b-y)_0$ colors of early-type stars. One employs the observed c_1 values to obtain the first approximation to the $(b-y)_0$ colors via the equation $(b-y)_0=-0.116+0.097c_1$. The color excesses are computed in the usual manner. Crawford (1966) has shown that the color excesses in $(b-y)$ and c_1 are related by the equation $E(c_1)=0.2E(b-y)$. The observed c_1 values are corrected for reddening via the equation $c_0=c_1-E(c_1)$. The second approximation to the $(b-y)_0$ colors is obtained by substituting c_0 in place of c_1 in the first equation. One iteration was sufficient to calculate the individual $(b-y)_0$ colors and color excesses for those stars in Table II with *uvby* photometry. The twenty-three program stars, selected as cluster members, yielded a mean cluster color excess of $E(b-y)=0\overset{m}{.}075\pm0\overset{m}{.}015$. This value corresponds to $E(B-V)=0\overset{m}{.}107\pm0\overset{m}{.}021$.

The tabulation by Johnson (1963) of intrinsic $(B-V)_0$ and $(U-B)_0$ colors as a function of MK spectral type allows one to determine the $(B-V)_0$ colors for those program stars listed in Table III with MK spectral types. Individual color excesses were next computed. The fifteen program stars, selected as cluster members, yielded a mean cluster color excess of $E(B-V)=0\overset{m}{.}112\pm0\overset{m}{.}022$.

B. DISTANCE MODULUS OF NGC 2362

The absolute magnitudes of the program stars were also determined by four methods, namely, (1) the intrinsic $(B-V)_0$ colors, (2) Fernie's (1965) (β, M_v) calibration adopted by Strömgren (1966), (3) Crawford's (1970) (β, M_v) calibration, and (4) MK spectral types. These methods will be discussed in some detail in turn.

In all four methods, the visual apparent magnitudes were corrected for interstellar absorption via either the equation $V_0=V-3.0E(B-V)$, derived by Morgan, Harris, and Johnson (1953), or $V_0=V-4.3E(b-y)$, derived by Crawford (1966).

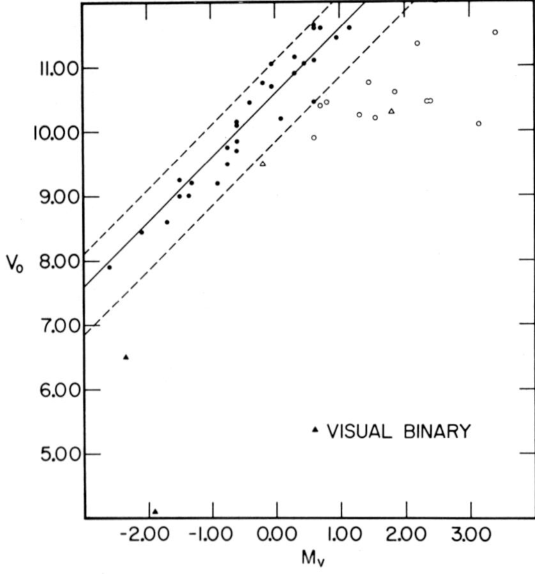

Fig. 2.   The $UBV(V_0,M_v)$ diagram for NGC 2362. Probable cluster members are denoted by filled circles, non-members by open circles. Visual binaries are indicated. The dashed lines are discussed in the text.

Following the example of Walker (1965), cluster membership was investigated by means of (V_0, M_v) diagrams in all four methods. Figure 2 illustrates the UBV (V_0, M_v) diagram for NGC 2362. Stars within the domain defined by the dashed lines are considered to be cluster members. These limits were set by assuming that no cluster member will lie more than $0^{m}\!.5$ above the line $V_0 = M_v + \text{constant}$ and that duplicity will not brighten a star more than $0^{m}\!.75$. The brightest members of a cluster may lie outside the membership domain because they are in the process of evolving away from the main sequence.

Blaauw (1963) has tabulated visual absolute magnitude as a function of $(B-V)_0$ for the zero-age main sequence. The twenty-seven program stars, selected as cluster members, in Table III with UBV photometry yielded a mean cluster distance modulus of $10^{m}\!.60 \pm 0^{m}\!.26$.

Figure 3 illustrates the UBV color-magnitude diagram for NGC 2362. The solid line traces Blaauw's ZAMS. The absolute magnitude scale on the right-hand side of the diagram is computed for a distance modulus of $10^{m}\!.60$.

Strömgren (1966) adopted Fernie's (1965) (β, M_v) calibration. The calibration was derived using various galactic clusters and associations with well-determined distance

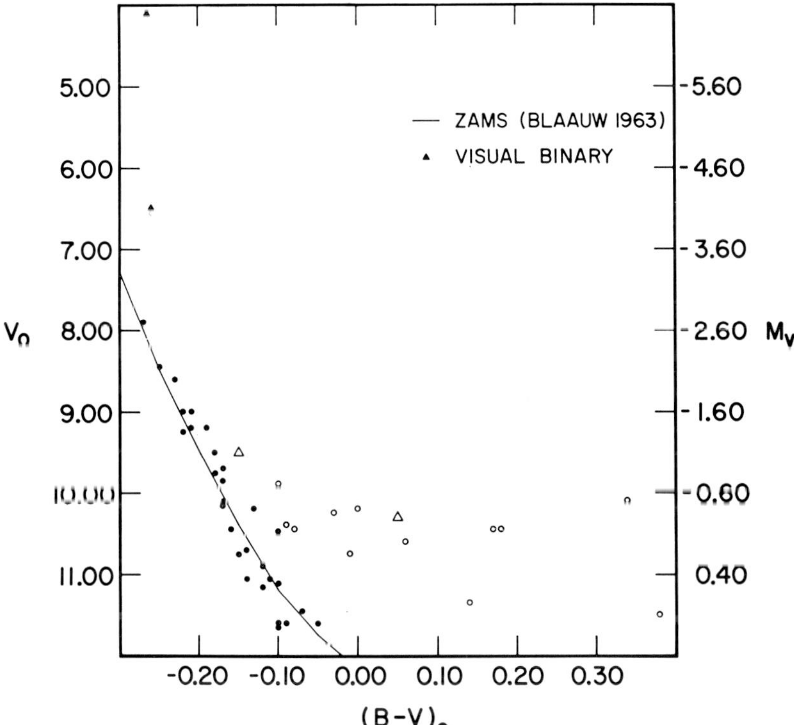

Fig. 3.   The UBV color-magnitude diagram for NGC 2362. Visual binaries are indicated. The solid line traces Blaauw's (1963) zero-age main sequence. The absolute magnitude scale on the right-hand side of the diagram is computed for a distance modulus of $10^{m}\!.60$.

moduli; the cluster members had measured hydrogen line intensities which Fernie reduced to the photoelectric β system. The nineteen program stars, selected as cluster members, in Table II with Hβ photometry yielded a mean cluster distance modulus of $11^{m}22 \pm 0^{m}24$.

Crawford (1970) derived his (β, M_v) calibration by fitting the shape of the (V_0, β) relations of various galactic clusters and associations; the zero point of the calibration was defined by the known distance moduli of the α Persei and Pleiades clusters. The eighteen program stars, selected as cluster members, in Table II with Hβ photometry yielded a mean cluster distance modulus of $11^{m}22 \pm 0^{m}26$.

The tabulation by Blaauw (1963) of absolute magnitude as a function of MK spectral type also allows one to determine the absolute magnitudes of those stars in Table III with MK spectral types. The fifteen program stars, selected as cluster members, yielded a mean cluster distance modulus of $11^{m}27 \pm 0^{m}30$.

4. Discussion

Table IV includes a summary of the results from the four determinations of the mean cluster color excess and corrected distance modulus of NGC 2362 made in this paper. The adopted values are averages of the four determinations weighted according to their mean errors, with the following exception. Because the (β, M_v) calibrations of Fernie and Crawford are quite similar, only one value of $V_0 - M_v = 11^{m}22$ was included in the weighted average. It should be pointed out that the color excess and distance modulus determined from the MK spectral types is not strictly compatible with the values determined by the three other methods. The former determination refers to stars found within the main-sequence band while the three latter determinations refer to the ZAMS. In addition, Table IV lists the previous determinations by Johnson, and Johnson and Morgan. In both of these papers, the authors assumed that interstellar reddening in the cluster was negligible.

Other investigators, including Johnson and Hiltner (1956), Sandage (1957), Johnson (1957), Johnson and Iriarte (1958), and Blaauw (1963) have derived estimates of the color excess and corrected distance modulus of the cluster by the main-sequence fitting

TABLE IV

Summary of reddening and distance determinations

Source	Type	$\bar{E}(B-V)$	No.	$\overline{V_0 - M_v}$	No.
Johnson (1950)		$0^{m}00$		$10^{m}75$	
Johnson and Morgan (1953)	UBV	0.00		11.9	
Perry's Method No. 1	MK	0.112 ± 0.022	15	11.27 ± 0.30	15
Perry's Method No. 2	UBV	0.100 ± 0.020	29	10.60 ± 0.26	27
Perry's Method No. 3	$uvby - H\beta$	0.129 ± 0.031	19	11.22 ± 0.24	19
Perry's Method No. 4	$uvby - H\beta$	0.107 ± 0.021	23	11.22 ± 0.26	18
Perry's Adopted Mean		0.109		11.02	

procedure. Their estimates cluster about the values of $E(B-V)=0\overset{m}{.}11$ and $V_0 - M_v = 10\overset{m}{.}8$.

As stated above, the question of cluster membership was investigated by means of (V_0, M_v) diagrams. The *UBV* analysis indicates that 14 of the 45 program stars (Johnson No. 64, $-24°2197$, $-24°2201$, $-24°2208$, Johnson No. 65, $-24°2220$, $-24°2221$, $-24°2226$, $-24°2235$, $-24°2240$, $-24°2246$, $-24°2250$, $-24°2253$, and $-24°2254$) are not cluster members. The other three analyses almost always confirm this conclusion.

Acknowledgements

The writer expresses his thanks to Dr V. M. Blanco, Director of the Cerro Tololo Inter-American Observatory, for permission to use the observatory facilities. He also acknowledges the helpful assistance of Sr Carlos Bolelli at the telescope. Last but not least, the writer thanks Dr Rudolph Schild for his spectral classifications used in this paper.

This research was supported by a grant from the National Science Foundation.

References

Blaauw, A.: 1963, in K. Aa. Strand (ed.), *Basic Astronomical Data*, University of Chicago Press, Chicago, p. 383.
Crawford, D. L. and Strömgren, B.: 1964, *Vistas in Astronomy* **8**, 149.
Crawford, D. L.: 1966, in K. Lodén, L. O. Lodén, and U. Sinnerstad (eds.), 'Spectral Classification and Multicolor Photometry', *IAU Symp.* **24**, 170.
Crawford, D. L.: 1970, in A. Slettebak (ed.), *Stellar Rotation*, D. Reidel, Dordrecht, Holland, p. 114.
Fernie, J. D.: 1965, *Astron. J.* **70**, 575.
Jeffers, H. M. and van den Bos, W. H.: 1963, *Publ. Lick Observ.* **21**, 7.
Johnson, H. L.: 1950, *Astrophys. J.* **112**, 240.
Johnson, H. L.: 1957, *Astrophys. J.* **126**, 121.
Johnson, H. L.: 1963, in K. Aa. Strand (ed.), *Basic Astronomical Data*, Univesity of Chicago Press, Chicago, p. 204
Johnson, H. L. and Morgan, W. W.: 1953, *Astrophys. J.* **117**, 313.
Johnson, H. L. and Hiltner, W. A.: 1956, *Astrophys. J.* **123**, 267.
Johnson, H. L. and Iriarte, B.: 1958, *Lowell Obs. Bull.* **4**, 47.
Morgan, W. W., Harris, D. L., and Johnson, H. L.: 1953, *Astrophys. J.* **118**, 92.
Morgan, W. W. and Harris, D. L.: 1956, *Vistas in Astronomy* **2**, 1124.
Perry, C. L. and Hill, G.: 1969, *Astron. J.* **74**, 899.
Sandage, A.: 1957, *Astrophys. J.* **125**, 435.
Schild, R.: private communication.
Strömgren, B.: 1966, in L. Goldberg (ed.), *Ann. Rev. Astron. Astrophys.* **4**.
Walker, G. A. H.: 1965, *Astrophys. J.* **141**, 660.

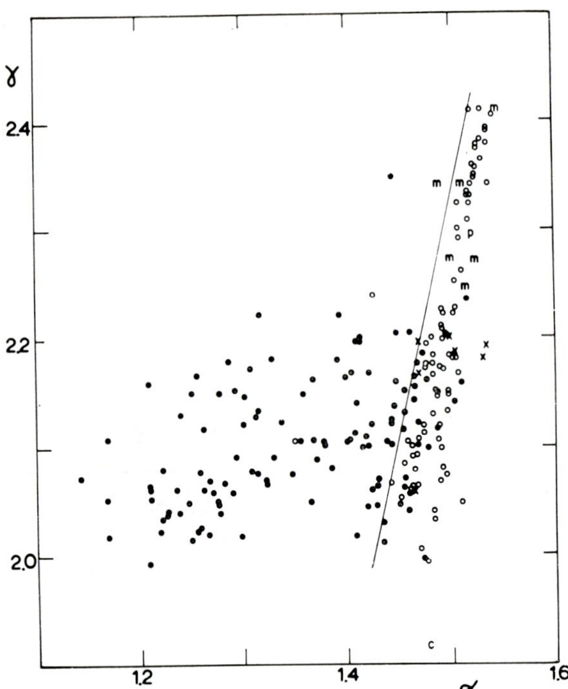

Fig. 2.   The (α,γ) diagram for all the observed stars.

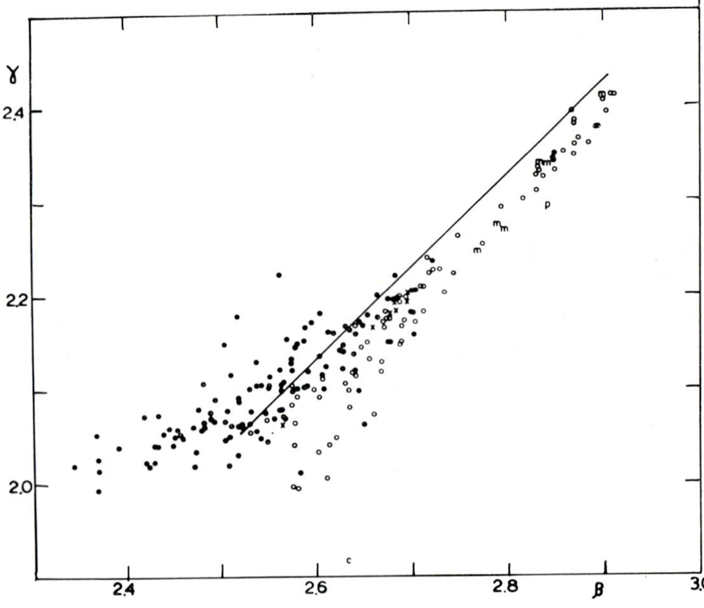

Fig. 3.   The (β,γ) diagram for all the observed stars.

PHOTOELECTRIC MEASURES OF Hα, Hβ AND Hγ
IN EARLY-TYPE STARS

A. FEINSTEIN

Observatorio Astrónomico, Universidad Nacional de La Plata, Argentina

Abstract. The results of over 200 stars measured with interference filters in Hα, Hβ and Hγ are given here. 113 Be stars from the northern and southern hemispheres were included. The comparison of 60 of them, in common with the observations made by Mendoza about 15 years ago, shows that nearly half of them have differences larger than $0^\text{m}030$ in Hβ (Crawford system). On the other hand the comparison with the observations made recently by Crawford indicate variations in a short time for some active Be stars.

During 1970 the Hα, Hβ and Hγ lines in over 200 stars in both hemispheres were measured with interference filters. These observations made at the Kitt Peak National Observatory and at the Cerro Tololo Inter-American Observatory were carried out using the photomultiplier 1P21 and with a combination of a narrow and a wide interference filter, both centered at each line. Thus, the same procedure was employed that Crawford and associates do for their measures of the Hβ line.

The β index related to the measures of the Hβ line is in the system established by Crawford and Mander (1966), since their standard stars were measured. For the Hα and Hγ lines all the observations were reduced to a standard system defined by these same stars, which give the α and γ indices.

The data included observations of 81 standard stars (Crawford and Mander, 1966; Crawford *et al.*, 1970), 118 Be stars (Mendoza, 1958; Feinstein, 1968), 7 helium-weak stars (Jaschek *et al.*, 1969), 6 metallic-line stars, one peculiar star (α Circinis), and the cepheid *l* Carinae.

For all observed stars, the Figures 1, 2 and 3 show the relation between the three indices: α, β and γ. A clear separation between the standard and the Be stars results from these three diagrams, as smaller numbers mean that the line is in emission. However, some stars classified as Be are contained in the same part of the diagrams as the standards, which suggests that at the epoch they were measured, the hydrogen lines had no emission.

From these three diagrams we can obtain the condition which an early-type star must satisfy in order to be an emission-line star. Nearly all the Be stars have smaller values of the indices: α, β and γ, in comparison with normal B stars. Then, from (α, β) and (α, γ) diagrams we may conclude that any star with $\alpha < 1.44$ is a Be star. This condition includes nearly all the observed Be stars, as only very few stars with larger α values also have emission. There is also a condition for the Hβ line, since from the (α, β) and (β, γ) diagrams we see that any star with an index $\alpha < 2.50$ is a Be star, as was already pointed out by Haug (1970). However, this is not fulfilled by approximately half of the Be stars. For the γ index it does not seem possible to establish a

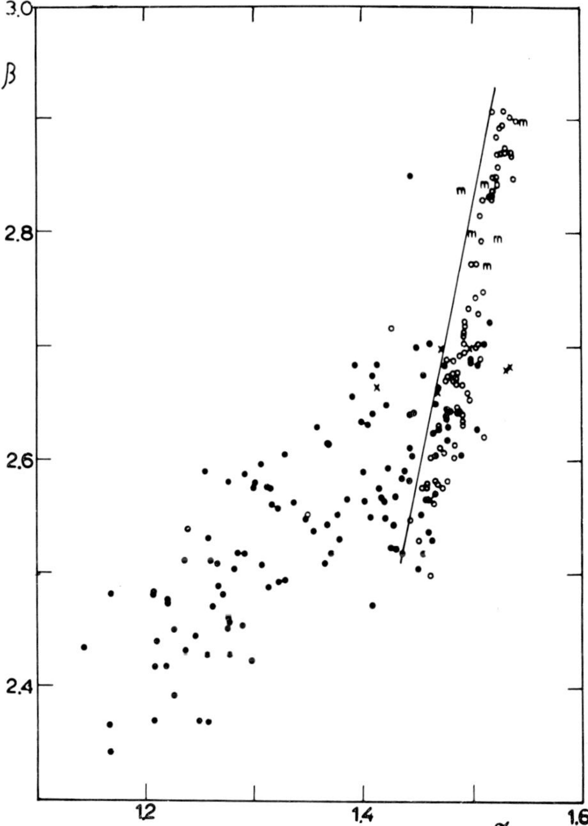

Fig. 1.   The (α,β) diagram for all the observed stars. Open circles denote standard stars, dots Be stars, crosses helium-weak stars, *m* metallic-line stars, *p* the peculiar star α Circinis, and *c* the cepheid *l* Carinae.

condition, as the emission in this line is always small, much smaller than in Hα and in Hβ.

Figure 4 presents the diagram $(\alpha, \beta-\gamma)$ for the observed stars, which also shows very clearly the condition for the Be stars, that is $\alpha < 1.44$.

The comparison of the Hβ measures of the Be stars with observations obtained by Mendoza before 1958 (private communication) shows that of 63 stars in common, 25 of them display differences larger than $0^{m}.03$ (Table I). It seems to be no systematic differences as there are nearly equal number of positive and negative values. The data indicates that about 40% of the Be stars have variable emission in the Hβ line in a time interval of around 13 yr.

A similar comparison with the measures for southern Be stars obtained between 1963 and 1969 (Crawford *et al*, 1970) is plotted in Figure 5. It indicates that of 21 stars in common, 6 have differences larger than $0^{m}.03$, but although in this case the time interval is shorter, the percentage is nearly the same.

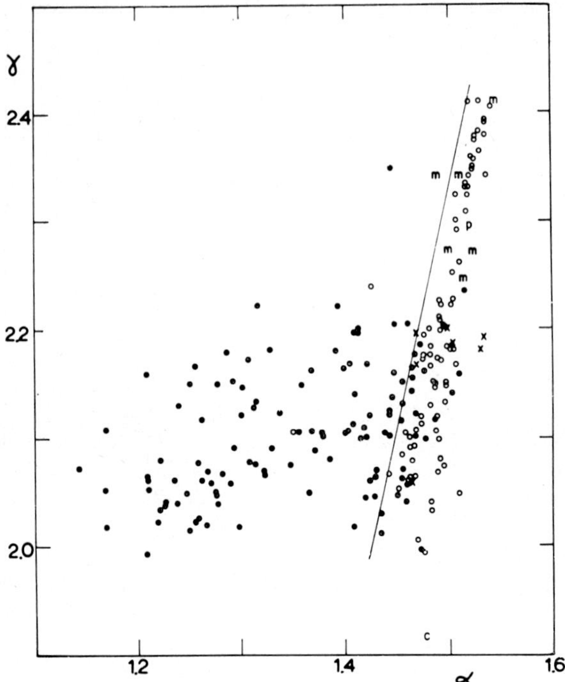

Fig. 2.   The (α,γ) diagram for all the observed stars.

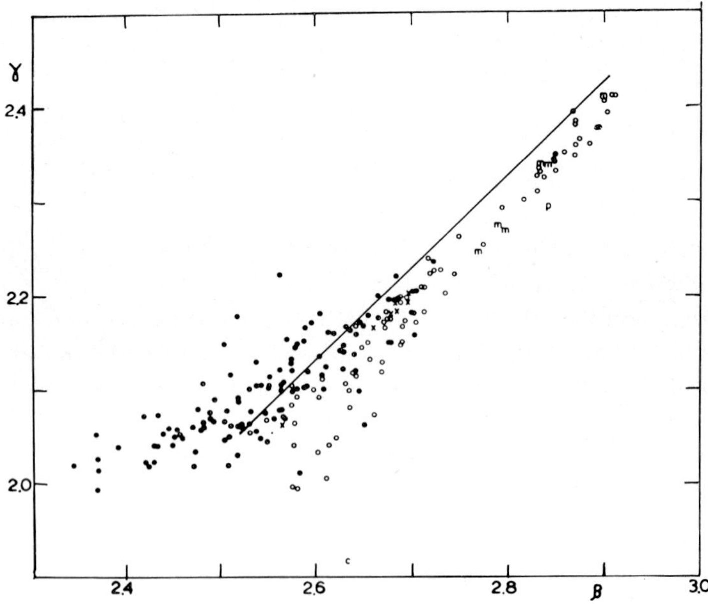

Fig. 3.   The (β,γ) diagram for all the observed stars.

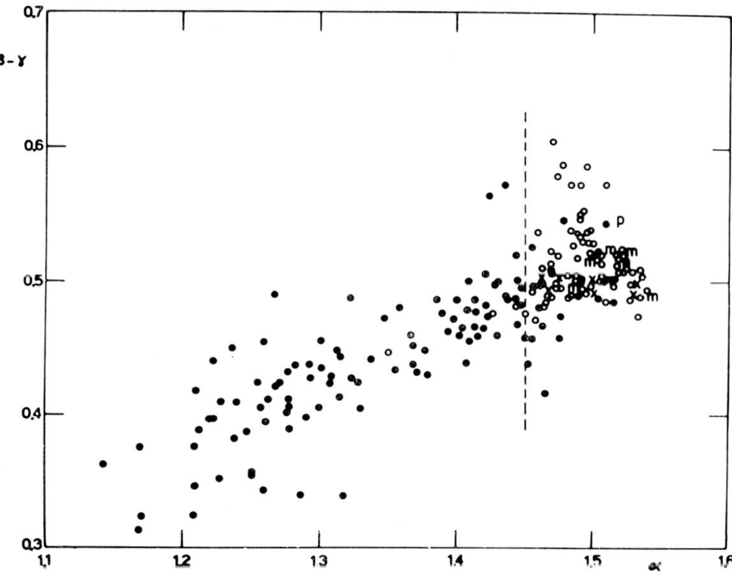

Fig. 4. The $(\alpha, \beta-\gamma)$ diagram for all the observed stars. The vertical line at $\alpha = 1.44$ separates the Be from the normal stars.

TABLE I

Comparison of β indices from measures made by
Feinstein and Mendoza

HD	β_F	β_M	$\beta_F - \beta_M$
5394	2.370	2.484	-0.114
11606	2.850	2.508	$+0.342$
19243	2.429	2.509	-0.080
20336	2.604	2.566	$+0.038$
21650	2.589	2.627	-0.038
22780	2.665	2.749	-0.084
23480	2.614	2.660	-0.046
28497	2.458	2.572	-0.114
32343	2.504	2.560	-0.056
33604	2.489	2.523	-0.034
41117	2.543	2.415	$+0.128$
43285	2.684	2.630	$+0.054$
45910	2.432	2.400	$+0.032$
58050	2.493	2.434	$+0.059$
60855	2.606	2.538	$+0.068$
83953	2.580	2.483	$+0.097$
184279	2.576	2.609	-0.033
187567	2.518	2.606	-0.088
187811	2.646	2.565	$+0.081$
189687	2.629	2.539	$+0.090$
191610	2.538	2.623	-0.085
200120	2.531	2.494	$+0.038$
206773	2.369	2.400	-0.031
212571	2.391	2.460	-0.069
217891	2.605	2.666	-0.061

Another interesting point to investigate is the Balmer decrement of the normal stars. Figure 6 shows the relation of the α vs the $(\alpha-\beta)$ indices for the standard stars, with the spectral type for each star given. Beginning with the star with earlier spectrum, around O9, it is found that the $(\alpha-\beta)$ values decrease with later spectral types whereas the α index increases slightly until around A0. At this point the relation takes the reverse direction, that is the $(\alpha-\beta)$ values increase with later spectral types. The two sequences are not superposed, as there is small difference between the α values. But a very noticeable fact is presented around spectral types F0 in one way, and B9 in the other. In this position there is a large gap with a few stars. This gap of $0^{\mathrm{m}}06$ wide in $(\alpha-\beta)$ is occupied by: 3 metallic-line stars, one δ Scuti star (HD 211336, F0IV), and just one normal star: HD 196867, B9V. Nearly the same gap is also presented in Figure 7, where the relation between the α and the $(\alpha-\gamma)$ indices is given.

In conclusion: by means of measuring photoelectrically the α and β indices, and

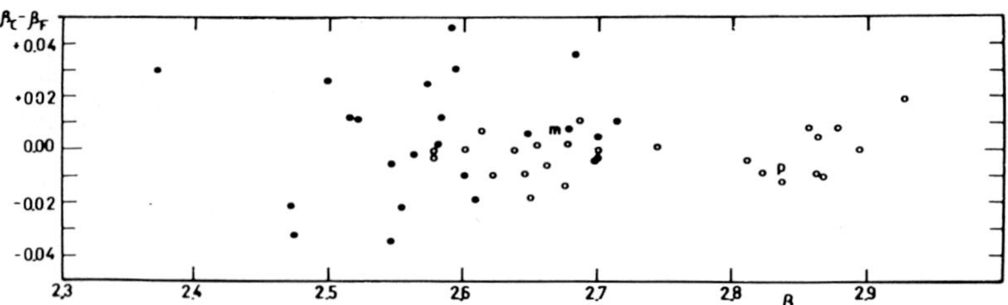

Fig. 5. The relation of the β index vs the difference $\beta_{\mathrm{C}} - \beta_{\mathrm{F}}$ for the stars in common with Crawford *et al.* (1970).

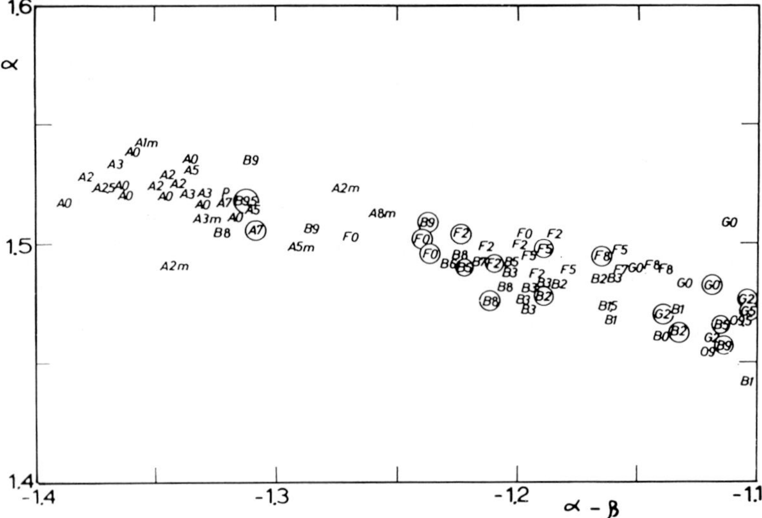

Fig. 6. The relation of the α vs the $\alpha-\beta$ indices for the standard stars. The position of each star is indicated by its spectral type. A circle around it means a supergiant or a giant.

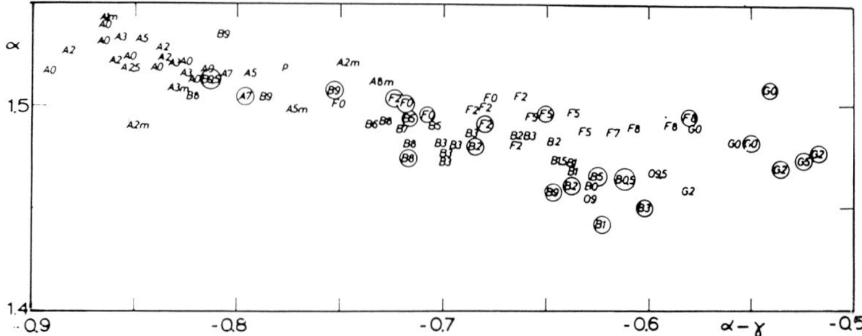

Fig. 7.   The relation of the α vs the $\alpha-\gamma$ indices for the standard stars.

eventually the γ index, of the first three Balmer lines, it is feasible to detect easily the early-type stars which have hydrogen emission. Also, from the comparison of the Hβ measures with those of other authors, this method seems very useful in order to follow the variations of the emission-line strengths in the Be stars.

Acknowledgements

I would like to express my appreciation for the use of Kitt Peak and Cerro Tololo Observatory facilities. A large part of this work was done with a John Simon Guggenheim Memorial Fellowship.

References

Crawford, D. L. and Mander, J.: 1966, *Astron. J.* **71**, 114.
Crawford, D. L., Barnes, J. V., and Golson, J. C.: 1970, *Astron. J.* **75**, 624.
Feinstein, A.: 1968, *Z. Astrophys.* **68**, 29.
Haug, U.: 1970, *Astron. Astrophys.* **9**, 453.
Jaschek, M., Jaschek, C., and Arnal, M.: 1969, *Publ. Astron. Soc. Pacific* **81**, 650.
Mendoza, E. E.: 1958, *Astrophys. J.* **128**, 207.

NARROW BAND PHOTOMETRY OF SUPERGIANT STARS*

E. E. MENDOZA V.

Instituto de Astronomía, Universidad Nacional Autónoma de México

Abstract. At the National Astronomical Observatory in San Pedro Mártir, B. C. (Mexico), we have performed photoelectric photometry for 31 stars in a narrow-band system. The system allows the measurements of total absorptions of neutral oxygen at 7774 Å through three interference filters (20–25 Å half-width), one for the OI lines and two for the continuum.

The preliminary results are very encouraging: Supergiant stars, ranging in spectrum from the later subdivisions of type B to early G, can be clearly separated from other luminosity classes. The total absorptions depend strongly on the stellar luminosity and, to a lesser degree on the effective temperature (spectral type) of the star. The system may be improved by using narrower filters and detectors with a higher quantum efficiency, already in existence.

* Published *in extenso* in *Bol. Obs. Tonantzintla y Tacubaya*, No. 37 (Dec. 1971).

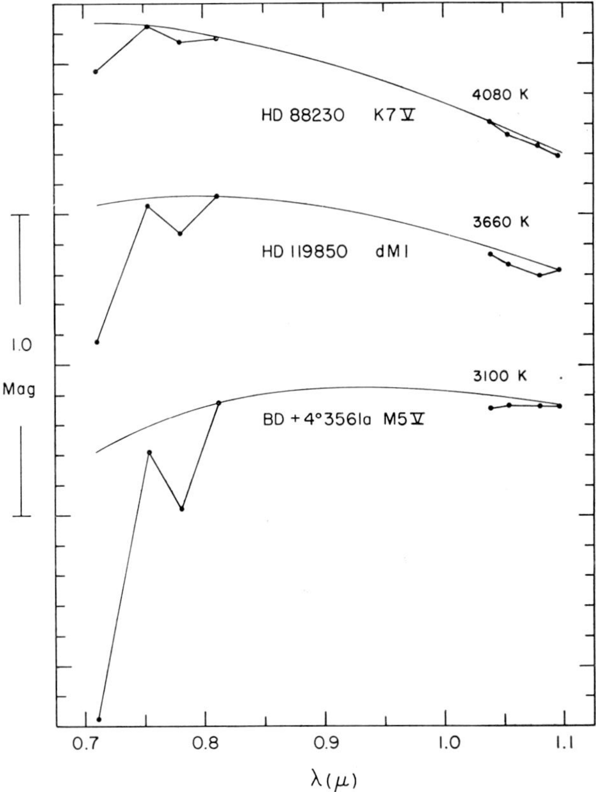

Fig. 2. Spectra on the 8–color system are shown for three late-type dwarfs. The depressions at the first and third filters are caused by TiO. Published spectral types appear after the star names; the revised types, from top to bottom, are K5.1V, M1.8V, and M3.7V. Each spectrum has been fitted with a blackbody curve of the indicated temperature.

their fitted blackbody continua. One of these is again Barnard's star, so that the data on the 8-color and 27-color systems may be compared directly. The numbers 3 through 8 accompanying the spectrum of Barnard's star in Figure 1 indicate the points corresponding most closely to measurements with filters 3 throught 8 on the 8-color system. Note in particular that the relatively weak TiO band measured by filter 3 is the strongest TiO band included on the scanner program. Filter 2, at 7544 Å, corresponds to a point that was added to the 27-color system after the scans of Figure 1 had been obtained.

The index of TiO strength used for classification purposes is defined as the depression at filter 1 with respect to the fitted blackbody continuum, in magnitudes. One advantage of defining the index in this manner is that it is completely uncoupled from the color of the star and hence is independent of reddening (intensity estimates or equivalent width measurements from spectrograms share this characteristic, but most photometric indices do not). It can be seen in Figure 2 that the TiO index is easily measurable in the K7V star with a value of 0.17 mag., and that it grows to 0.90 mag. in Barnard's star. Since the uncertainties in the photometry (as judged from the results for standard stars) normally amount to about ± 0.015 mag. in the magnitudes and ± 0.010 mag. in the colors, the spectral type interval separating HD 88230 from Barnard's star is about 70 times the error of measurement; thus spectral types may meaningfully be given to tenths of subclasses.

Temperatures of unreddened stars are derived from the 8-color photometry by finding the blackbody curve that passes through one point in each group of four filters and above all the other points (see Figure 2). In most kinds of cool stars, including M dwarfs, it is known from the 27-color photometry (Wing, 1967) that these high points are satisfactorily close to the true continuum. The wavelength baseline is such that an error of 0.01 mag. in the photometry at either continuum point introduces an error of about 0.01 in θ, the reciprocal temperature $(5040.T^{-1})$. At the temperature of Barnard's star, this translates into an uncertainty of ± 20 K. Likewise, for each 0.01 mag. of error in the absolute calibration between the two continuum points there results a systematic error in all the temperatures amounting to 0.01 in θ. The actual temperatures given in Figure 2 agree well with the work of Johnson (1965); he does not give temperatures for individual stars, but he gives the mean temperatures for types K7V, M1V, and M5V as 4160, 3680, and 3120 K.

The TiO band strengths are plotted against the reciprocal temperature θ in Figure 3.

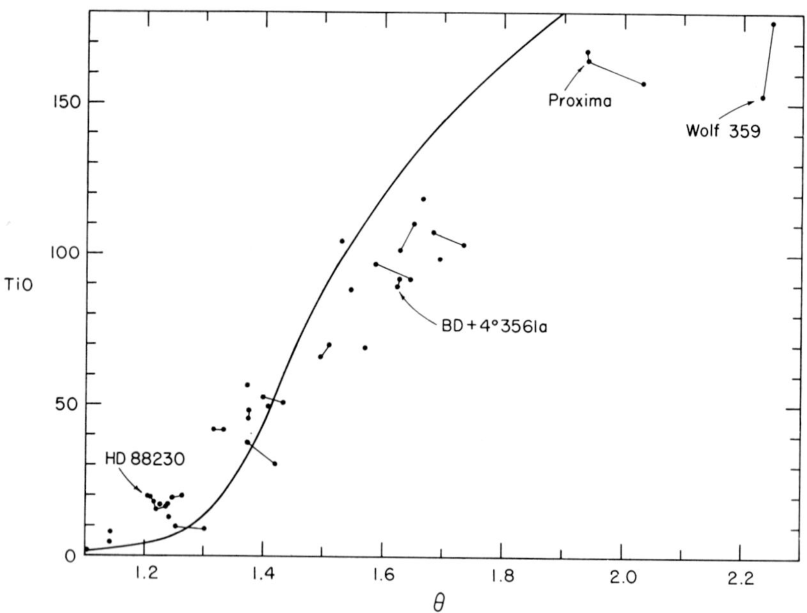

Fig. 3. The TiO index, defined as the depression at filter 1 in units of 0.01 mag., is plotted against the reciprocal temperature θ, obtained from a blackbody fit to the photometry. All observations of dwarfs are plotted, and observations of the same star are connected. The smooth curve is the mean relation for luminosity class III giants.

All observations of dwarfs later than K4 obtained to date – 45 observations of 26 different stars – are plotted, and different observations of the same star are connected by straight lines. The solid curve is the relation for luminosity class III giants, based upon a much larger number of observations. The dwarf relation differs significantly from it, being more nearly linear and also showing more scatter.

Some of the scatter shown by the dwarfs may be due to variability, but the changes shown by stars observed twice are as often perpendicular to the general trend as parallel to it; this suggests that the changes are due mainly to observational scatter, and that the lengths of these line segments may be used to estimate the total observational uncertainty. The probable errors found in this manner are about ± 0.03 or 0.04 mag., as compared with ± 0.01 mag. for the colors and indices of standard stars. The difference simply reflects the faintness of the dwarfs, for which the accuracy is limited by photon noise rather than sky quality. The two observations of Wolf 359, the faintest and reddest dwarf observed, agree well at seven of the eight filters but differ by more than 0.2 mag. at filter 1 where the TiO index is obtained; when we consider that in this case we are trying to measure the deepest part of the strongest absorption band in a star of visual magnitude 13.6, the scatter at filter 1 is understandable. This is the worst case, and even here the uncertainties in the resulting spectral types are acceptably small: the two individual classifications of Wolf 359, on the scale discussed below, are M5.3 and M5.9.

An effort has been made to identify and observe dwarfs that are as cool and as late in type as possible, and several stars previously classified dM5 or later are among those plotted in Figure 3. Most of these stars fall quite close to Barnard's star in the middle of the diagram, while two stars – Proxima Centauri and Wolf 359 – stand out as having much stronger bands and redder colors than the others. Although a few stars are known to exist which have even fainter absolute magnitudes than Wolf 359 (for example, the companion to $BD+4°4048$, which has not yet been observed on this program) there is at present no evidence for the occurrence of stronger TiO bands in dwarfs than those shown by Proxima and Wolf 359. The upper border of Figure 3 corresponds to type M6.0 on the scale used here.

The supposition that the color difference between Proxima and Wolf 359 represents a real difference in temperature can be supported by three arguments: (1) an infrared spectrogram of Wolf 359 taken by Wing and Ford (1969) shows no molecular absorption that could seriously depress the shortward continuum point; (2) the derived temperature for Wolf 359, 2250 K, agrees well with the value 2230 K obtained from scanner measurements made at the Lick 120-in. telescope by Spinrad and Wing, who used a different shortward continuum point (Wing, 1967); and (3) Wolf 359 is 1.7 mag. fainter that Proxima in M_V and hence would be expected to be substantially cooler. If this temperature difference is accepted, then the equality of the band strengths strongly suggests that an upper limit exists to the TiO band strength attainable in dwarfs. It will be important to determine whether a continuous sequence of star exists between Proxima and Wolf 359, or whether the latter is really a lone wolf.

The differences between the mean relations for dwarfs and giants in Figure 3 can probably be accounted for qualitatively in the following manner. In the early M stars, where only a small fraction of the titanium atoms is in molecular form, the dwarfs have stronger bands than giants of the same temperature as a direct result of their higher gas pressures and molecular concentrations. At the later types, however, when molecule formation is considerable in both dwarfs and giants, the band strengths may be governed mainly by the opacity in the continuum; if so, we would infer from Figure 3 that, with decreasing temperature, the opacity either increases in the dwarfs or decreases in the giants, so that the mean relations for the two luminosity classes cross each other. If there exists an upper limit to the TiO band strength in dwarfs, it could be imposed either by increasing opacity or by the formation of titanium dioxide, which should be much more important in dwarfs than in giants.

The observations presented here were obtained with two telescopes: the Perkins 72-in. telescope of the Ohio State and Ohio Weslyan Universities located near Flagstaff, Arizona, and the 60-in. telescope of the Cerro Tololo Inter-American Observatory in Chile. Many of the stars lie in the equatorial region, and several have been observed from both hemispheres.

4. The Spectral Types

The reasons for deciding to classify dwarfs on the scale established for giants by Wing and Keenan (1973) have already been stated. Thus, types for dwarfs can be obtained simply by entering the TiO index, as measured on the 8-color system, into the calibration already derived. Two points, however, need to be discussed.

The first is that type K7, which has been used on the MK system for dwarfs but not for giants, will no longer exist for any luminosity class on the new unified system. Type M0.0 follows immediately after type K5.9, and the types K4.0, K5.0, M0.0, and M1.0 have been defined to represent equal intervals in TiO band strength (and hence nearly equal intervals in the inverse temperature θ).

The second point is that absorption by the CN molecule is nearly continuous throughout the red and infrared, and in particular the region of filter 1 contains fairly strong CN absorption. In order to obtain spectral types that are independent of the CN strength and that are truly on the same scale for all luminosity classes, corrections must be applied for the contribution of CN to the depression of filter 1. Fortunately, the appropriate correction for each star can be derived from the 8-color photometry alone (see White, 1971), since filters 4 and 8 provide explicit CN measurements. Since Wing and Keenan (1973) used CN-corrected TiO indices in deriving the calibration for giants, no further corrections need to be applied here, because the effects of CN are negligible in K and M dwarfs. If the giants had not been corrected, however, a systematic error of about half a subclass between dwarfs and giants would have been introduced.

The accuracy of the spectral types depends upon the photometric accuracy (already discussed) and the width of the spectral type intervals in terms of the TiO index. In the interval K4.0–M1.0, as defined by Wing and Keenan (1973), the CN-corrected TiO index increases by precisely 0.01 mag. per spectral type interval of 0.1 subclass, so that tenths of subdivisions may meaningfully be used for the means of repeated observations of non-variable stars or for single observations of bright stars made on good nights. With advancing type, the relation between band strength and spectral type becomes steeper, so that for example each 0.1 subclass has a width of 0.04 mag. in the range M4.0–M6.0. In progressing down the main sequence, the increasing width of these intervals tends to compensate for the fainter apparent magnitudes with which we are forced to contend.

The spectral types for dwarfs obtained to date are given in the last column of Tables I and II, under the heading 'revised MK type'. It is hoped that these types will not only be considered to be MK types but will be adopted as the definition of the MK scale for dwarfs. The new types are compared with types given by Kuiper (1942) and Joy (1947), and with those found in the literature expressed in the MK notation.

Table I gives the mean types for stars observed more than once (n is the number of observations), and it is these stars that are recommended for use as standards of spectral type. The list is too short to give a really adequate distribution in both right ascension and spectral type, but most of the stars are observable from both hemispheres and additional standards will be made available in the near furture. Some of these stars undergo flare activity, but none is known definitely to be variable in spectral type or in the infrared continuum. The degree of agreement of the different observations of these stars can be seen by inspection of Figure 3. From the considerations discussed earlier, the mean types in Table I are judged to have probable errors of ± 0.1 subclass and maximum errors of ± 0.2 subclass. In Table II are given the types obtained for stars observed only once; their uncertainties depend somewhat upon the circumstances of the particular observation, but estimates for typical cases indicate probable errors of ± 0.2 subclass and maximum errors of ± 0.3 subclass.

TABLE I

Spectral types for standard stars

Star	α(1900)	δ(1900)	Kuiper (1942)	Joy (1947)	Published MK type	n	Revised MK type
HR 753 B	2^h30^m6	$+ 6°25'$	M6	–	–	2	M4.1V
HD 33793	5 07.7	−44 59	M0	–	M0V	2	K4.1V
BD −12°2918	9 26.4	−13 03	M4	–	–	2	M2.9V
HD 88230	10 05.3	+49 58	K8	–	K7V	6	K5.1V
Wolf 359	10 51 6	+ 7 37	M8	dM6e	–	2	M5.6:V
Ross 128	11 42.6	+ 1 23	M5+	dM5	–	2	M4.1V
HD 111631	12 45.6	− 0 13	K8	–	M0.5V	2	K5.3V
BD +11°2576	13 24.9	+10 55	M1	dM1	M1V	2	M0.8V
HD 119850	13 40.7	+15 26	M1+	–	M4V	2	M1.8V
Proxima Cen	14 22.8	−62 15	–	–	–	3	M5.6V
BD +4°3561a	17 52.9	+ 4 25	M5+	sdM4.5	M5V	2	M3.7V
BD −15°6290	22 47.9	−14 47	M5	dM4.5	–	2	M3.8V
HD 217987	22 59.4	−36 26	M0+	–	M2V	2	M1.5V
HD 225213	23 59.5	−37 51	M3	–	M4V	2	M2.1V

TABLE II

Spectral types for additional stars

Star	$\alpha(1900)$	$\delta(1900)$	Kuiper (1942)	Joy (1947)	Published MK type	Revised MK type
BD $+5°1668$	7^h22^m0	$+ 5°31'$	M4+	dM4	–	M4.1V
Wolf 358	10 45.8	$+ 7\ 22$	M5	dM5	–	M4.0V
BD $-11°3759$	14 28.9	$-12\ 06$	M4	–	–	M4.5V
HR 5568 A	14 51.6	$-20\ 58$	K4	–	K5V	K3.5V
B			M0+	–	M2V	M2.0V
BD $-7°4003$	15 14.2	$- 7\ 21$	M4	–	–	M3.0V
BD $-12°4523$	16 24.7	$-12\ 25$	M4	dM4.5	–	M3.6V
HD 154363	16 59.8	$- 4\ 54$	–	–	K5V	K4.2V
BD $-4°4226$	17 00.0	$- 4\ 55$	–	–	M3.5V	M2.4V
HD 157881	17 20.8	$+ 2\ 14$	K7	–	K7V	K5.0V
61 Cyg A	21 02.4	$+38\ 15$	K3	–	K5V	K3.5V
B			K5	–	K7V	K4.7V

5. Plans for Further Work

The observations presented here represent approximately the first 20% of a program intended to furnish two or more observations of each of about 100 M dwarfs. The observing list includes the several brightest single stars of each spectral type, with emphasis on the latest types. The 8-color photometry will provide color temperatures and $I(104)$ magnitudes in the infrared continuum, as well as spectral types. Discussion of the following topics is planned, but will be deferred until the observational data are complete:

(1) By combining the 8-color photometry with trigonometric parallaxes, a color-magnitude diagram involving the absolute $I(104)$ magnitude and the reciprocal temperature θ will be constructed in order to discuss once again such topics as the width and slope of the lower main sequence, and the existence of subdwarfs.

(2) New values for the mean colors, effective temperatures, and bolometric corrections as a function of spectral type will be derived on the basis of the wideband photometry published by Johnson (1965), simply by regrouping the stars he observed according to the new and more self-consistent spectral types.

(3) Algebraic relations will be derived whereby the bolometric magnitude and effective temperature are expressed as functions of the $I(104)$ magnitude, the reciprocal color temperature θ, and the spectral type, all of which are given directly by the 8-color photometry.

Acknowledgements

Support from the National Science Foundation is gratefully acknowledged. I would like to thank the Director and staff of the Cerro Tololo Inter-American Observatory for their hospitality and for observing time on the 60-in. telescope, with which most of the data discussed here were obtained.

References

Hayes, D. S.: 1970, *Astrophys. J.* **159**, 165.
Johnson, H. L.: 1965, *Astrophys. J.* **141**, 170.
Johnson, H. L. and Morgan, W. W.: 1953, *Astrophys. J.* **117**, 313.
Joy, A. H.: 1947, *Astrophys. J.* **105**, 96.
Keenan, P. C.: 1963, in K. Aa. Strand (ed.), *Basic Astronomical Data*, University of Chicago Press, Chicago, p. 78.
Kuiper, G. P.: 1942, *Astrophys. J.* **95**, 201.
Morgan, W. W.: 1938, *Astrophys. J.* **87**, 589.
White, N. M.: 1971, Dissertation, The Ohio State University.
White, N. M. and Wing, R. F.: 1973, in preparation.
Wing. R. F.: 1967, Dissertation, University of California, Berkeley.
Wing, R. F.: 1971, in G. W. Lockwood and H. M. Dyck (eds.), *Proc. of the Conference on Late-Type Stars*, Kitt Peak National Observatory Contribution No. 554, p. 145.
Wing, R. F. and Ford, W. K., Jr.: 1969, *Publ. Astron. Soc. Pacific* **81**, 527.
Wing, R. F. and Keenan, P. C.: 1973, in preparation.

THE IDENTIFICATION OF T TAURI-LIKE STARS BY MULTICOLOUR PHOTOMETRY

K. NANDY

Royal Observatory, Edinburgh, Scotland

Abstract. A method of distinguishing between T Tauri-like stars and reddened early type field stars by the application of *UBVRI* photometry is described, and the results for T Tauri-like variables in NGC 2264 are presented. The plates have been measured using the new measuring machine GALAXY. $(V-R)$ colours, which are less affected by spectral peculiarities, are used to determine infrared and ultraviolet excesses. The range of variability in *UBVRI* colours of these varables has also been studied.

1. Introduction

The young clusters and associations contain a large number of T Tauri-like variables which populate the elevated main sequence. These stars are known to belong to the contracting phase of the main sequence. It has become evident from the work of Mendoza (1966, 1968) and Low *et al.* (1970) that these stars are intrinsically brighter in the infrared and ultraviolet than expected from their spectral types. Because of their intrinsic spectral peculiarities, the separation of these stars from reddened field stars presents a problem. In this paper, a method distinguishing between T Tauri-like stars and reddened early type field stars by the application of *UBVRI* photometry is described, and the results for T Tauri-like variables in NGC 2264 as regards their infrared and ultraviolet excesses, and the range of variability in *UBVRI* colours are presented.

2. Observations and Reductions

The observational material was obtained with the 40/60/150 cm Schmidt telescope of the Royal Observatory, Edinburgh's outstation at Monte Porzio during the period from March 1968 to February 1969. For U, B, V, R and I magnitudes the following emulsion filter combinations are used:

$$
\begin{array}{ll}
U & \text{IIaO} + \text{UG2 (2 mm Schott)} \\
B & \text{IIaO} + \text{GG13 (2 mm Schott)} \\
V & \text{IIaD} + \text{GG14 (2 mm Schott)} \\
R & \text{IaE } + \text{Ilford colour filter 204} \\
I & \text{IN } + \text{Ilford colour filter 207}
\end{array}
$$

U, B and V plates are calibrated by photoelectric sequences of Walker (1956). For R and I magnitudes photographic calibration has been achieved by using a pair of doubly refracting calcite plates in conjunction with a polaroid (Brück *et al.*, 1969). (For future work photoelectric R and I magnitudes are now being obtained with the Observatory's Twin telescope). The plates have been measured using the Observatory's new measuring machine GALAXY. The calibration curve obtained from the GALAXY

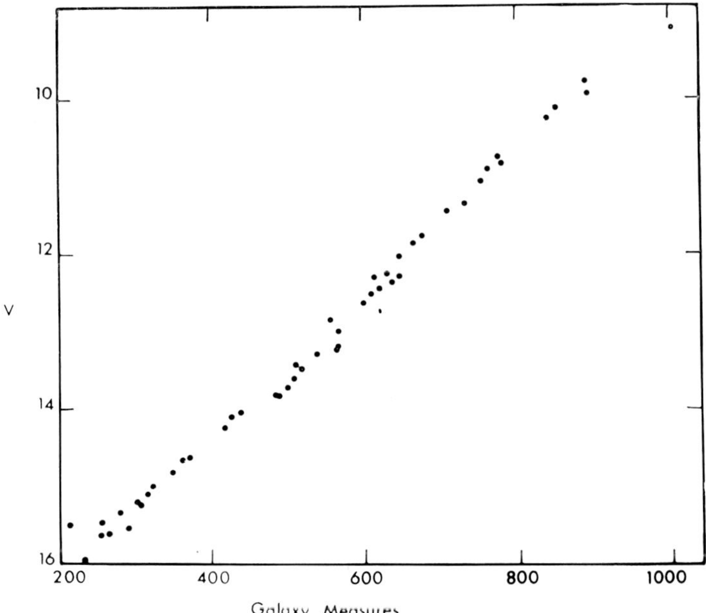

Fig. 1.   The calibration curve obtained from the GALAXY measures.

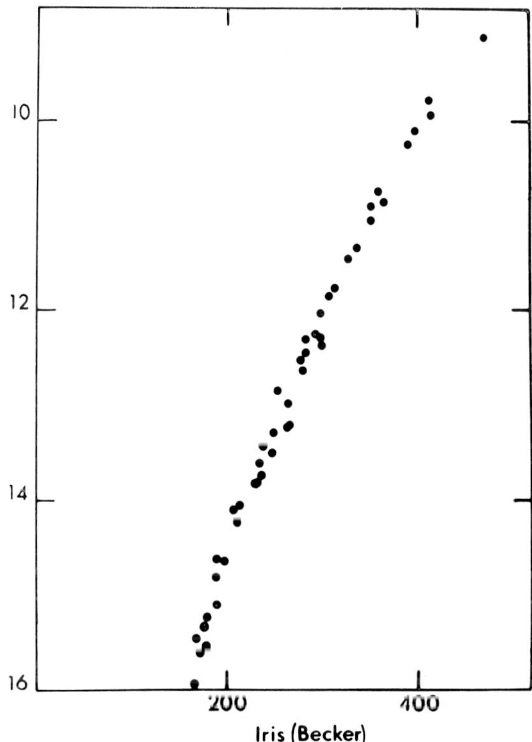

Fig. 2.   The calibration curve derived from the iris photometer readings.

measures is shown in Figure 1 (For comparison, Figure 2 shows the calibration curve derived from the conventional iris photometer readings, the same plate being used). It is to be noted (cf. Figure 1) that the GALAXY measures are linear with magnitudes ($-2.5 \log I$) over the magnitude range from $V = 8^{m}.0$ to the plate limit.

3. Identification of T Tauri-like Stars

The separation of T Tauri-like stars from reddened field stars can be achieved in the following way: if the stars are intrinsically brighter in the ultraviolet and infrared than normal for their spectral type as indicated by their absorption lines, they occupy the same region of the $(U-B, B-V)$ diagram as highly reddened early type stars. However, the $(U-V)$, $(V-R)$ and $R-I)$ colours can be predicted for the early type stars from the parameter Q defined as $Q = (U-B) - 0.70 (B-V)$ (Q determines the spectral type and the extinction law being known (Nandy, 1968) the colour excesses E_{B-V}, E_{V-R}, and E_{V-I} can be computed). In the case of reddened early type stars the predicted colours would agree with the observed $(V-R)$ and $(V-I)$ colours. On the other hand, if the difference between the predicted and observed colours is much larger than the errors of observations, the stars are likely to be of later types with anomalous colours characteristics of T Tauri stars. For these stars $(V-R)$ colours corrected for interstellar reddening, which are less affected by peculiarities in the stellar spectra, are found necessary to estimate their spectral types, and infrared and ultraviolet excesses can then be directly computed. The spectral types of T Tauri-like stars determined from $UBVRI$ photometry and from low dispersion objective prism spectra as faint as $V = 12^{m}.5$ agree within $\frac{1}{2}$ spectral class.

4. Sources of Error and Range of Variability

The main sources of error is the variability of the stars, since U, B, V, R and I magnitudes have not been obtained simultaneously. During the period of observations a considerable number of plates in each colour have been obtained. The range of variability (in amplitude) of T Tauri-like stars in each colour is estimated by the intrinsic rms dispersion of the mean magnitudes. It is found that the rms dispersion in U magnitudes of the T Tauri-like variables is much larger than that in I magnitudes (Nandy and Pratt, 1972). The mean rms dispersion in $UBVRI$ colours obtained for the known T Tauri-like stars in the cluster NGC 2264 is as follows:

	$\bar{\sigma}_U$	$\bar{\sigma}_B$	$\bar{\sigma}_V$	$\bar{\sigma}_R$	$\bar{\sigma}_I$
T Tauri-like stars	0.6 ± 0.4	0.4 ± 0.3	0.2 ± 0.1	0.15 ± 0.1	0.1 ± 0.1
Photometric error as determined from rms. error of standard stars	± 0.08	± 0.06	± 0.04	± 0.06	± 0.07

Only those plates which were taken in a small interval of time (1 week) have been used to determine $(V-R)$ colour, and therefore the standard error of mean $(V-R)$ colour is not more than $\pm 0^{m}\!.2$.

5. Ultraviolet and Infrared Excesses

The ultraviolet, blue and infrared excesses are defined as follows:

$$\delta(U) = (U-V)_{\text{obs}} - (U-V)_{\text{predicted}} - E_{U-V}$$
$$\delta(B) = (B-V)_{\text{obs}} - (B-V)_{\text{predicted}} - E_{B-V}$$
$$\delta(I) = (V-I)_{\text{predicted}} - (V-I)_{\text{obs}} + E_{V-I}.$$

The variables, in particular, the Hα emission stars in NGC 2264 exhibit infrared and ultraviolet excesses and there appears to be a positive correlation between them (Nandy, 1970). These stars, as Mendoza (1968) pointed out, are more luminous than indicated by their colour magnitude diagram. Their mean infrared excesses are comparable to the T Tauri-like stars with $L_{\text{total}}/L_{\text{sp}} \sim 3$. Masses and ages of these variables computed on this assumption (Nandy, 1970) are shown in Figure 3. It is found that their masses range from one to two solar masses, and the average age is $\sim 10^{6}$ yr, in agreement with the estimated age of the B stars of the cluster.

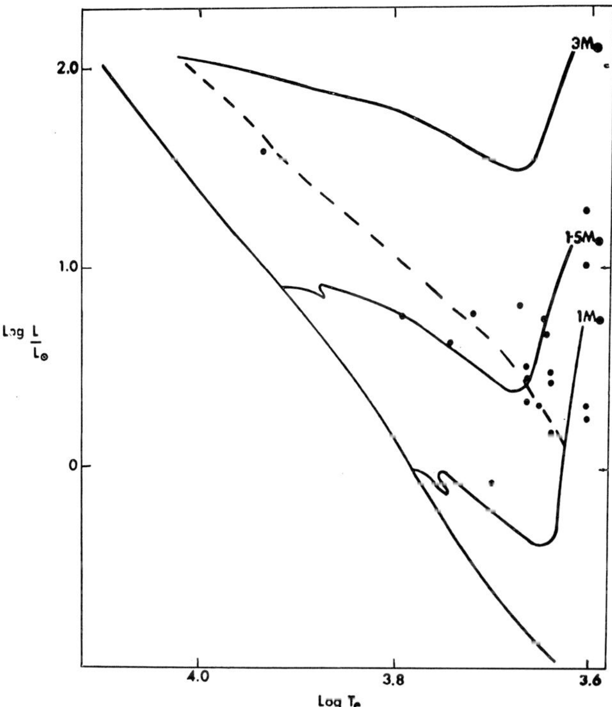

Fig. 3.   H-R diagram for T Tauri-like stars in NGC 2264 as studied here. The dashed line indicates the locus of stars 10^{6} yr old.

6. Dust Model for T Tauri-Like Stars

The indication that the range in variability in I is much less than that in U, and the infrared excesses and ultraviolet excesses are correlated, favour the model of circumstellar dust clouds for T Tauri-like stars (Low *et al.*, 1966) in which grains absorb the visual radiation and radiate in the infrared.

The extinction law for the circumstellar dust cloud around T Tauri-like stars may be very much different from those for the general interstellar medium, since the physical conditions in the dense cloud favour the growth of large particles. Hartmann (1970) has suggested that planetisimal particles (iron grains coated with silicates) may be formed in these dense clouds. The theoretical extinction curves for such particles have been computed for various sizes of mantles (Nandy and Wickramasinghe) and it is found that for large sizes of mantles the extinction in ultraviolet becomes nearly non-selective (Figure 4). The simultaneous existence of infrared and ultraviolet excesses may be explained by the fact that large planetisimal particles are relatively more transparent in the ultraviolet (c.f. Figure 4) and radiate more efficiently in the infrared. However, until we get more observational data as regards ultraviolet and infrared excesses of T Tauri-like stars in cluster and associations, these theoretical results remain tentative.

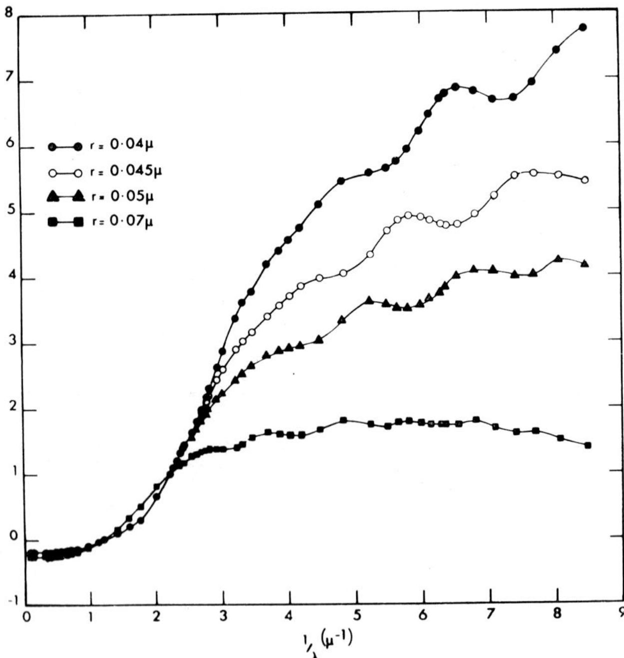

Fig. 4.   Theoretical extinction curves for planetisimal particles
(iron grains coated with silicates).

References

Brück, M. T., Caprioli, G., Nandy, K., and Smiglio, F.: 1969, *Astrophys. Space Sci.* **4**, 213.

Hartman, K. W.: 1970, 'Évolution stellaire avant la séquence principale', *Soc. Roy. Sci. Liège, 5e série* **19**, Institut d'Astrophysique, Cointe-Ougrée, Belgique.

Low, F. J., Johnston, H. L., Kleinman, D. E., Latham, A. S., and Geisal, S. L.: 1970, *Astrophys. J.* **160**, 531.

Low, F. J. and Smith, B. J.: 1966, *Nature* **212**, 675.

Mendoza, E. E.: 1966, *Astrophys. J.* **143**, 1010.

Mendoza, E. E.: 1968, *Astrophys. J.* **151**, 977.

Nandy, K.: 1968, *Publ. Roy. Obs. Edinburgh* **6**, 169.

Nandy, K.: 1971, *Publ. Roy. Obs. Edinburgh* **7**, 47.

Nandy, K. and Pratt, N. M.: 1972, *Astrophys. Space Sci.* **19**, 219.

Nandy, K. and Wickramasinghe, N. C.: 1973, *Astrophys. Space Sci.*, in press.

Walker, M. F.: 1956, *Astrophys. J. Suppl.* **2**, 365.

CLASSIFICATION OF POPULATION II STARS
IN THE *RGU* SYSTEM

U. W. STEINLIN

Astronomisches Institut der Universität Basel, Binningen, Switzerland

Abstract. The theoretical treatment of the blanketing effect and the first photometric results had shown that the separation of stars of different metal content should be better in the *RGU* system than in the *UBV* system. A three color photometry of the globular cluster M5 is undertaken to show the position of these stars in the *RGU* two-color diagram, to derive the conversion formulae between the two photometric systems for these stars and to use the results for the classification of population II field stars in statistical investigations of the galactic halo.

This work starts from the long known facts that the *RGU*-system allows, other than the *UBV*-system, the separation of late type dwarfs and giants (Becker and Steinlin, 1956) and that the separation of stars with different metal content is better in *RGU* (increased amount of blanketing and greater angle between blanketing lines and main sequence curve in the two-color diagrams) (Smith and Steinlin, 1964). Using the system for Population II stars those advantages, however, may partly be imperiled by the possibility that metal-poor giants might just be moved to the places of metal-rich dwarfs. But altogether *RGU* photometry should be more apt to classify late type stars according to populations. This was the basis for starting an extensive photographic photometry program in high galactic latitude fields in the course of the halo program of the Basel Observatory (Becker, 1965). In this program the densities of Population II stars in different directions in the halo should be determined.

To check the behaviour of Population II stars in the *RGU*-system Palomar Schmidt plates were taken for one of these fields (which contains the globular cluster M5) in the five colors *RGUBV* with limiting magnitude around $19^{m}.5$ in *V*. In M5 a photo-electric and photographic *UBV*-sequence is established (Arp, 1962) and the cluster itself is well known so that the position of the stars in the two-color diagram for *UBV* is reliable. The photometry of cluster and field stars in both three-color systems (the *U*, of course, being common to both) should also provide a set of transformation equations to convert photoelectric *UBV* standards among Population II stars in *RGU* magnitudes for calibration purposes, as they were previously only known for Population I stars (Steinlin, 1968).

At this moment measurements and reductions are not yet finished: only three out of six plates in each color are measured and final corrections for field and scale errors etc. are not yet calculated and applied. The scatter is therefore still large and the results can only be discussed in a qualitative way.

In Figure 1 the two-color diagrams with the main results are given. In *UBV* (left side) the photographic measurements fall into a stripe which is well centered on Arp's cluster sequence (broken line); only for the faintest stars – at the top – the new measurements seem to lie slightly higher, in the sense of a better separation between giant

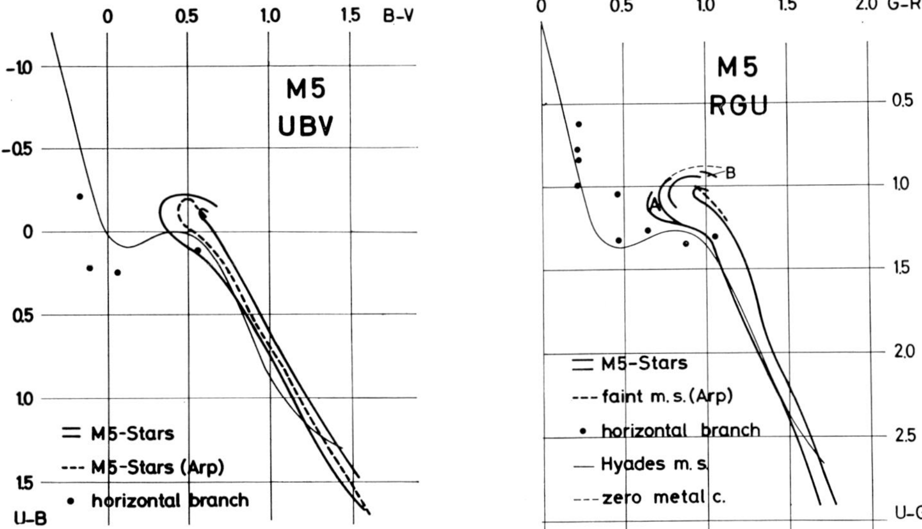

Figs. 1a–b.   Two–color diagrams in (a) *UBV* and (b) *RGU* for photographic photometry of the globular cluster M5.

branch and beginning main sequence. But as we reach just at this point the limiting magnitude of the Schmidt plates, not too much weight can be given to this fact. (As the scatter of the individual stars in these diagrams is, in this preliminary stage, still large, no single stars are marked, but only – to obtain a clearer picture – the stripe in which the majority of them falls is indicated.)

In *RGU* (right side) the giant-supergiant branch indeed coincides with the normal (metal-rich) main sequence. But already from $U-G=1.50$ upward the globular cluster stars are well above this curve and permit a neat distinction between them and ordinary field stars. The fainter stars, as expected, turn farther away from the main sequence curve than in the left-hand picture; regarding the normal scatter of photographic magnitudes this difference may be decisive for the possibility of a statistical separation of the two populations. The broken line indicates where the cluster main-sequence stars, assuming the same metal deficiency (as determined in *UBV* by Arp for M5) would lie. As far as one can judge from the fact that only a few uppermost stars of the main sequence are measured in *RGU*, they seem in this diagram as well as in the left one to lie somewhat higher (that is: to show a somewhat lower metal content) than in Arp's paper.

Three groups of stars will need further consideration in the final evaluation. The positions of two of them in the color-magnitude diagram are schematically marked in Figure. 2 (regions A and B). At first glance stars in these regions seem either to be an increasing number of faint field stars or just indicating a large scatter in the color indices. Their mean position is also marked in Figure 1. The group at A is certain'y not a group of Population I field stars in this direction, but definitely a group of

Population II stars. Whether there are halo field stars in this number to be expected can only be judged when the photometry of an adjoining field of about 2000 stars (now under way) will give numbers of stars per square degree at this magnitude range; however it seems more plausible from the number of stars in group A that they are mostly cluster members.

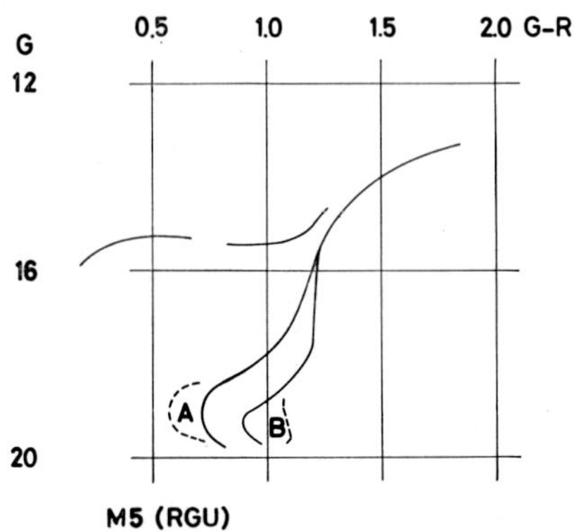

Fig. 2.   Color–magnitude diagram of the globular cluster M5 with star groups A and B (schematic).

More surprising is the position of the group B stars in Figure 1. They are definitely at the upper edge of the distribution, that is they seem to indicate a number of cluster stars at the inside of the knee of the cluster diagram with slightly lower metal content than the bulk of the cluster stars.

The third group with unexpected behavior are the horizontal branch stars. In *UBV* they are well below the main sequence curve (confirming Arp's results), but in *RGU* they are right on the main sequence line (each point in Figure 1 represents a mean value of several stars closely together in the cluster color-magnitude diagram). Only a more detailed inspection of the spectral intensity distribution of these stars can show the reason for this behavior.

The conclusions that can be drawn are:

(a) The *RGU* system offers somewhat better possibilities for a statistical separation of stars of the two populations, above all for main sequence stars.

(b) On the other hand is it clear that with this kind of problem the limits of the possibilities of a pure three-color photometry are reached – even more so, when we consider the increased complexity of the diagrams in the case of even only moderate interstellar absorption. This is not surprising, as we are in fact trying to classify the stars in three dimensions (temperature, luminosity, metal content), with interstellar absorption as a possible fourth dimension, with only two independent measured values (color indices). Only in simple and favorable cases this may work; in the present problem we go beyond these limits.

(c) The *RGU* system, however, may give some indications in which direction one may have to look for an additional fourth color to add a new possibility of discrimination between different groupes. A theoretical approach to this problem is under way at the Basel Observatory.

References

Arp, H.: 1962, *Astrophys. J.* **135**, 311.
Becker, W.: 1965, *Z. Astrophys.* **62**, 54.
Becker, W. and Steinlin, U.: 1956, *Z. Astrophys.* **39**, 188.
Smith, L. and Steinlin, U.: 1964, *Z. Astrophys.* **58**, 253.
Steinlin, U.: 1968, *Z. Astrophys.* **69**, 276.

PHOTOMETRIC CLASSIFICATION
OF BLUE HORIZONTAL-BRANCH STARS

A. G. DAVIS PHILIP*

Dudley Observatory and State University of New York at Albany

Abstract. The Strömgren four-color system has been used to classify blue horizontal-branch stars in globular clusters and in the general field. A relation between $\delta \log g$ ($\log g_{\text{main sequence}} - \log g_{\text{star}}$) and δc_1 ($c_{1\,\text{observed}} - c_{1\,\text{main sequence}}$) has been determined which gives $\log g$ to within ± 0.2 for stars with $\log g$'s between 4.4 and 2.0. Newell (1970) has determined a relation between $(B-V)_0$ and θ_e. Thus the gravities and effective temperatures of blue horizontal-branch stars can be determined from photometric measures alone. Five globular clusters have been studied at the present time. The BHB stars in the globular cluster M3 and M13 have lower surface gravities than BHB stars in the other clusters studied. Iben and Rood (1970) have recently suggested that the BHB stars of M3 may have lost more mass than BHB stars in more metal poor clusters.

1. Introduction

The purpose of this preliminary report is twofold. First, it will be shown how blue horizontal-branch (BHB) stars can be classified by photometric means and, second, the relationship of the field horizontal-branch (FHB) stars to the BHB stars will be indicated.

In the course of a study of galactic structure at high galactic latitudes, finding lists of early-type stars in a number of selected regions (this volume, p. 295) have been compiled. Among the fainter stars in each list, a natural group of stars has been found with the common characteristics of large Balmer jumps and low metal abundances. Since these qualities are similar to blue stars found in globular clusters, a program of measuring a number of BHB stars in a number of globular clusters with a large range in metal abundance has been started. Sufficient observations have been made of BHB and FHB stars that certain conclusions can now be drawn.

2. Field Horizontal-Branch Stars

Before discussing the four-color photometry of BHB stars, it would be useful to point out some relations derived among temperature, color, luminosity, and other parameters. In the range of temperature, $0.56 < \theta_e < 0.72$ it has been shown (Newell, 1970) that horizontal-branch stars have a different relation in the θ_e, $(B-V)_0$, plane than do Population I stars, in the sense that for the same $(B-V)_0$, the BHB stars have lower effective temperatures. The relation for BHB stars is shown in Figure 1 (Figure 3 from Newell, 1970) and is the one used in this study. Mihalas (1970) has computed the strength of the Balmer jump from LTE model atmospheres as a function of effective

---

* Visiting Astronomer, Kitt Peak National Observatory and Cerro Tololo Inter-American Observatory, which are operated by the Association of Universities for Research in Astronomy, Inc., under contract with the National Science Foundation.

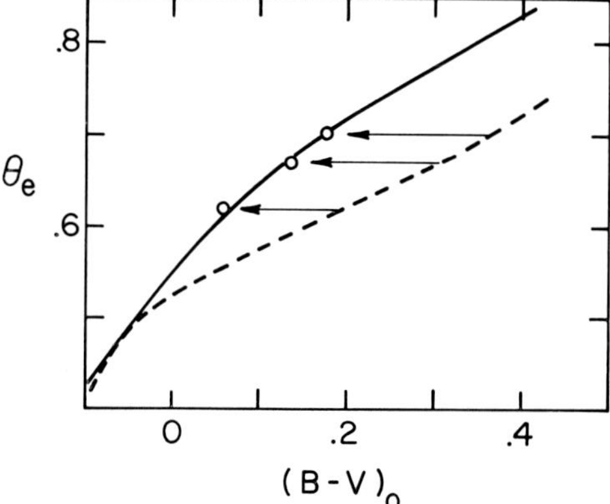

Fig. 1. θ_e vs $(B-V)_0$ (Figure 3 from Newell, 1970). The solid line represents the θ_e, $(B-V)_0$ calibration for BHB stars from Newell *et al.* (1969). The dashed line represents the Morton and Adams
(1968) main sequence.

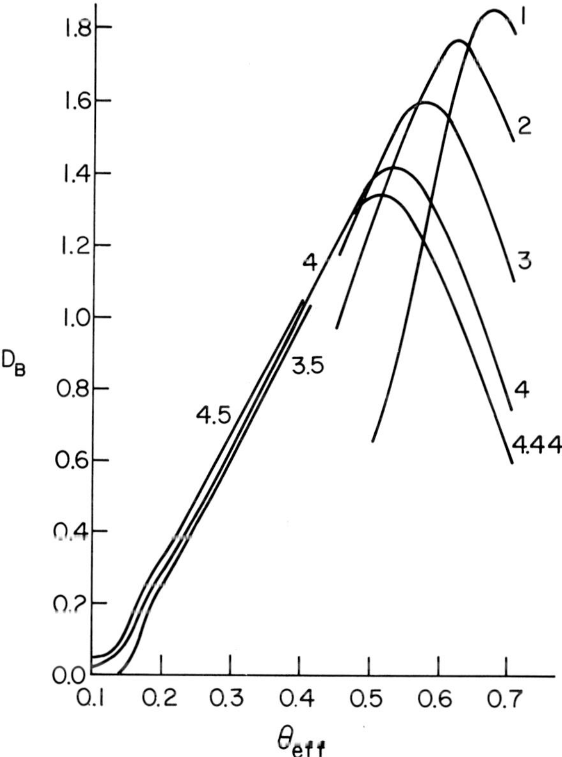

Fig. 2. Balmer groups computed from LTE model atmospheres, as a function of effective temperature and gravity. (Figure 6–3 from Mihalas, 1970). The Balmer jump, in magnitudes, is plotted vs the
effective temperature. The curves are labeled with $\log g$.

temperature and gravity. For $0.56 < \theta_e < 0.72$ there appears to be a linear relationship between changes in $\log g$ (between $\log g = 4.4$ and 3.0) and D_B, the magnitude of the Balmer jump. This relationship is shown in Figure 2. (Figure 6–3 from Mihalas, 1970). The c_1 index $[c_1 = (u-v)-(v-b)]$ is a measure of the Balmer jump and thus should be a good measure of $\log g$ in this temperature range, if the temperature is first determined by some other method. The m_1 index $[m_1 = (v-b)-(b-y)]$ has been shown to be a measure of the metal abundance in the range of effective temperatures being considered. The u and v filters are affected by blanketing while the b and y filters are relatively unaffected. Thus a metal-rich star of the same effective temperature as a metal-poor star will have a different m_1 index.

Since θ_e, $\log g$, and a measure of the metal abundance can be obtained from a study of the four-color indices, the Strömgren four-color system is an excellent system for the study of BHB stars.

The first stars to be measured were four prototype FHB stars, HD 2857, 86986, 109995, and 161817, listed by Oke *et al.* (1966). Matching photoelectric spectrum scans to atmospheric models they obtained estimates of $\log g$ and θ_e ($\log g = 3.0$, $\theta_e = 0.6$) which indicated that these stars were FHB stars. Three of these stars (86986, 109995, 161817) have been analyzed for chemical abundance (Kodaira, 1964; Wallerstein and Hunziker, 1964; and Kodaira, 1969) and were found to have metal abundances, Fe/H, which were down by factors greater than 10 from Fe/H values for Population I stars. Kodaira *et al.* (1969) also showed that these stars have large negative V velocities and high W velocities. These stars have been set up as secondary standards in the four-color and Hβ systems. The standards values are shown in Table I along with the ZAMS values for an A0 star for comparison.

For early A stars, it can be seen that the c_1 index of a FHB star is about 0.2 higher, the m_1 index about 0.04 lower, and the Hβ index about 0.1 lower than a main sequence star. Some of these differences can be seen by inspection of the spectra shown in Figure 3 (Figure 5 of Philip, 1968). Spectra, at a dispersion of 128 Å mm^{-1} of two of the prototype FHB stars are shown at the top of the figure. In the middle are spectra of two Population I stars and on the bottom are spectra of two FHB stars found at high galactic latitudes. The Balmer lines of the FHB stars are sharper than the Balmer lines in the Population I spectra and more Balmer lines can be seen in the FHB spectra. On spectra taken in the UV a stronger Balmer jump can be seen.

TABLE I

Four–color and Hβ indices for prototype field horizontal–branch stars

Star	$b-y$	c_1	m_1	n	β	n
Population I A0 star	0.000	1.070	0.171		2.943	
HD 109995	0.048	1.286	0.134	48	2.823	46
HD 86986	0.089	1.266	0.123	30	2.822	25
HD 161817	0.129	1.207	0.112	39	2.670	31
HD 2857	0.135	1.213	0.114	36	2.684	29

Fifty FHB stars have been identified in areas at high galactic latitudes (Philip, 1967, 1968, 1969a, b, and 1970a). In five of the regions under study, sufficient A stars have been measured so that the color excess as a function of distance modulus has been determined, and thus the intrinsic colors of FHB stars found in these regions are easily calculated. Forty-five stars, in five regions (NGP, SGP, 1 HLF 2, 1 HLF 3, and

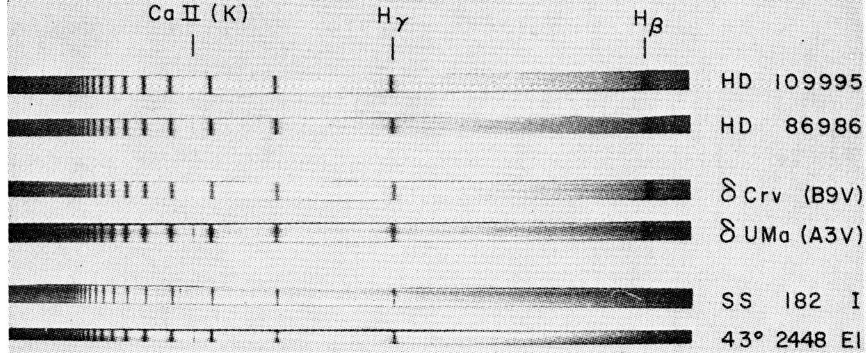

Fig. 3. Spectra of six stars. (Figure 5 from Philip, 1968). The upper two are of field horizontal-branch stars (Oke *et al.*, 1966). The middle two are of two Population I main sequence stars. The bottom two spectra are of two field horizontal-branch stars (Philip, 1968).

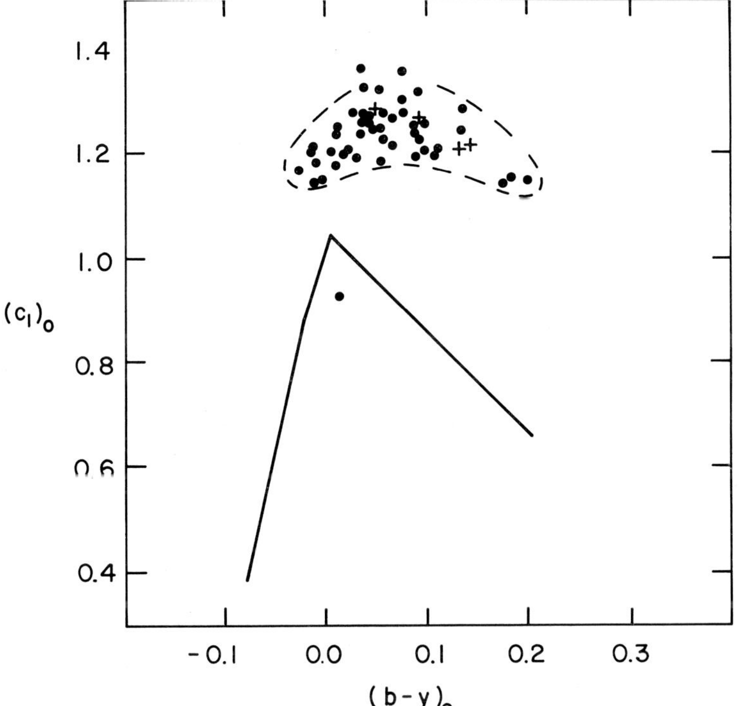

Fig. 4. The $(c_1)_0$ index vs $(b-y)_0$ for 45 FHB stars in regions at high galactic latitudes. The Solid line represents the zero age main sequence. The dotted line indicates the area in which the FHB stars fall. The crosses represent the indices of the four prototype FHB stars.

4 HLF 6) have been measured in the four-color system and their intrinsic color indices
are plotted in Figures 4 and 5. The $(c_1)_0$ index is plotted against $(b-y)_0$, in Figure 4.
The solid lines indicate the zero age main sequence for B stars (Slettebak *et al.*, 1968)
and A stars (Crawford, 1970). The indices for the four prototype FHB stars are plotted
as crosses. The point falling below the ZAMS represents SS 182 I, a star identified by
Slettebak and Stock (1959) as a FHB star. The rest of the stars fall in a common area.
A dotted line has been drawn around these points to mark the area occupied by the
natural group.

The $(m_1)_0$ index is plotted against $(b-y)_0$ in Figure 5. Again, the indices for the
four prototype FHB stars are plotted as crosses. Dotted lines have been drawn around
the area occupied by the natural group, an area nearly perpendicular to the zero age
main sequence, indicated by the solid line. The two points that fall outside the area

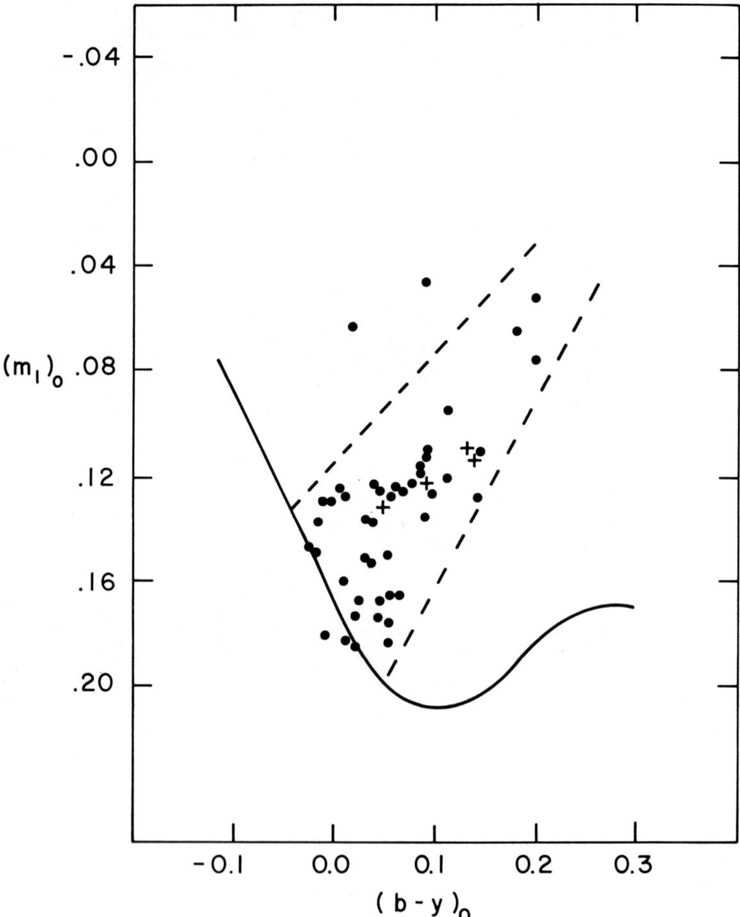

Fig. 5.   The $(m_1)_0$ index vs $(b-y)_0$ for 45 FHB stars in regions at high galactic latitudes. The symbols are the same as in Figure 4.

marked by the dotted lines are indices representing SS 182 I and another star with very high c_1 and very low m_1 indices.

The Hβ index is plotted against $(b-y)_0$ for the group of stars in Figure 6. Four crosses represent the prototype FHB stars. The Hβ indices of the FHB stars generally fall below the zero age main sequence. In Figures 4, 5, and 6, the indices representing the four prototype and the FHB stars found in high galactic latitude areas fall in the same areas. The position occupied by the points in Figure 4 is an effect of higher luminosity or lower surface gravity. The u magnitude is depressed due to the larger Balmer jump and the v magnitude is slightly greater due to the smaller blanketing, both of which lead to a higher c_1 index. In Figure 5, the position occupied by the points representing FHB stars is caused by two effects. One is a luminosity effect due to the Hδ line which falls in the middle of the passband of the v filter. The other is the effect of a lower metal abundance in FHB stars which decreases the blanketing thus causing a brighter v magnitude and a smaller m_1 index. In Figure 6, the position of the points representing the Hβ indices of FHB stars is in the same sense as the variations in the Balmer lines seen in Figure 3.

Plots of $(m_1)_0$ vs $(b-y)_0$ could be better indicators of metal abundance variations were it possible to reduce the scatter in a diagram such as Figure 2. The presence of Hδ in the v passband introduces difficulties of interpretation which has led to the introduction of a new v^1 filter to be added to the Strömgren system. The v^1 filter has a central wavelength of 4220 Å and a half width of 45 Å. A new m_1 index will be computed using the v^1 filter which should be a better measure of the metal abundance.

3. Radial Velocities of Field Horizontal Branch Stars

Radial velocities have been measured for 33 FHB stars, 18 by Philip (1969c, 1970b) and 15 by Rodgers (1971). The velocity measures are presented in Table II.

The velocity dispersion for the group in Table II, for which $V > 10.0$, is ± 113 km s^{-1}. Woolley and Stewart (1967) observed stars in the NGP region and found $\sigma = \pm 18$ km s^{-1} for A stars with $8 < V < 10$. For brighter A stars at the NGP, $\sigma = \pm 8$ km s^{-1}, thus there is a rapid increase in σ with increasing Z.

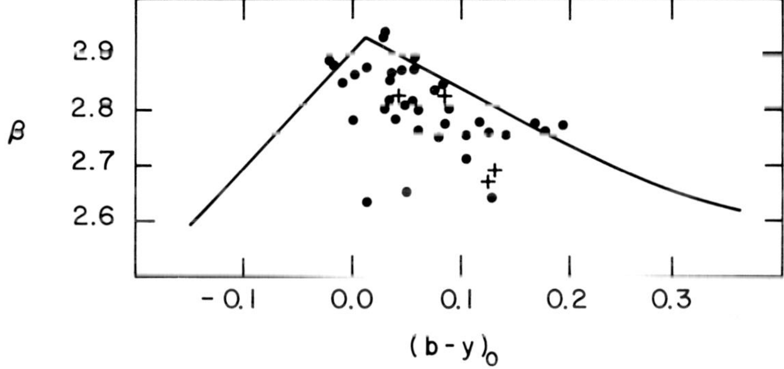

Fig. 6.   The β index vs $(b-y)_0$ for 36 FHB stars. The symbols are the same as in Figure 4.

TABLE II

Radial velocities of field horizontal branch stars

Star		Area	V km s^{-1}	$W(K)$ in Å	Source[a]
SS 182	I	NGP	− 143		1
SS 287	I	NGP	123		1
SS 194	II	NGP	36		1
SS 197	II	NGP	150		1
SS 202	II	NGP	123		1
SS 227	II	NGP	83		1
SS 229	II	NGP	− 90		1
50°2236 El		1 HLF 3	− 89		1
47°2324 Nl		1 HLF 3	− 56		1
43°2448 El		1 HLF 5	− 230		1
40°2179 El		4 HLF 3	70		1
40°2325 Nl		4 HLF 5	83		1
S 10		1 HLF 2	− 160		2
13 110		1 HLF 2	− 123		2
16 111		1 HLF 2	− 253		2
17 24		1 HLF 2	76		2
17 136		1 HLF 2	− 212		2
18 21		1 HLF 2	− 53		2
PS 1	II	SGP	− 64	< 0.4	3
PS 5	II	SGP	− 132	< 0.4	3
PS 6	II	SGP	− 51	< 0.4	3
PS 16	II	SGP	− 215	0.7	3
PS 23	II	SGP	92	1.2	3
PS 25	II	SGP	− 32	0.7	3
PS 27	II	SGP	− 100	2.9	3
PS 35	II	SGP	− 51	0.6	3
PS 36	II	SGP	106	1.7	3
PS 45	II	SGP	− 2	0.4	3
PS 50	II	SGP	147	0.5	3
PS 52	II	SGP	− 34	< 0.4	3
PS 53	II	SGP	82	< 0.4	3
PS 57	II	SGP	6	3.3	3
PS 59	II	SGP	26	0.4	3

[a] Sources: 1, Philip (1969c); 2, Philip (1970b); 3, Rodgers (1971).

4. Equivalent Width of the Calcium K Line

Rodgers (1971) published spectrographic measures of the faint A stars at the SGP from the list of Philip and Sanduleak (1968). One of the measures was the equivalent width of the calcium K line. $W(K)$ is plotted against $(B-V)$ for the faint A stars measured by Rodgers (1971, his Figure 2) in the top part of Figure 7. The main point of his graph was that there are a number of faint A stars far below the galactic plane with normal calcium abundance. In the lower graph the figure is replotted this time showing only the stars from Table II, which have all been classified as FHB stars on the basis of four-color and Hβ photometry. The great majority of these stars have much smaller equivalent widths than is normal for a Population I star. This evidence, along with the high velocity dispersion, tends to confirm that FHB stars are members of Population II.

5. Stellar Densities at the Galactic Poles

Since the samples of early-type stars at the galactic poles have been selected in an unbiased way and are complete to a limiting visual magnitude fainter than 13th mag., the stellar density of these stars can be compiled, knowing the volume of the area searched. The number of stars per 10^6 pc^3 is plotted against the perpendicular distance from the galactic plane in kiloparsecs in Figure 8. The densities of A0 and A2–A7 stars from Upgren (1963) are plotted as solid lines in the upper part of the diagram. The line across the bottom of the diagram represents the density of RR Lyrae stars at the NGP found by Kinman *et al.* (1966). The densities of the FHB stars are plotted as circles (NGP) and crosses (SGP). For distances less than two kiloparsecs accurate densities can not be calculated because the volume of the cone being investigated is small and the density low. For distances greater than five kiloparsecs the magnitudes

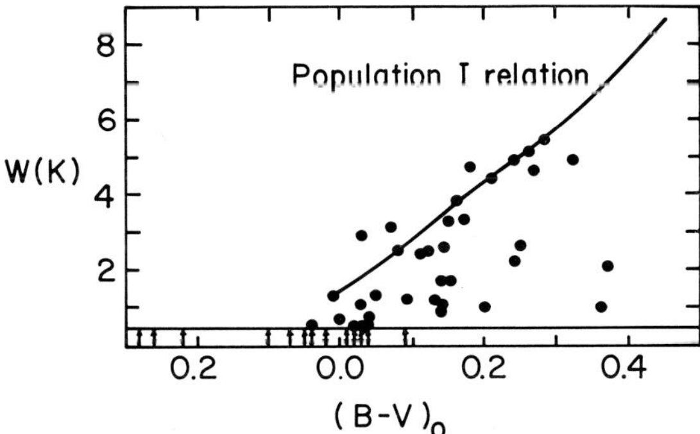

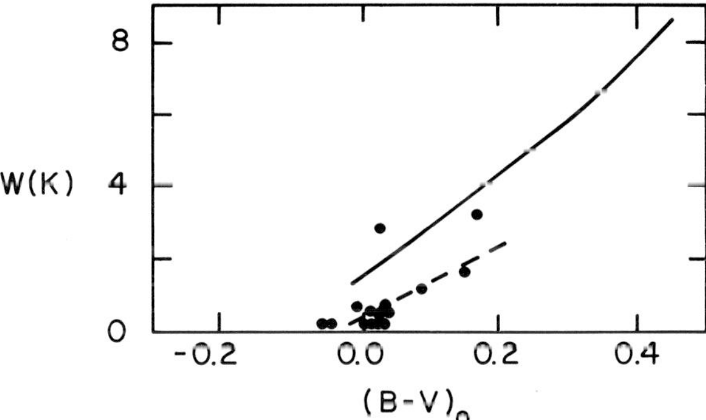

Fig. 7. $W(K)$, the equivalent width of the calcium K line, vs $(B-V)_0$. The upper figure is Figure 2 from Rodgers (1971). The lower figure presents the indices for FHB stars at the SGP.

of the stars would be below the plate limit and thus the data are incomplete. In between these two limits, where a complete count is being made of the blue halo stars, the space density of the FHB stars is approximately 10 times that of the RR Lyrae stars, a situation similar to that found in many globular clusters.

6. Blue Horizontal Branch Stars in Globular Clusters

Observations have been made of BHB stars in the globular clusters M3, M4, M13,

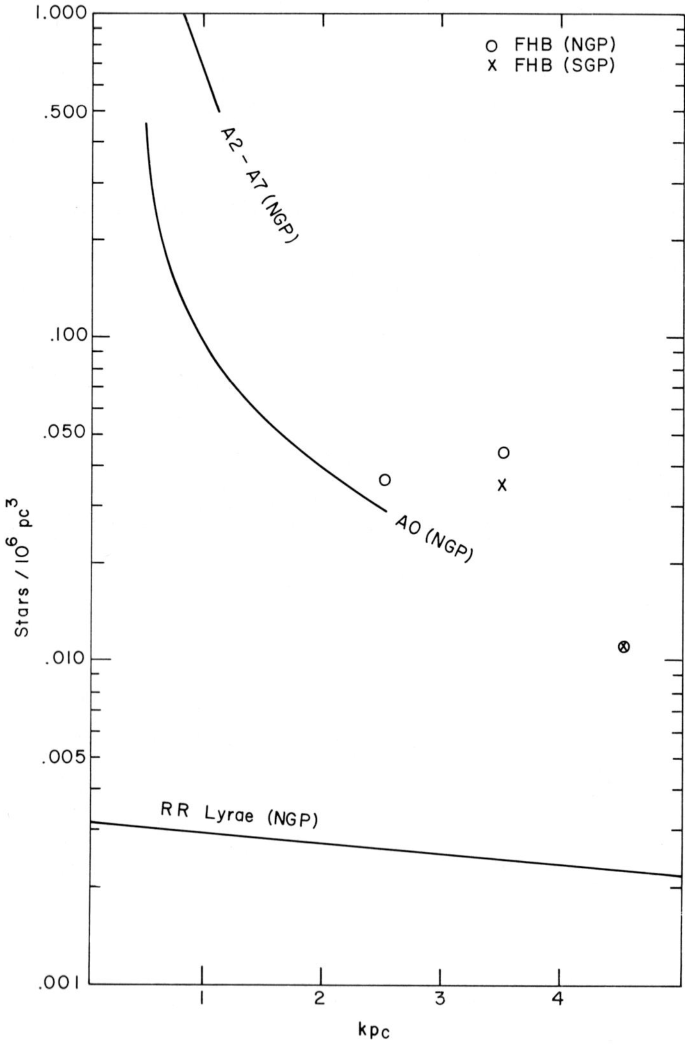

Fig. 8.   The stellar density (stars per 10^6 pc^3) as a function of the distance from the galactic plane in kiloparsecs. At 3–4 kpc, the FHB stars have densities approximately ten times that of the RR Lyrae stars.

M55, M92, and NGC 6397. These stars are quite faint, reaching a V magnitude of 16.2, and thus the precision with which the four-color indices can be measured is less than that of the FHB stars, which are not fainter than $V = 14$. In spite of this handicap, certain similarities can be seen between the FHB and BHB stars, as is shown in the following graphs. The Balmer jump index, $(c_1)_0$ is plotted against $(b-y)_0$ in Figure 9. The solid line indicates the zero age main sequence, the dotted line indicates the area occupied by the FHB stars. Stars in two of the clusters studied, M3 and M13, of intermediate metal abundance, are represented by circles; stars in other clusters are represented by dots. The distribution of globular cluster BHB stars is quite similar to that for the FHB stars.

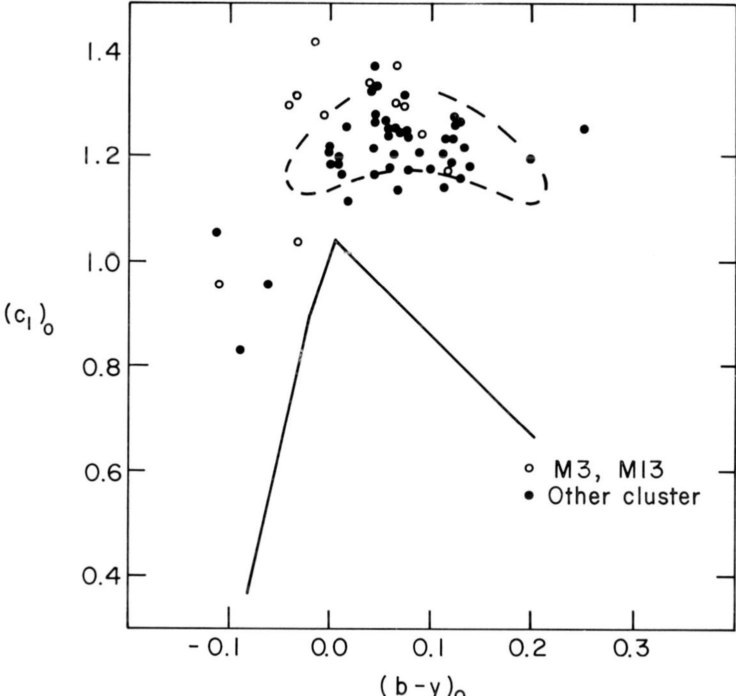

Fig. 9. The $(c_1)_0$ index vs $(b-y)_0$ for BHB stars in globular clusters. The $(c_1)_0$ indices for BHB stars in M3 and M13 are larger, near $(b-y)_0 = 0.00$, than $(c_1)_0$ indices for other clusters.

There are some bluer BHB stars which fall outside the area marked by the dotted lines; no stars this blue were found in the field (except possibly SS 182 I). At $(b-y)_0 =$ $= 0.00$ there are some points representing stars in M3 and M13 which fall higher than the dotted line. This effect can be seen more clearly in Figure 10, where mean lines have been drawn through all the points representing the two types of clusters. The higher $(c_1)_0$ index of the intermediate metal abundance clusters seems quite real.

The metal index, $(m_1)_0$ is plotted against $(b-y)_0$ in Figure 11. The solid and dotted lines have the same meaning as in Figure 9. There does not seem to be any difference between the two types of clusters in this diagram; all the cluster points fall in the same area as the FHB stars. As noted earlier, the m_1 index is complicated by the presence of Hδ in the v passband. The index does not discriminate between stars of low metal abundance but works best between a solar metal abundance and one tenth solar abundance.

7. The log g, θ_e Relation

A search of the literature has yielded 34 early-type stars* for which log g and θ_e have been determined by means of high dispersion spectra or photoelectric spectrum scans

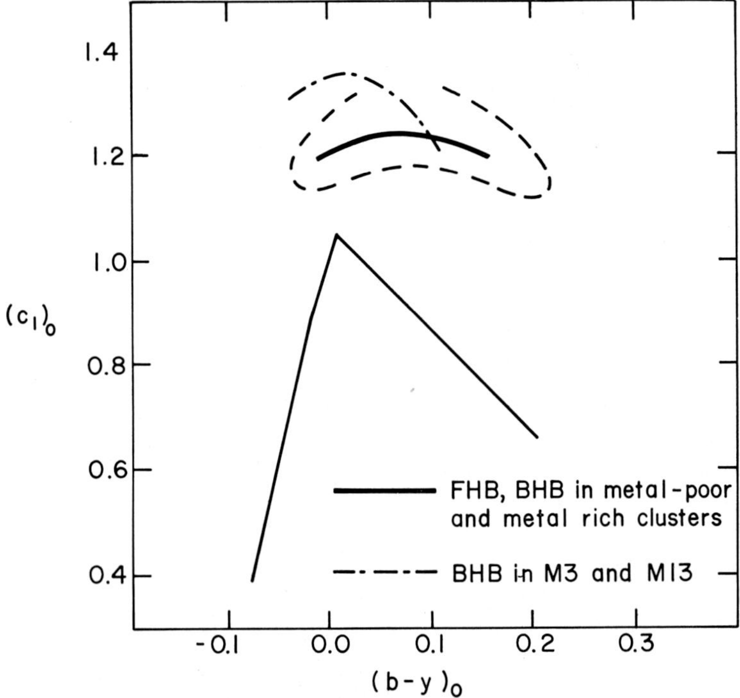

Fig. 10.   The mean $(c_1)_0$ index as a function of $(b-y)_0$ for intermediate metal abundance clusters and other clusters.

matched to atmospheric models. Four-color data exists for these stars; either in the Strömgren-Perry Catalogue (1962) or in the author's unpublished catalogue of FHB stars. In Figure 12, (Figure 1 of Philip, 1972) δ log g is plotted against δc_1 for each of the 34 stars. The dotted line represents a theoretical relation derived by Strom (1970) between δ log g and δc_1. The observational points fall about this relation and indicate that theory and observation agree quite well. The δ log g can be estimated from the

* A list of these stars will be published elsewhere (Philip, 1972).

δc_1 with a probable error of ± 0.2 in $\log g$. Using this relation, one can compute $\log g$ for the BHB stars in the globular clusters so far investigated.

In Figure 13 (Figure 2 of Philip, 1972) the solid line represents the mean relation between $\log g$ and θ_e found for NGC 6397 (Newell *et al.*, 1969), while the symbols represent the results obtained here. The small symbols represent BHB stars in M4, M55, and M92 (a metal-rich and two metal-poor clusters). These points scatter about

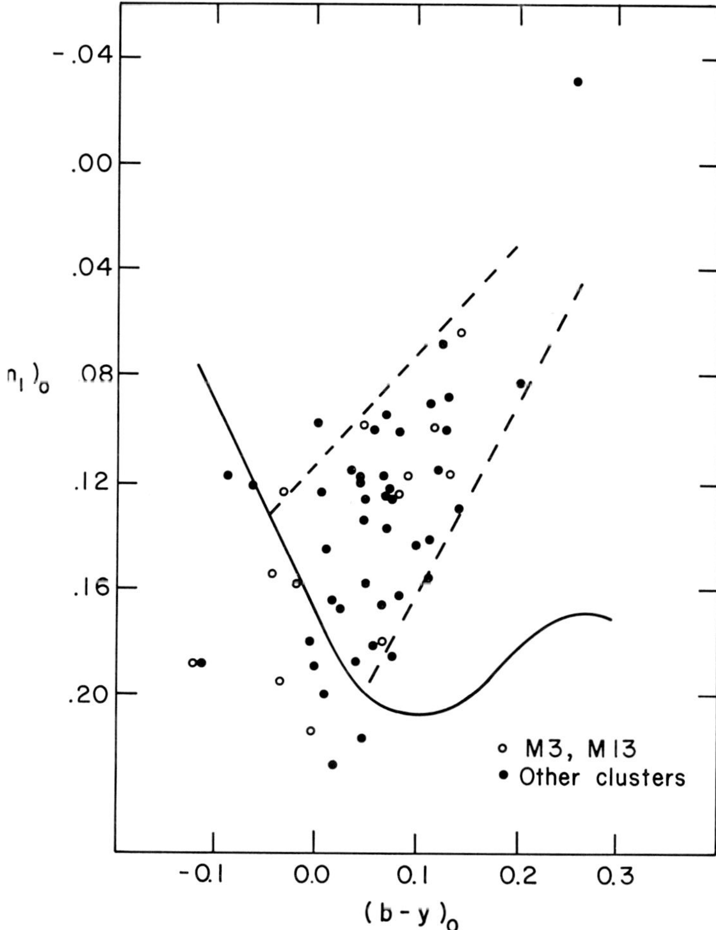

Fig. 11.   The $(m_1)_0$ index vs $(b-y)_0$ for BHB stars in globular clusters.

the mean relation for NGC 6397. Large symbols, representing BHB stars in M3 and M13, fall above the mean relation (in the same sense as the higher $(c_1)_0$ indices in Figure 10) and scatter about a (dotted) line which is about 0.3 higher than the mean relation. This behavior would be expected if the BHB stars in intermediate metal abundance clusters have lost 0.2 $M_\odot$ and the radii have changed by 20%.

8. Conclusion

The FHB stars have been shown to be metal-poor, to have a high velocity dispersion, and to have similar colors to BHB stars in globular clusters. The hypothesis that the FHB stars are indeed BHB stars in the general field seems to be confirmed in light of all these data. More observations of additional stars and more observations per star are needed to confirm the low surface gravities of BHB stars in globular clusters of inter-

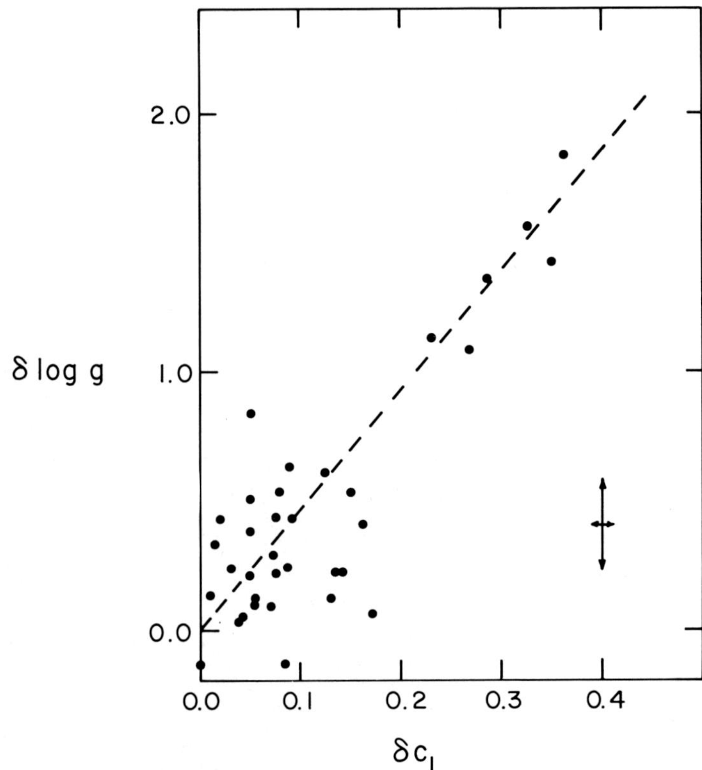

Fig. 12.   log g vs δc_1 for 34 early type stars for which θ_e and log g have been obtained by means of high dispersion spectra or photoelectric scans.

mediate metal abundance. If these additional data agree with the preliminary data, they will be an interesting confirmation of the hypothesis of Iben and Rood (1970) that BHB stars in clusters of intermediate metal abundance have lost more mass than BHB stars in metal-poor clusters.

Acknowledgements

The Directors of the Cerro Tololo Inter-American Observatory, Kitt Peak National Observatory, and Steward Observatory are thanked for making time available on the

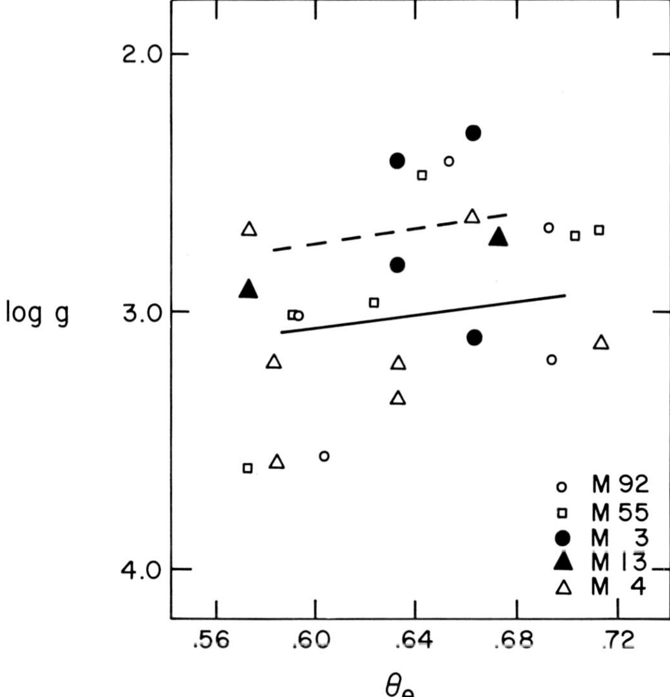

Fig. 13. Log g vs θ_e for BHB stars in five globular clusters. The line represents the mean relation for NGC 6397 (Newell, Rodgers, and Searle, 1969). The intermediate metal abundance clusters M3 and M13 have BHB stars with log g's about 0.3 smaller than the mean relation.

60-in., 36-in., 84-in., and 90-in. telescopes for this project. The National Science Foundation and the SUNY Research Foundation have partially supported this project.

References

Crawford, D. L.: 1970, in A. Slettebak (ed.), *Stellar Rotation*, D. Reidel Publ. Co., Dordrecht, Holland, p. 204.
Iben, I and Rood, R. T.: 1970, *Astrophys. J.* **161**, 587.
Kinman, T. D., Wirtanen, C. A., and Janes, K. A.: 1966, *Astrophys. J. Suppl.* **13**, 379.
Kodaira, K.: 1964, *Z. Astrophys.* **59**, 139
Kodaira, K., Greenstein, J. L., and Oke, J. B.: 1969, *Astrophys. J* **155**, 525.
Mihalas, D.: 1970, *Stellar Atmospheres*, W H Freeman and Co., San Fransisco, p. 190.
Newell, E. B.: 1970, *Astrophys. J.* **159**, 443.
Newell, E. B., Rodgers, A. W., and Searle, L.: 1969, *Astrophys. J.* **156**, 597.
Oke, J. B., Greenstein, J. L., and Gunn, J.: 1966, in R. F. Stein and A. G. W. Cameron (eds.), *Stellar Evolution*, Plenum Press, New York.
Philip, A. G. D.: 1967, *Astrophys. J.* **148**, L143.
Philip, A. G. D.: 1968, *Astrophys. J.* **152**, 1107.
Philip, A. G. D.: 1969a, *Astron. J.* **74**, 209.
Philip, A. G. D.: 1969b, *Astron. J.* **74**, 812.
Philip, A. G. D.: 1969c, *Astrophys. J. Letters* **158**, 113.
Philip, A. G. D.: 1970a, *Astron. J.* **75**, 957.

Philip, A. G. D.: 1970b, *Astron. J.* **75**, 246.
Philip, A. G. D.: 1972, *Astrophys. J.* **171**, L51.
Philip, A. G. D. and Sanduleak, N.: 1968, *Bol. Obs. Tonantzintla y Tacubaya* **4**, 253.
Rodgers, A. W.: 1971, *Astrophys. J.* **165**, 581.
Slettebak, A. and Stock, J.: 1959, *Astron. Abh. Hamburg*, No. 5.
Slettebak, A., Wright, R. R., and Graham, J. A.: 1968, *Astron. J.* **73**, 152.
Strom, S.: 1970, private communication.
Strömgren, B. and Perry, C. L.: 1962, unpublished.
Upgren, A. R.: 1963, *Astron. J.* **68**, 475.
Wallerstein, G. and Hunziker, W.: 1964, *Astrophys. J.* **140**, 214.
Woolley, R. and Stewart, J. O.: 1967, *Monthly Notices Roy. Astron. Soc.* **136**, 329.

SOME BLUE STARS OF PECULIAR TYPE IN THE
REGION OF THE SOUTH GALACTIC POLE

J. A. GRAHAM

Cerro Tololo Inter-American Observatory, Chile*

and

A. SLETTEBAK

Perkins Observatory, The Ohio State and Ohio Wesleyan Universities, U.S.A.

Abstract. *uvby* photometric observations have been used in conjunction with slit spectra to classify 90 stars which were noted as peculiar by Slettebak and Brundage in a recent objective prism survey of the South Galactic Pole region. In this paper, we review the photometric classification criteria and identify in the Slettebak-Brundage list, 8 subdwarf O stars, 10 subdwarf B stars, 10 horizontal branch stars, 1 white dwarf star and 26 late subdwarf stars. Three stars with outstanding peculiarities are SB (Slettebak-Brundage) 58 which is a helium subdwarf O star, SB 319 (CD$-38°245$), a late type star with extremely weak metal lines and SB 845 (BD $-13°6465$), an A type star with a very small Balmer discontinuity.

1. Introduction

Intermediate band photometry can be used very effectively as a means of classifying stars of many different types on the basis of color measurements alone. The increased spectral resolution over broad band photometry allows quantitative measures to be made of such features as the color gradient, the Balmer discontinuity and of some lines and bands in stellar spectra. These photometric indices can be used in turn as classification parameters. In this paper, we report on an application of such a procedure to a group of 90 peculiar high latitude blue stars. These stars were recently found during the course of an objective prism survey for early type stars which was carried out by Slettebak in collaboration with R. K. Brundage (Slettebak and Brundage, 1971). We have obtained in addition slit spectra for the purpose of checking and further refining some of the photometric classifications.

In the Slettebak-Brundage investigation, a $4\frac{1}{2}°$ ultraviolet transmitting prism was used with the Curtis Schmidt telescope at Cerro Tololo Inter-American Observatory to obtain spectra at a dispersion of 580 Å mm^{-1} (at Hγ) over an area of 840 sq deg centered on the South Galactic Pole. The limiting magnitude of the plates is between 14 and 15 for the purpose of spectral classification. During the course of the survey, a number of stars with unusual spectral characteristics were noted. These were specifically, (i) all O and B type stars since at the faint limit of the survey, these stars are likely to be subluminous with respect to the main sequence of early type stars, (ii) stars with unusually broad, weak or strong spectral lines for their spectral type. Some of the stars included by the second criterion had spectra later than type F0 which was

* Operated by the Association of Universities for Research in Astronomy, Inc., under contract with the National Science Foundation.

the limit for listing normal stars, however, these objects were retained because of their special interest as possible faint subdwarf stars.

2. Photometry and Photometric classification Procedure

To date, Graham has obtained photometry for 90 stars which were noted as peculiar in the Slettebak-Brundage survey. The intermediate band *uvby* system described by Strömgren (1963) and by Crawford and Barnes (1970) has been used for this purpose. Elsewhere, Graham (1970) has outlined a classification scheme which is based on the properties of the *uvby* system. This scheme in turn derives from the detailed study of a number of high latitude blue stars by Sargent and Searle (1968). Since the Graham (1970) paper was written, a number of small modifications have been made to the classification criteria. Table I shows the scheme as presently formulated. Note the addition of a separate group to include the late type subdwarf stars. We emphasise here, however, that a photometric classification scheme does have limitations. For example, no distinction can be made between a B type main sequence star and a B type horizontal branch star which has lower mass and lower luminosity but rather similar surface gravity. The differences seen on slit spectrograms such as the variation in helium line intensities have not as yet been detected photometrically. Complications also appear in the classification procedure when heavy interstellar reddening is present. At high galactic latitudes, however, these effects are not serious. Generally speaking, we find that the spectral types estimated from the *uvby* photometry are more accurate than those derived from objective prism plates but that the most precise classifications of all follow from the inspection of slit spectrograms especially when complementary photometric data are also available.

TABLE I

Criteria for *uvby* photometric classification

Type	Classification criteria
sd O subdwarf O	$(b-y)$ very blue, < -0.120, $m_1 < +0.100$ showing Balmer lines are weak if present. $c_1 < 0.000$.
sd B subdwarf B	$(b-y) < -0.100$, $0.100 < m_1 < +0.150$ indicating strong Balmer lines for color, $c_1 < 0.000$.
B	Colors of unreddened main sequence B star. Category includes stars with weak helium lines.
Aw Weak line A	$(b-y)$ and c_1 of a main sequence A star, $m_1 < +0.125$.
AFhb horizontal branch	$c_1 > 1.200$ or at least 0.200 greater than c_1 given by standard Hyades relation (Crawford and Perry, 1966). This is indicative of the low surface gravity. $(b-y) > 0.000$. RR Lyrae stars are included in this group.
DA hydrogen line white dwarf	$m_1 > 0.150$ indicative of very strong hydrogen lines. $c_1 < 0.000$, $(b-y) < +0.200$.
sd FG late subdwarf	$(b-y) > +0.180$, c_1 normal but m_1 less than standard Hyades value by at least 0.045.

3. Slit Spectroscopy

Slit spectrograms have been obtained for 41 of the stars of peculiar type in the Slettebak-Brundage list. These were taken with the Cassegrain spectrograph at the 60-in. telescope of the Cerro Tololo Inter-American Observatory. 14 of the brightest stars were observed at a dispersion of 39 Å mm^{-1}. These are the stars numbered in the Slettebak-Brundage list (SB) 11, 80, 94, 134, 158, 253, 290, 294, 483, 527, 707, 735, 815, 939. The spectrograms were examined in detail by Slettebak. Notes were made about the differences between these and those of standard stars taken under similar observing conditions. Measurements of radial velocity were also made of each of the 39 Å mm^{-1} spectrograms. 27 fainter stars were examined also with spectrograms of dispersion 200 Å mm^{-1}. These included several stars whose colors had been particularly difficult to interpret. Again, Slettebak made a detailed comparison between these stars and a number of standard stars and his notes were used in conjunction with the uvby data to classify the stars into the groups listed in Table I. These low dispersion spectrograms were not suitable for radial velocity measurement but a special note was made if the spectrum seemed to be red or blue shifted by a significant amount with respect to the iron arc comparison lines.

4. The Peculiar Stars in the Slettebak-Brundage List of Early Type Stars

With the combination of the objective prism, photometric, and slit spectroscopic data, we have classified in Table II the 90 stars for which we have observations. A number of these stars have colors of main sequence A and F stars and these are included as a separate group.

8 subdwarf O stars are listed. These include one very unusual object, SB 58. Our spectrogram of this star is shown in Figure 1. Strong lines of helium, both neutral and ionised are seen. The Balmer lines are weak and barely visible. A possible large blue shift was noted on the 200 Å mm^{-1} spectrogram.

10 stars are considered as probable subdwarf B stars. A study of the data from all three sources indicates that the helium lines are almost invariably weak for the color. A typical example of this class, SB 459 (CD $-33°$ 417) is shown in Figure 1. The characteristic broadening of the hydrogen lines with respect to those found in main sequence stars is clearly seen in this spectrogram when a comparison is made with the standard B2 star γ Peg. Two stars, SB 7 and SB 744, are unusual in that they have very red $(b-y)$ indices for their class. This suggests that either the stars are composite (however there was no sign of the necessary red companions on our spectrograms) or that the effect is caused by a circumstellar dust shell. Reddening by interstellar material is unlikely because in no single case did any of the more numerous A and F stars in our list show any such color anomaly. In our spectrogram of SB 7 there are traces of a weak K-line but we cannot be sure that this is not a plate flaw.

The group of 14 B stars contains a mixture of types. Some of the brighter stars appear indeed to be normal main sequence stars (SB 80, SB 158). SB 253 and SB 463

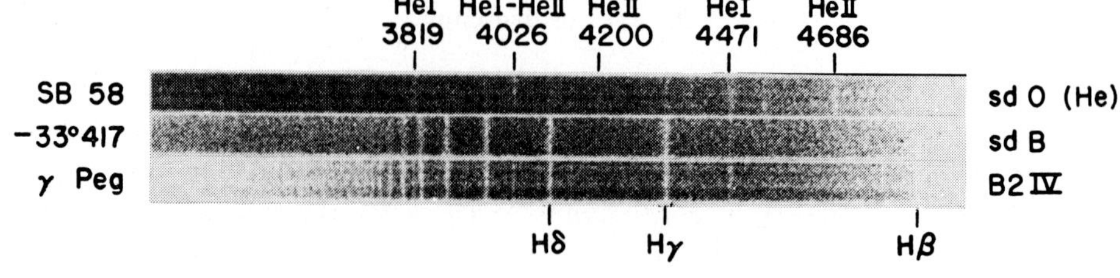

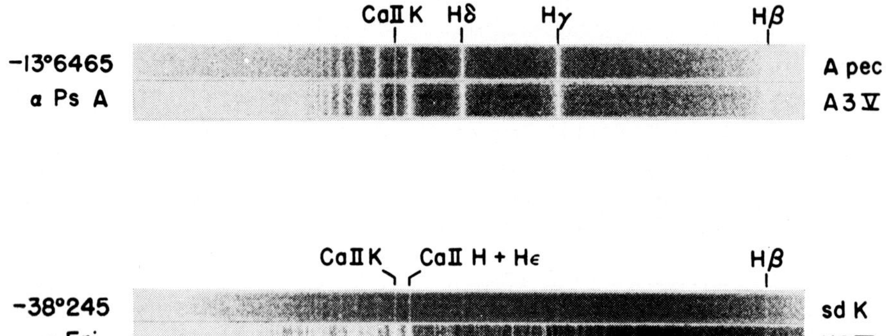

Fig. 1. Spectrograms of peculiar high latitude stars compared with spectrograms of bright standard stars.

have normal spectra but probable high velocities and are most likely runaway stars from the galactic plane. SB 395 and SB 460 appear to be B stars with weak helium lines and similar to those classified as Bw by Sargent and Searle (1968). SB 527 with high velocity, narrow lines and spectral type B8 is a probable horizontal branch star with lower surface gravity than a main sequence star of similar type.

One at first surprising result of our work is the very small number of horizontal branch A and F stars that are found. Of the 10 stars classified as horizontal branch AF stars, 8 are probable RR Lyrae stars. 5 of these are variables which were discovered prior to our study and it is interesting to note that all 8 probable RR Lyrae stars are identifiable as horizontal branch stars on the basis of color measurements alone and without knowledge of their variability. When average colors are plotted on a $(b-y)$-c_1 diagram, it is found that the RR Lyrae variables fall on the continuation of the sequence occupied by non-variable horizontal branch stars. The almost complete absence of the A-type horizontal branch stars in our list of peculiar types can be explained by the fact that these stars do not show conspicuous spectroscopic anomalies at the dispersion used in the objective prism survey. This was deduced after examining in detail a sample of 17 stars classed as (normal) early A type. These stars

TABLE II

Classification of 90 blue stars of peculiar type in the Slettebak-Brundage list

Type	Slettebak-Brundage number	No. of stars
sdO	38 58[a,f] 147 169 705 884 931[b] 933	8
sdB	7[c,d] 8[c] 290[c] 360 410[c] 459[c] 485[c] 707[c] 744[cd] 815[c]	10
B	21 80[e] 158[e] 171 253[e,f] 357[f,g] 395[c,f] 446 460[c,f] 463[e,f] 527[f,h] 735 774 939	14
AFhb	31[i] 107 338[i] 405[i] 406[i] 453[i] 714[i] 830 900[i] 909[i]	10
sdFG	26 44 68 178 264 273 274 319 330 342 421 442[f] 469 501 606 623 683 730 919 941 961[f] 962 963 965 966 967	26
DA	702	1
AF	11[j] 51 94[k] 115 134[l] 189 260 282[l] 294[m] 349 419 483[l] 495 540 553[f] 712 718 719 720 803	20
Peculiar	845	1

[a] SB 58. Slit spectrogram shows very strong He I and He II but very weak Balmer lines. Helium subdwarf O star. Colors red for a sd O star.

[b] SB 931. $(b-y)$ red for a sd O star. No spectrogram available.

[c] Probably weak helium lines.

[d] $(b-y)$ unusually red for classification which is based on slit spectrogram.

[e] No unusual features in spectrum. Normal spectral type.

[f] Possible high velocity. Slight red or blue shift noted on 200 Å mm^{-1} spectrum.

[g] SB 357. Balmer lines rather weak. $H\beta$ not visible suggesting presence of line emission.

[h] SB 527. Slit spectrum indicates B8 but Balmer lines are sharp. Horizontal branch star?

[i] RR Lyrae variable.

[j] SB 11. Peculiar A-star of the silicon type.

[k] SB 94. Strong Sr II, similar to late Am or β CrB star.

[l] Metallic line A-star.

[m] SB 294. λ Bootis star, weak metal lines, low velocity, moderate rotation.

have apparent magnitudes close to 13. *uvby* observations have been made of these stars and it is found that of the 17, 7 show the definite characteristics of horizontal branch stars. In the Slettebak-Brundage list, there are 159 early A stars of 13th mag. The small sample observed suggests that between 50 and 80 of these are horizontal branch stars. It is clear, therefore, that if *uvby* observations had been made of the entire Slettebak-Brundage list, many more horizontal branch A stars would have been detected.

26 late type subdwarf stars seem to be present in our list. Perhaps the most interesting of these is SB 319 (CD $-38°245$) which may be a very cool, extreme member of the class. The 200 Å mm^{-1} spectrogram that we have is reproduced in Figure 1. Only the H and K lines are prominent. The Balmer lines are very sharp and weak and no other lines are visible. The *uvby* photometry ($y: 12.00$, $(b-y): +0.580$, $m_1: +0.070$, $c_1: +0.409$) is suggestive of a very metal deficient K star. The comparison with the spectrogram of the K2 standard star in Figure 1 indicates how extreme this deficiency may be. A second very peculiar star in our list is SB 845 (B.D. $-13°6465$). We are unable to classify this star into any of the categories in Table I. It appears as a peculiar

A star with weak Balmer lines at objective prism dispersion. The *uvby* photometry is itself peculiar (y: 10.74, $(b-y)$: $+0.106$, m_1: $+0.123$, c_1: $+0.352$). The $(b-y)$ and m_1 indices are normal for an early A star but the Balmer discontinuity is very small. Our slit spectrogram, reproduced in Figure 1, shows a weak but definite K-line (equivalent to type A2) and weak Balmer lines with no other lines visible. The very small Balmer discontinuity can be seen in Figure 1 from the comparison with the A3V star, α PsA. It does not seem likely that the star is composite, since no combination of two stars seems capable of producing the observed colors and spectrum. There is no sign of Balmer emission either line or continuum. The star is not faint and we feel that it would merit a very much more detailed investigation.

References

Crawford, D. L. and Perry, C. L.: 1966, *Astron. J.* **71**, 206.
Crawford, D. L. and Barnes, J. V.: 1970, *Astron. J.* **75**, 978.
Graham, J. A.: 1970, *Publ. Astron. Soc. Pacific* **82**, 1305.
Sargent, W. L. W. and Searle, L.: 1968, *Astrophys. J.* **152**, 443.
Slettebak, A. and Brundage, R. K.: 1971, *Astron. J.* **76**, 338.
Strömgren, B. 1963, in K. Aa. Strand (ed.), *Basic Astronomical Data*, University of Chicago Press, Chicago, p. 123.

QUELQUES SOUS-NAINES DANS LA DIRECTION
DU GRAND NUAGE DE MAGELLAN

M.-N. PERRIN

Observatoire de Lyon, 69230 – Saint Genis-Laval, France

Abstract. Six G and K type stars, chosen among Fehrenbach's stars, in the direction of the Large Magellanic Cloud, were recognized as metal deficient galactic stars, on ground of multicolour photometry (*UBV* system and Lick six-color system). Data concerning these stars are given in Tables I and II. The stars are plotted in a $U - B/B - V$, and in a $U - G/G - I$ diagram.

Three more of Fehrenbach's stars also appear in the Tables and Figures. They are: HD 30 229 and HD 268 957 – Thackeray's extreme and moderate subdwarfs – and HD 270 011, already recognized by Fehrenbach *et al.* as an extremely deficient galactic giant. This last star falls close to HD 122 563 in the $U - G/G - I$ diagram. The *I*-Colour of HD 30 229 is probably contaminated by nearby cooler stars.

Cette communication porte sur des observations photoélectriques faites au télescope de 1 m de diamètre de l'ESO à La Silla, en U, B, V et en six couleurs (système de Lick), sur des étoiles de type G et K, dans la direction du Grand Nuage de Magellan.

On sait que, dans cette direction ($l \sim 280°$, $b \sim -33°$), les grandes vitesses radiales observées sont le reflet de la rotation galactique et que la composante V de la vitesse spatiale (comptée positivement dans le sens de la rotation galactique) qui correspond à une vitesse radiale ϱ est approximativement égale à $-\varrho \cos b$.

Dans ces conditions, une vitesse radiale très grande (de l'ordre de 250–300 km s^{-1}) peut signifier: appartenance au Nuage ou appartenance au halo galactique, une vitesse moins grande pouvant signifier: étoile galactique de Population II plus ou moins extrême.

On s'attend donc à trouver des sous-naines galactiques parmi les étoiles de vitesse radiale grande ou moyenne, signalées par Fehrenbach et Duflot (1970).

Les deux premières sous-naines ainsi mises en évidence ont été HD 30229 et HD 268957, que nous appellerons sous-naines de Fehrenbach-Thackeray (Thackeray, 1966). J'ai essayé d'en trouver d'autres par la photométrie multicouleur.

Pour cela, j'ai trié, parmi des étoiles de type F avancé, G et K déjà observées en 1968, les étoiles les plus riches en ultra-violet (les supergéantes du Nuage, supposées pauvres en UV, étant étudiées par ailleurs).

Pour une quinzaine d'étoiles ainsi choisies (magnitude visuelle entre 9.5 et 12.5), j'ai eu à ma disposition des mesures en U, B, V (valeurs moyennes) et des mesures en six couleurs (généralement, 3 observations par étoile).

Les mesures photoélectriques (observations et réductions aux standards de l'hémisphère Nord) font partie d'un travail collectif effectué par les équipes de photométrie des Observatoires de Marseille et de Lyon.

Sur le matériel d'observation, j'ai d'abord procédé à divers contrôles, nécessaires à deux points de vue: les rattachements sont récents; les étoiles observées sont faibles pour la photométrie en six couleurs.

Un premier contrôle a porté sur la comparaison des deux systèmes photométriques. On peut constater que, pour les nombreuses étoiles G et K, naines et sous-naines, étudiées en six couleurs par Sears et Whitford (1969) et étudiées aussi en U, B, V la différence $D=(U-V)_{3c}-(U-G)_{6c}$ est constante.

Pour les étoiles de la Séquence Principale (Johnson et Knuckles, 1955; Sears et Whitford, 1969) D est égale à 0.82 ± 0.02 (rms).

Pour divers groupes d'étoiles plus ou moins déficientes en métaux, l'erreur quadratique moyenne est plus forte (0.05), mais la moyenne ne change pas. Je me suis assurée que, pour les observations dont je disposais, on retrouve cette même valeur moyenne.

Un second contrôle s'applique aux 6 couleurs: les couleurs moyennes de chaque étoile ont été portées dans un diagramme en 6 couleurs des Hyades, où une étoile déficiente en métaux, comparée à une étoile des Hyades de même indice $B-R$, fait apparaître un effet Code (Code, 1959; Rousseau, 1968 Figure 6b). Je considère comme sûre une étoile qui fait apparaître un effet Code normal (excès de I, de V et de U bien proportionnés).

Les contrôles m'ont fait rejeter quatre étoiles, pour lesquelles les mesures en six couleurs sont d'ailleurs discordantes. Ces étoiles ont été conservées seulement dans

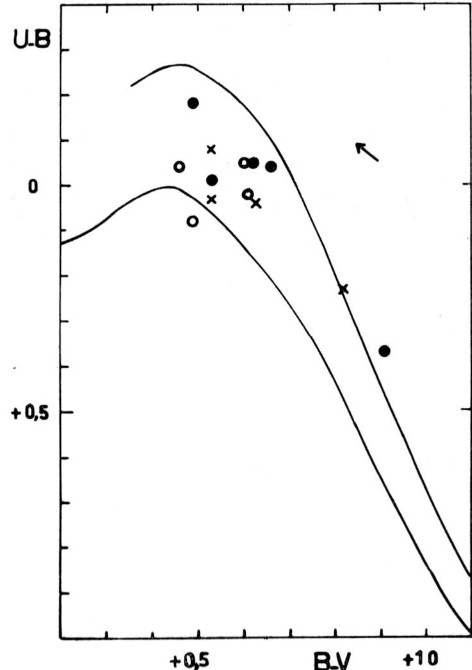

Fig. 1. Diagramme $U-B/B-V$. Séquence Principale des Hyades et enveloppe supérieure d'après Sandage (1969, Table IA, colonnes 1, 2 et 3). Points noirs: étoiles de la Table I (les plus sûres en 6 couleurs). Cercles: étoiles de la Table II (moins sûres). Croix: quatre étoiles qui ont été rejetées parce que douteuses en 6 couleurs. Ces étoiles ne sont pas portées dans la Figure 2. Ce sont: NC. 71, NC. 64, P 282, NC. 67 (n° OM). La flèche représente un rougissement de $A_v=0.15$.

le diagramme $U-B/B-V$, (Figure 1). De plus: parmi les objets que j'avais considérés comme galactiques, parce que riches en UV, il semble se trouver des amas et même des objets bleus présentant des caractères de supergéantes, dont la discussion dépasse la photométrie en 6 couleurs (G 480, G 234B et G 296).

Il reste alors 9 étoiles, situées dans la direction du Grand Nuage de Magellan ou de la Jonction Grand Nuage-Petit Nuage de Magellan, et qui comprennent:

(1) L'étoile G 446, classée G5III par Ardeberg *et al.* (1971) et déjà signalée par Fehrenbach *et al.* (1970) comme géante galactique très déficiente en métaux.

(2) Les deux sous-naines de Fehrenbach-Thackeray.

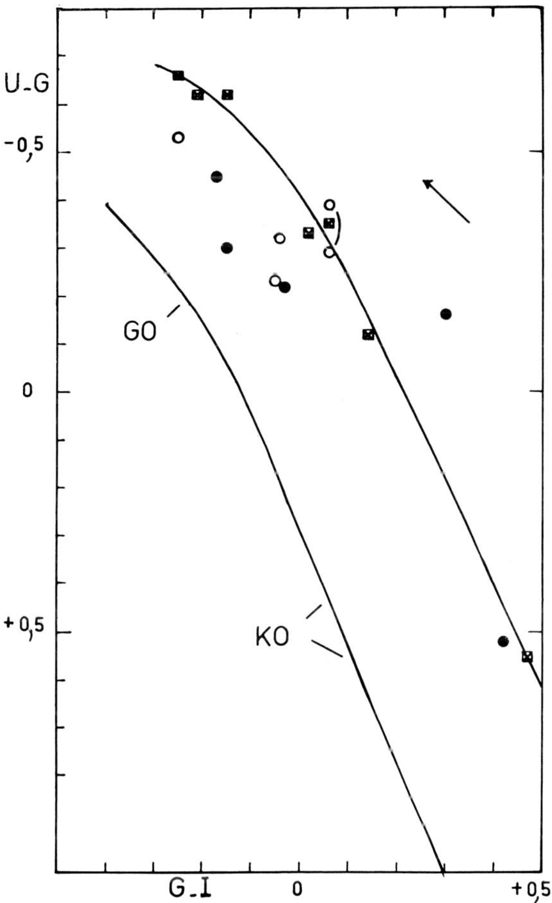

Fig. 2.   Diagramme $U-G/G-I$. Séquence Principale des Hyades, d'après Sears et Whitford (1969), Enveloppe supérieure tracée à partir de six sous-naines extrêmes (dont 4 sous-naines de Roman, 1954) et de HD 122 563 géante du halo (($G-I=+0.47$). Les sous-naines sont, par ordre de $G-I$ croissant: $+72°94$, HD 219 617, $+26°2606$, HD 64 090, $-13°3834$ et HD 195 636. Points noirs et cercles, mêmes notations que pour la figure 1. Pour l'étoile NC. 65, on donne la valeur observée et une valeur corrigée (cf. texte). La flèche représente un rougissement interstellaire de $A_v=0.15$.

(3) Quatre étoiles de vitesses radiales positives moyennes.

(4) Deux étoiles de vitesses radiales négatives.

Au sujet de ces deux dernières étoiles, rappelons que la théorie du Collapse de Eggen, Lynden-Bell et Sandage (1962) prévoit l'existence d'étoiles du halo ayant un grand moment angulaire (composante V de la vitesse spatiale, positive). De telles étoiles, observées dans la direction du L.M.C., auront des vitesses radiales négatives. (Pour les étoiles du halo à grand moment angulaire, voir en particulier Sandage (1969).)

Parmi les 9 étoiles, j'ai distingué: 5 étoiles sûres et 4 étoiles moins sûres. Ces étoiles sont portées dans les Tables I et II, par ordre de $G-I$ croissant. Elles sont portées aussi dans un diagramme $U-B/B-V$, et dans un diagramme $U-G/G-I$ (Figure 2).

Dans ce second diagramme, les lignes de blanketing sont verticales, dans la mesure où les effets de line-blocking et de backwarming se compensent pour l'indice $G-I$. D'après Cayrel (1968, Table III) la compensation se produit entre les types spectraux G0 et K0, et l'effet total reste faible pour les types considérés ici.

De plus, d'après ce qui a été dit plus haut, le diagramme est équivalent au diagramme $(U-V)_{3c}/G-I$, les deux diagrammes étant superposables, aux erreurs de mesures près.

La Figure 2 met donc en évidence les corrections:

$$\Delta(U-G)_{6c} = \Delta(U-V)_{3c} = \Delta(B-V) + \Delta(U-B).$$

Autrement dit: les corrections $\Delta(B-V)$ et $\Delta(U-B)$, qui se contrarient dans la Figure 1 (effet de guillotine), s'ajoutent dans la Figure 2.

1. Explication des tables

Colonne 1: numéro HD ou HDE.

Colonne 2: numéro de l'étoile dans les listes de Ardeberg *et al.* (1971).

Colonne 3: numéro de l'Observatoire de Marseille.

(1) Les étoiles précédées des lettres G et P sont extraites des listes G et P de Fehrenbach et Duflot (1970).

(2) Les étoiles de la Jonction sont précédées de NC et sont extraites des listes de Carozzi (1971).

Colonnes 4 et 5: ascension droite et déclinaison 1950.

Colonnes 6–10: résultats de la photométrie U, B, V, et indices $G-I$ et $U-G$ de la photométrie 6 couleurs.

Colonne 11: différence $D=(U-V)_{3c}-(U-G)_{6c}$; D doit être égal, en moyenne, à 0.82.

Colonnes 12 et 13: vitesse radiale au Prisme Objectif, et vitesse radiale obtenue à partir de spectres pris au spectrographe RV Cas. (Ardeberg *et al.*, 1971).

Colonne 14: types spectraux de Ardeberg *et al.* (1971).

Colonnes 15 et 16: pour huit étoiles, Magnitudes absolues approximatives et dis-

TABLE I

Étoiles les plus sûres

HD ou HDE	No. A	No. OM	α 1950	δ 1950	V	$B-V$	$U-B$	$G-I$	$U-G$	D	ϱ_{PO}	ϱ_A	Types spectraux	M_v	r_{pc}
(1)	(2)	(3)	(4)	(5)	(6)	(7)	(8)	(9)	(10)	(11)	(12)	(13)	(14)	(15)	(16)
269221	171	G 237	5^h13^m07	$-71°00\!.3$	11.25	$+0.50$	-0.18	-0.17	-0.45	0.76	$+149$	$+150$	G0–F0	5.3	155
270668	607	P 99	4 42.39	-69 10.4	11.37	$+0.53$	-0.00	-0.15	-0.30	0.82	-101			5.4	156
269764	632	P 1173	5 34.63	-69 15.2	11.02	$+0.61$	-0.05	-0.03	-0.22	0.79	$+129$			6.0	101
30229 (1)	604	'C1'	4 40.90	-65 24.5	9.40	$+0.66$	-0.04	$+0.30$	-0.16	0.78		$+307$	G2–F0	7.3	26
270011	693	G 446	5 42.99	-70 15.0	11.25	$+0.92$	$+0.36$	$+0.42$	$+0.52$	0.76	$+267$	$+262$	G5III	(-0.7)	(2500)

TABLE II

Étoiles moins sûres

HD ou HDE	No. A	No. OM	α 1950	δ 1950	V	$B-V$	$U-B$	$G-I$	$U-G$	D	ϱ_{PO}	ϱ_A	Types spectraux	M_v	r_{pc}
		NC 51	1^h14^m40	$-73°47\!.8$	10.73	$+0.46$	-0.04	-0.25	-0.53	0.95	-72	-88		4.3	193
268957 (2)	632	P 474	5 02.15	-70 07.2	9.78	$+0.49$	$+0.08$	-0.04	-0.32	0.89	$+179$	$+173$	G0–F2	5.9	60
		NC 59	2 16.43	-74 17.9	11.58	$+0.60$	-0.05	-0.05	-0.23	0.78	$+161$	$+173$	G2V–V1	5.85	140
(3)		NC 65	2 42.06	-74 1.2	12.52	$+0.61$	$+0.02$	$+0.06$	(-0.39) -0.29	(1.02) 0.92	$+101$			6.4	167

(1) Sous-naine extrême de Fehrenbach-Thackeray.
(2) Sous-naine modérée de Fehrenbach-Thackeray.
(3) Colonnes 10 et 11 – entre parenthèses: valeur observée – Au-dessous: valeur corrigée.

tances r, d'après les données de Cayrel (1968, Table III), dans l'hypothèse où les étoiles sont des sous-naines.

Pour la géante G 446, j'ai adopté la magnitude absolue attribuée à HD 122563 par Wallerstein *et al.* (1963).

Si l'indice $G-I$ de HD 30229 est rendu trop positif par un compagnon rouge, l'éclat et la distance de cette étoile sont sous-estimés.

2. Commentaires sur quelques étoiles

P 474 (la sous-naine modérée de Fehrenbach-Thackeray) est l'étoile la plus douteuse de la Table II, pour les 6 couleurs. Ses observations U, B, V paraissent erronées, elles aussi.

Dans la Figure 2, l'étoile NC 65 se place un peu au-dessus de l'enveloppe supérieure. Mais, dans le diagramme en 6 couleurs, l'étoile présente un excès de U exagéré (par rapport aux excès de I et de V). De plus la différence D, trop grande, confirme que le U des 6 couleurs pourrait être trop brillant de 0.20. Si on corrige $U-G$ de 0.10, on obtient un point situé sur l'enveloppe supérieure. Si on faisait la correction entière de 0.20, l'étoile aurait encore les couleurs d'une sous-naine très déficiente en métaux (à peu près celles de 85 Peg).

L'étoile la plus remarquable de la Figure 2, est la sous-naine de Fehrenbach-Thackeray (Table I), qui se situe nettement en dehors de l'enveloppe. Cette fois, tout paraît correct dans les observations. La position anormale de HD 30229 pourrait provenir d'un compagnon rouge: une étoile M pourrait rendre l'indice $G-I$ trop positif de 0.15, sans presque modifier l'indice $U-G$. On peut se demander si les étoiles B et C signalées par Thackeray n'ont pas contaminé les couleurs de l'étoile principale dans l'infra-rouge, malgré la distance angulaire (21").

L'étoile la plus froide de la Table I (la géante G III de Fehrenbach *et al.*, 1970), se place, dans la Figure 2, au voisinage de HD 122563, la géante de déficience extrême de Wallerstein *et al.* (1963), étudiée aussi par Cayrel et Fringant (1964). Les six couleurs des deux étoiles sont très voisines. HD 122563 (située au voisinage du Pôle Galactique Nord) n'est pas rougie. Si G 446 est elle-même peu rougie, elle peut être aussi une géante du halo de déficience extrême, ce qui est compatible avec sa vitesse radiale de $+262$ km s^{-1}.

3. Conclusion

La recherche d'étoiles galactiques de Population II plus ou moins extrême, est une application intéressante des mesures de vitesses radiales de Fehrenbach et Duflot dont l'objet principal est la mise en évidence des objets appartenant au Grand Nuage de Magellan.

Reconnaissances

Je remercie l'equipe des vitesses radiales de l'Observatoire de Marseille, pour m'avoir communiqué des vitesses avant publication. Je rappelle que le présent travail n'est qu'une partie isolée du travail des équipes photométriques de Marseille et de Lyon.

Je remercie R. Canavaggia, qui m'a encouragée dans la rédaction de la présente communication.

Bibliographie

Ardeberg, A., Brunet, J.P., Maurice, E., et Prévot, L.: 1972, *Astron. Astrophys. Suppl. Ser.* **6**, 249.

Carozzi, N.: 1971, Communication privée.

Cayrel, G.: 1968, *Astrophys. J.* **151**, 997.

Cayrel, G. et Fringant, A-M.: 1964, *Compt. Rend. Acad. Sci. Paris* **258**, 3195.

Code, A.: 1959, *Astrophys. J.* **130**, 473.

Eggen, O. J., Lynden-Bell,D., et Sandage, A.: 1962, *Astrophys. J.* **136**, 748.

Fehrenbach, Ch. et Duflot, M.: 1970, *Astron. Astrophys., Special Supplement Series,* No. 1.

Fehrenbach, Ch., Brunet, J-P., Maurice, E., et Prévot, L.: 1970, *Compt. Rend. Acad. Sci. Paris,* **271**, 470.

Johnson, H. L. et Knuckles, C. F.: 1955, *Astrophys. J.* **122**, 209.

Roman, N. G.: 1954, *Astron. J.* **59**, 307.

Rousseau, J.: 1968, *Ann. Astrophys.* **31**, 413.

Sandage, A.: 1969, *Astrophys. J.* **158**, 1115.

Sears, R. L. et Whitford, A. E.: 1969, *Astrophys. J.* **155**, 899.

Thackeray, A. D.: 1969, *Observatory* **86**, 60.

Wallerstein, G., Greenstein, J. L., Parker, R., Helfer, H. L., et Aller, L. H.: 1963, *Astrophys. J.* **137**, 280.

TWO DIMENSIONAL SPECTRAL CLASSIFICATION
OF EARLY TYPE STARS BY LOW DISPERSION
SPECTROPHOTOMETRY

A. GUTIÉRREZ-MORENO and H. MORENO

Departamento de Astronomía, Universidad de Chile

Abstract. Spectrophotometric data of early type stars (spectral intensity distributions from 3100 to 6000 Å and equivalent widths of Hβ, Hγ and Hδ) have been analysed, looking for the best parameters for a two dimensional spectral classification. It was found that the best correlation with MK classification is given by the equivalent width of Hβ and the Balmer discontinuity.

The classification scheme is well defined for stars from B0 to A1 of luminosity classes V to III. No stars of luminosity class II were observed. Luminosity class I is clearly separated from the less luminous stars, but the number of supergiants observed is too small to give a clear separation of spectral types.

The usefulness of quantitative measurements of the Balmer discontinuity and other hydrogen parameters for stellar classification is a well known fact. The photoelectric spectrum scans obtained by Gutiérrez-Moreno *et al.* (1967, 1968, 1972) and Moreno (1972) extend from 3000–6000 Å and thus allow the determination of the size and position of the Balmer discontinuity, and the measurement of equivalent widths of Hβ, Hγ and Hδ. The observations were made at the Cerro Tololo Inter-American Observatory using a spectrum scanner with a slit 60 Å wide. The parameters measured are shown in Figure 1, where D and λ_D are the size and position of the Balmer discontinuity as defined by Barbier and Chalonge (1939) and Chalonge and Divan (1952). Besides, ΔD is a measure of the emission in the Balmer continuum, present in some stars and already detected by Slettebak and Stock (1957) on objective prism plates; and UD is an ultraviolet deficiency, which exists in some stars and which was measured for its possible physical significance. The blue-green and ultraviolet gradients, ϕ_{BG} and ϕ_{U} were also measured.

All these data were tested for a two dimensional spectral classification. The first attempt was made following Barbier and Chalonge (1939) and Chalonge and Divan (1952). A comparison of our data with those from Paris showed that the measurements of the Balmer discontinuity are in close agreement:

$$2.5\, D_{\mathrm{Paris}} = 0.04 + D_{\mathrm{Chile}}. \tag{1}$$

Nevertheless, a systematic dependence on the equivalent widths of the hydrogen lines was found for the relation between the Paris position of the Balmer discontinuity, λ_1, and ours, λ_D. For example:

$$\lambda_D - \lambda_1 = 7.545 + 0.692\,(\lambda_D - 3700) - 2.366\, W_{\mathrm{H}\beta}. \tag{2}$$

Similar relations hold if $W_{\mathrm{H}\gamma}$ or $W_{\mathrm{H}\delta}$ are used. Relation (2) reproduces λ_D with a dispersion of $\pm 5\,\text{Å}$.

Ch. Fehrenbach and B. E. Westerlund (eds.), Spectral Classification and Multicolour Photometry, 258–266. All Rights Reserved.
Copyright © 1973 by the IAU.

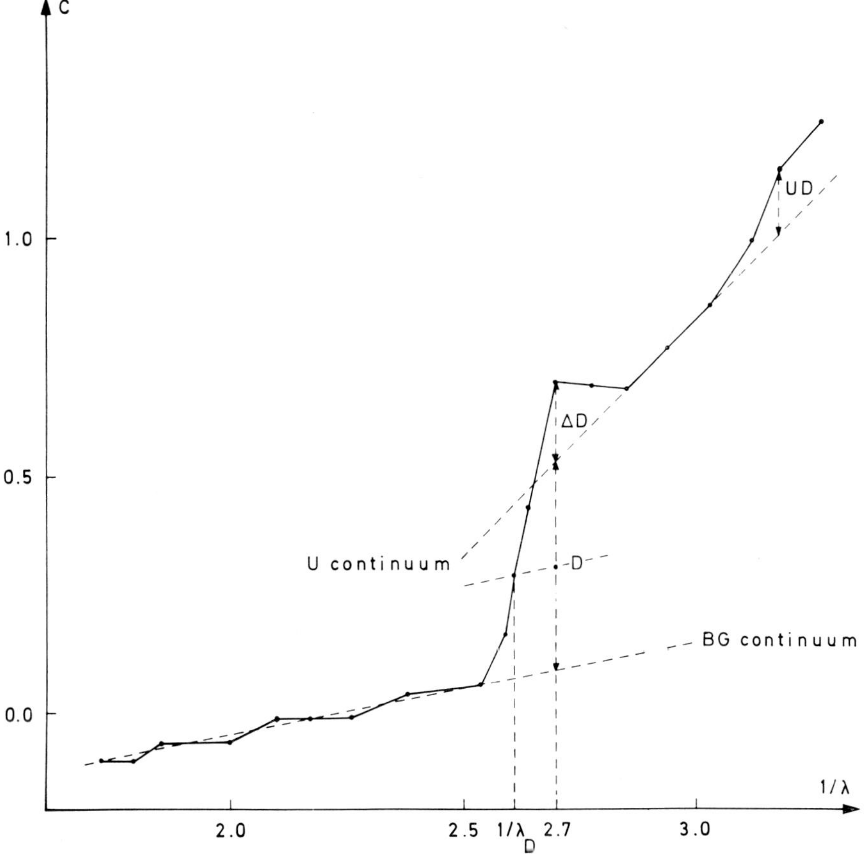

Fig. 1.   Schematic diagram showing the different parameters measured.

May be due to this systematic effect, a graph of D vs λ_D failed to give a positive result for a two dimensional spectral classification.

The analysis of all the remaining data showed that the closest agreement with the MK system is obtained by using the relation between the size $D_1 = D + \Delta D$ of the Balmer discontinuity (spectral type indicator) and the equivalent width of Hβ (luminosity indicator). It was also found that D_1 could be replaced by the color index:

$$c_D = m_{3695} - m_{3930},\qquad(3)$$

which is, numerically, almost identical to D_1 and has advantages from the practical point of view. The magnitudes in relation (3) have been corrected by atmospheric extinction, but not for interstellar absorption. Nevertheless, no misclassifications were detected due to this fact. On the other hand, a comparison of D, D_1 and c_D with the broad-band parameter Q, which is independent of reddening (Johnson, 1958), gives the best agreement for the (Q, c_D) relation. An attempt to make a correction for reddening, by using a color difference defined in a similar way to the parameter Q showed no improvement whatsoever.

Up to B7, the standard stars used to establish the system have been taken only from Lesh (1968) or from Hiltner *et al.* (1969). Preliminary attempts made independently with each of these authors gave coincident classification schemes and, consequently, both lists were considered to form a homogeneous classification and were used to derive the empirical diagram shown in Figure 2. From B8 on, the system is not so homogeneous, since the classifications were taken from different authors (Jaschek and Jaschek, 1966; Jaschek *et al.*, 1964). The stars used as standards are listed in Table I. The references for this Table and for Table II are given at the end of Table II.

Since the lines separating spectral types and luminosity classes in Figure 2 have been obtained empirically, they have a certain amount of uncertainty; nevertheless, the separations between luminosity classes V, IV and III are fairly well defined, mainly for the earlier types. No stars of luminosity class II were included in the observations and, consequently, the lower limit for luminosity class III is rather arbitrary. Luminosity class I stars are well detached. The sample of these stars is small, but they show that their classification is difficult due to the effect of luminosity on the Balmer discontinuity, which produces a crowding effect of spectral types towards small c_D values. As an example, we show the star HD 111613, which is an A1Ia star. Figure 2 suggests the convenience of using the $(c_D, W_{H\beta})$ diagram to separate the high luminosity stars, and then use for them a classification scheme based on different parameters.

A group of 75 stars, considered as program stars, was classified according to the scheme in Figure 2. In general, the discrepancies with respect to Lesh (1968) and Hiltner *et al.* (1969) are small as defined by Jaschek and Jaschek (1966): only 5 stars, of a total of 46 in common with these authors differ from their classification by more than one sub-class in spectral type, or more than one class in luminosity. For these 5 stars, a classification in better agreement with ours is given in the La Plata Catalogue (Jaschek *et al.*, 1964). The average index of discrepancy, defined also by Jaschek and Jaschek (1966) is $\bar{e} = 0.55$, which implies that the average uncertainty of a single classification is less than 0.55 luminosity classes and 0.55 spectral subclasses. The internal accuracy of our classification, determined by the method of Butler and Thackeray (1940) is:

$$\sigma_{ST} = \pm\, 0.2\,\text{subclasses}, \qquad \sigma_L = \pm\, 0.5\,\text{classes}.$$

σ_{ST} is the mean of three determinations, made by comparison with different authors.

There are, nevertheless, nine stars for which the discrepancies with respect to other authors are not small. They were not included in the previous analysis, since they also show rather large discrepancies between the classifications made by other authors. They are listed in Table II. The table also shows the broadband spectral types S_Q. In most of the stars listed, S_Q is closer to the spectrophotometric types than to the MK classifications given by other authors. This shows that these stars have some peculiarities, even though these peculiarities may not be apparent in the spectrum.

An attempt to detect peculiar stars or emission stars was made by plotting the equi-

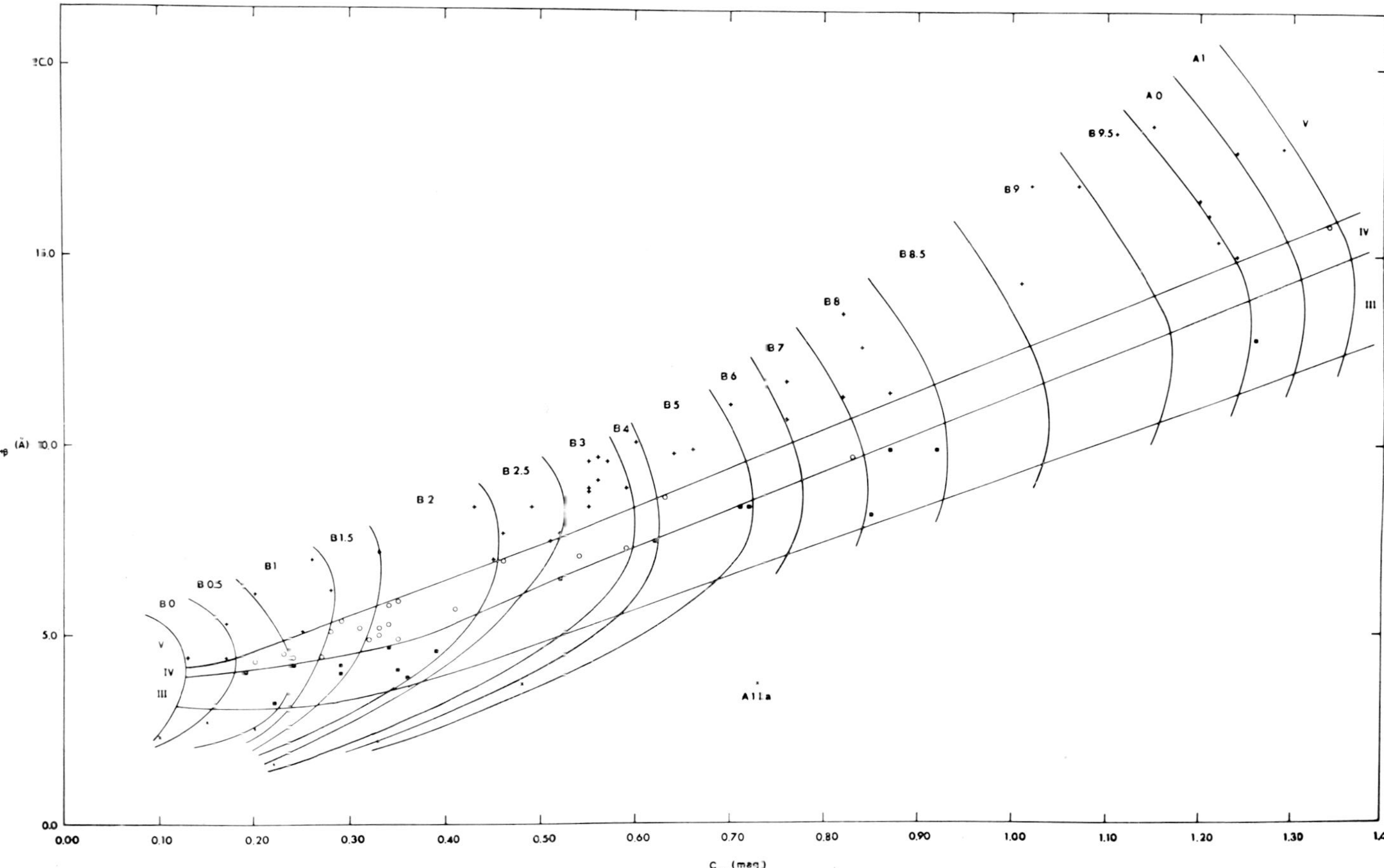

Fig. 2. Empirical diagram for obtaining MK spectral types on the basis of c_D, $W_{H\beta}$ measures.

A. GUTIÉRREZ-MORENO AND H. MORENO

TABLE I

Standard Stars

BS	HD	Name	c_D	$W_{H\beta}$	MK	Ref.
1855	36512	υ Ori	0.13	4.4	B0V	[1]
6165	149438	τ Sco	0.17	4.4	B0V	[1]–[2]
1903	37128	ε Ori	0.10	2.3	B0Ia	[1]–[2]
1887	36960	$-6°$ 1234	0.17	5.3	B0.5V	[1]
5953	143275	δ Sco	0.20	4.3	B0.5IV	[1]–[2]
1756	34816	λ Lep	0.23	4.5	B0.5IV	[1]
4853	111123	β Cru	0.19	4.0	B0.5III	[2]
7446	184915	κ Aql	0.22	3.2	B0.5III	[1]
2004	38771	κ Ori	0.15	2.7	B0.5Ia	[1]–[2]
1892	37018	42 Ori	0.20	6.1	B1V	[1]
1789	35439	25 Ori	0.25	5.1	B1V	[1]
1886	36959	$-6°1233$	0.26	7.0	B1V	[1]
1868	36695	VV Ori	0.28	6.2	B1V	[1]
5056	116658	α Vir	0.24	4.4	B1IV	[1]
5132	118716	ε Cen	0.24	4.2	B1III	[2]
4133	91316	ϱ Leo	0.20	2.5	B1Iab	[1]
1890	37017	$-4°1183$	0.33	7.2	B1.5V	[1]
6527	158926	λ Sco	0.27	4.4	B1.5IV	[2]
5695	136298	δ Lup	0.28	5.1	B1.5IV	[2]
1933	37481	$-6°1275$	0.29	5.4	B1.5IV	[1]
6247	151890	μ^1 Sco	0.31	5.2	B1.5IV	[2]
3468	74575	α Pyx	0.29	4.2	B1.5III	[2]
5469	129056	α Lup	0.29	4.0	B1.5III	[2]
1848	36430	$-6°1207$	0.43	8.4	B2V	[1]
5285	122980	χ Cen	0.45	7.0	B2V	[2]
1679	33328	λ Eri	0.32	4.9	B2IV	[1]
39	886	γ Peg	0.33	5.2	B2IV	[1]
6252	151985	μ^2 Sco	0.33	5.0	B2IV	[2]
5776	138690	γ Lup	0.34	5.8	B2IV	[2]
6508	158408	υ Sco	0.34	5.3	B2IV	[2]
779	16582	δ Cet	0.35	5.9	B2IV	[1]
5395	126341	τ^1 Lup	0.35	4.9	B2IV	[2]
5576	132200	κ Cen	0.41	5.7	B2IV	[2]
1790	35468	γ Ori	0.34	4.7	B2III	[1]
1463	29248	ν Eri	0.35	4.1	B2III	[1]
1552	30836	π^4 Ori	0.36	3.9	B2III	[1]–[2]
1567	31237	π^5 Ori	0.39	4.6	B2III	[1]
7029	172910	$-35°12876$	0.46	7.7	B2.5V	[2]
1891	37016	$-4°1184$	0.49	8.4	B2.5V	[1]
6875	168905	$-44°12569$	0.51	7.5	B2.5V	[2]
5812	139365	τ Lib	0.52	7.7	B2.5V	[2]
3886	84816	$-44°5846$	0.46	7.0	B2.5IV	[2]
801	16908	35 Ari	0.55	9.6	B3V	[1]
4848	110956	$-55°5215$	0.55	8.9	B3V	[2]
3925	85980	$-44°5987$	0.55	8.8	B3V	[2]

Table I (continued)

BS	HD	Name	c_D	$W_{\mathrm{H}\beta}$	MK	Ref.
4573	103884	$-61°2829$	0.55	8.4	B3V	[2]
5035	116087	$-60°4627$	0.56	9.7	B3V	[2]
3415	73390	$-57°1590$	0.56	9.1	B3V	[2]
4638	105937	ϱ Cen	0.57	9.6	B3V	[2]
3539	76161	$-47°4460$	0.59	8.9	B3V	[2]
2159	41753	ν Ori	0.54	7.1	B3IV	[1]
1320	26912	μ Tau	0.59	7.3	B3IV	[1]
3663	79447	$-61°1201$	0.52	6.5	B3III	[2]
2653	53138	o^2 CMa	0.22	1.6	B3Ia	[2]
4549	103079	$-64°1724$	0.60	10.1	B4V	[2]
6938	170523	δ^2 Tel	0.62	7.5	B4III	[2]
4940	113703	$-47°8088$	0.64	9.8	B5V	[2]
1839	36267	32 Ori	0.66	9.9	B5V	[1]
5292	123335	$-58°5383$	0.63	8.7	B5IV	[2]
1735	34503	τ Ori	0.71	8.4	B5III	[2]
5217	120908	$-52°6805$	0.72	8.4	B5III	[2]
3940	86440	ϕ Vel	0.48	3.7	B5Ib	[2]
2827	58350	η CMa	0.33	2.2	B5Ia	[2]
5026	115823	$-52°6405$	0.70	11.1	B6V	[2]
338	6882	ζ Phe	0.76	11.7	B7V	[2]
5625	133937	$-42°10050$	0.76	10.7	B7V	[2]
6934	170465	δ^1 Tel	0.83	9.7	B7IV	[2]
	37151	$-7°1131$	0.82	13.5	B8V	[4]
674	14228	ϕ Eri	0.82	11.3	B8V	[3]
1038	21364	ξ Tau	0.84	12.6	B8V	[4]
1088	22203	τ^5 Eri	0.87	11.4	B8V	[4]
2451	47670	ν Pup	0.85	8.2	B8III	[3]
8353	207971	γ Gru	0.87	9.9	B8III	[4]
4662	106625	γ Crv	0.92	9.9	B8III	[4]
1806	35640	$-5°1247$	1.01	14.3	B9V	[4]
126	2884	β^1 Tuc	1.02	16.9	B9V	[3]
806	16978	ε Hyi	1.07	16.9	B9V	[4]
3665	79469	θ Hya	1.11	18.3	B9.5V	[4]
8781	218045	α Peg	1.22	15.4	B9.5V	[4]
1570	31295	π^1 Ori	1.15	18.5	A0V	[4]
1544	30739	π^2 Ori	1.20	16.5	A0V	[4]
191	4150	η Phe	1.21	16.1	A0V	[4]
3410	73262	δ Hya	1.24	15.0	A0V	[4]
3981	87887	α Sex	1.26	12.8	A0III	[4]
7950	198001	ε Aqr	1.24	17.8	A1V	[3]
4359	97633	θ Leo	1.29	17.9	A1V	[4]
3685	80007	β Car	1.34	15.8	A1IV	[4]

TABLE II

Stars for which the classification disagrees

BS	HD	Name	c_D	$W_{H\beta}$	S.T.	S_Q	MK	Ref.
	37061	$-5°1325$	0.30	4.4	B1.5III	B1	B0V	[5]
							B1V	[6]
							B1V, B8	[7]
2294	44743	β CMa	0.23	5.2	B1V	B0.5	B1II–III	[1]
							B0.5II–III	[8]
							B1II	[9]
							B8V	[10]
2657	53244	γ CMa	0.75	8.6	B6III	B6	B8II	[3]
							B8III	[11]
2845	58715	β CMi	0.98	10.9	B8.5III	B8	B8V	[12]
3860	83979	ζ Cha	0.58	9.5	B3V	B4	B5V	[2]
							B5IV	[13]
4773	109026	γ Mus	0.58	8.3	B3V	B4	B5V	[2]
							B5IV	[14]
4823	110335	$-59°4393$	0.85	4.2	?	B6	B6IV	[2]
							B7IV	[13]
5528	130807	o Lup	0.53	8.7	B3V	B4	B5IV	[2]
							B6III	[15]
							B6V	[16]
							B8V	[14]
7129	175362	$-37°12982$	0.48	8.9	B2.5V	B3	B8IV	[17]
							B7V	[18]
							B9III	[19]

References for Tables I and II:

[1] Lesh, 1968.
[2] Hiltner *et al.*, 1969.
[3] Jaschek and Jaschek, 1966.
[4] Jaschek *et al.*, 1964.
[5] Divan, 1954.
[6] Strand, 1958.
[7] Wenzel, 1951.
[8] Underhill, 1947.
[9] Buscombe, 1962.
[10] Crawford, 1958.
[11] Woods, 1958.
[12] Bappu *et al.*, 1962.
[13] Morris, 1961.
[14] Cape Mimeogram No. 12, 1961.
[15] Walraven and Walraven, 1960.
[16] Eggen, 1961.
[17] Buscombe and Morris, 1960.
[18] Hoffleit, 1956.
[19] Feast *et al.*, 1957.
All the References from [5] on are taken from Jaschek *et al.*, 1964.

valent width of Hβ vs that of Hγ or Hδ since Hβ is more sensitive to emission effects than Hγ or Hδ. Figure 3 shows a graph of $W_{H\beta}$ against $W_{H\delta}$. The dots represent the stars used as standards for the classification, which we consider as 'normal' stars; the crosses are stars with emission and the squares are peculiar stars. Even though peculiar stars are not discriminated in this graph, emission stars are more or less clearly separated, depending on the amount of emission.

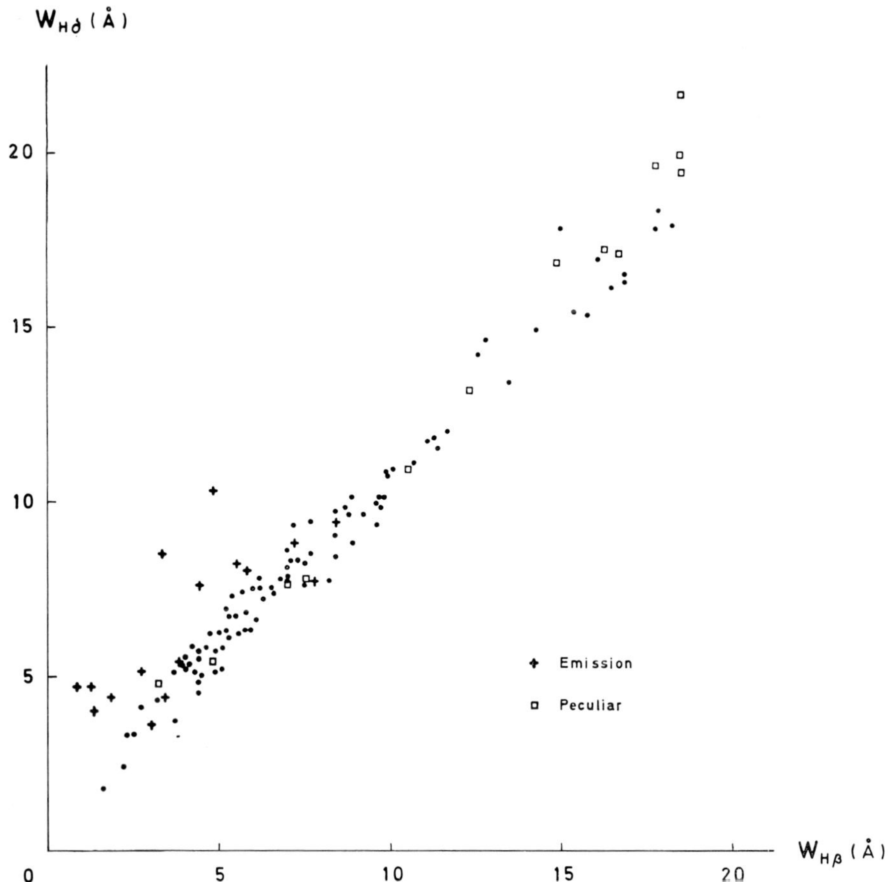

Fig. 3. $W_{H\beta}$, $W_{H\delta}$ relation. This diagram allows the separation of emission stars, indicated by crosses. The squares represent peculiar stars.

Future plans for this work include an attempt to segregate peculiar stars by using other parameters, as UD or ΔD. We also intend to continue the observations, with special emphasis in luminosity classes I and II and late B stars, in order to fill in the gaps and obtain a more precise classification scheme. If possible, the classification will be extended to later types.

References

Barbier, D. and Chalonge, D.: 1939, *Ann. Astrophys.* **2**, 254.

Butler, H. E. and Thackeray, *A.* D.: 1940, *Monthly Notices Roy. Astron. Soc.* **100**, 450.

Chalonge, C. and Divan, L.:1952, *Ann. Astrophys.* **15**, 201.

Gutiérrez-Moreno, A., Moreno, H., and Stock, J.: 1967, *Publ. Dept. Astron. Univ. Chile* **1**, No. 4, 49.

Gutiérrez-Moreno, A., Moreno, H., and Stock, J.: 1968, *Publ. Dept. Astron. Univ. Chile* **1**, No. 8, 127.

Gutiérrez-Moreno, A., Moreno, H., and Stock, J.: 1972, *Publ. Dept. Astron. Univ. Chile*, in press.

Hiltner, W. A., Garrison, R. F., and Schild, R. E.: 1969, *Astrophys. J.* **157**, 313.

Jaschek, C. and Jaschek, M.: 1966, in K. Lodén, L. O. Lodén and U. Sinnerstad (eds.), 'Spectral Classification and Multicolor Photometry', *IAU Symp.* **24**, 6 and 30.

Jaschek, C., Conde, H., and de Sierra, A. C.: 1964, *Obs. Astron. Univ. La Plata, Serie Astronómica*, **28**, No. 2.

Johnson, H. L.: 1958, *Lowell Bull.* **4**, No. 90, 37.

Lesh, J. R.: 1968, *Astrophys. J. Suppl.* **17**, No. 151, 371.

Moreno, H.: 1972, in preparation.

Slettebak, A. and Stock, J.: 1957, *Z. Astrophys.* **42**, 67.

LOW DISPERSION SPECTROPHOTOMETRY
OF LATE TYPE STARS

H. MORENO

Departamento de Astronomía, Universidad de Chile

Abstract. Low dispersion (slit 60 Å wide) spectrophotometric observations of late type stars have been made. The sample of stars observed cover from K0 to M8 and luminosity classes V to I.

A selection of these stars including each luminosity class and spectral type is being analyzed. The observed intensity distributions have been corrected for atmospheric extinction and have been normalized to a common maximum intensity. The resulting instrumental intensity distributions are being studied for the best parameters for spectral classification.

The object of this communication is to report on some observations and preliminary results concerning K and M stars. The observations were made at the Cerro Tololo Inter-American Observatory using a spectrum scanner with a slit 60 Å wide, attached to the 36- and 60-in. telescopes. The scanner uses a plane grating blazed for 3800 Å in the first order and a refrigerated 1P21 photomultiplier. The spectrum scans cover from 3000–6000 Å. Each spectrum was scanned twice in succession in opposite directions of wavelength, in order to eliminate systematic errors (Gutiérrez-Moreno *et al.*, 1966). A complete observation in both directions takes 24 min and it is registered in a Brown recorder with a speed of one inch per minute and a reciprocal dispersion of 10 Å mm^{-1}.

Each night, conditions permitting, at least one pair of standard stars was observed twice. These observations allow accurate determination of atmospheric extinction as well as the reduction of the data to a uniform system. Sky scans were obtained when necessary. The observations include 92 stars up to $m_v = 9.0$. The distribution of these stars in spectral types and luminosity classes is given in Table I. Most of the stars were observed on three nights.

No complete determinations of energy distributions have been made up to the moment. It has been preferred to make a preliminary analysis of a sample of stars covering all the MK types observed, for determining which are the points worth mea-

TABLE I

Stars observed

ST LC	K0	K1	K2	K3	K4	K5	K6	K7	K9	M0	M1	M2	M3	M4	M5	M6	M8
I	1		1	3		1				1	1	1					
II			1	2	2	1				1			1		2		
III	5	4	3	2	3	3	1			1	2	1	2	2	1	3	1
IV	3	3	2											1			
V	4	3	3	5	1	6		2	1	2		2					1

Remark: the spectral types were taken from the La Plata Catalogue (Jaschek *et al.*, 1964).

suring for later use in spectral classification. The instrumental energy distributions observed have been corrected for atmospheric extinction using mean values for the observing period. Amplification differences were eliminated from each spectrum and the resulting distributions were normalized to equal maximum intensity, to make the comparisons easier. The spectra of the faintest stars have lost some detail due to the large fluctuations of the signal.

Figures 1 to 4 show the variation of the energy distributions with spectral type and with luminosity class for K and M stars. Some identifications of absorption lines and bands have been made in the spectra. The identifications do not include all the lines present in a given blend, but only the most important ones. Some of the features already used for spectral classification are readily recognized in the curves. These include the CN bands at 3855 Å and 4180 Å; the G band at 4308 Å; the Ca I line at 4226 Å; the K line of Ca II at 3934 Å; and, for the M stars, the TiO bands.

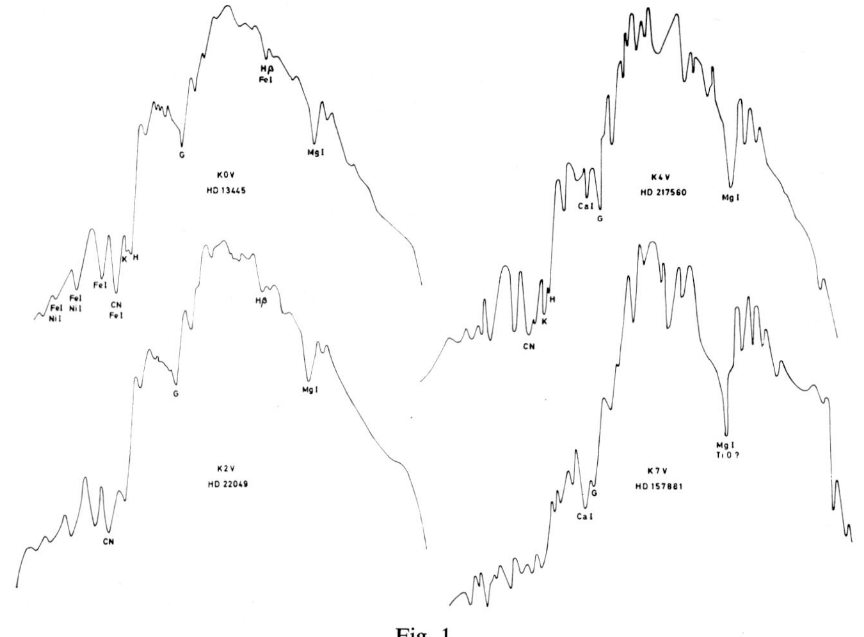

Fig. 1.

Classifications based on the schemes previously used by different authors (Lindblad and Stenquist, 1936; Elvius, 1956; Westerlund, 1951–1953; Strömgren and Gyldenkerne, 1955) have been attempted. Nevertheless, several difficulties were found, being the most important one the blends produced by the low resolution of the spectra. This makes difficult to decide what has to be measured. For example, if we measure an 'equivalent width' or the depth of a feature with respect to the apparent continuum, it is difficult to decide which continuum must be used, since this continuum keeps changing with varying spectral type or luminosity class due to the change of

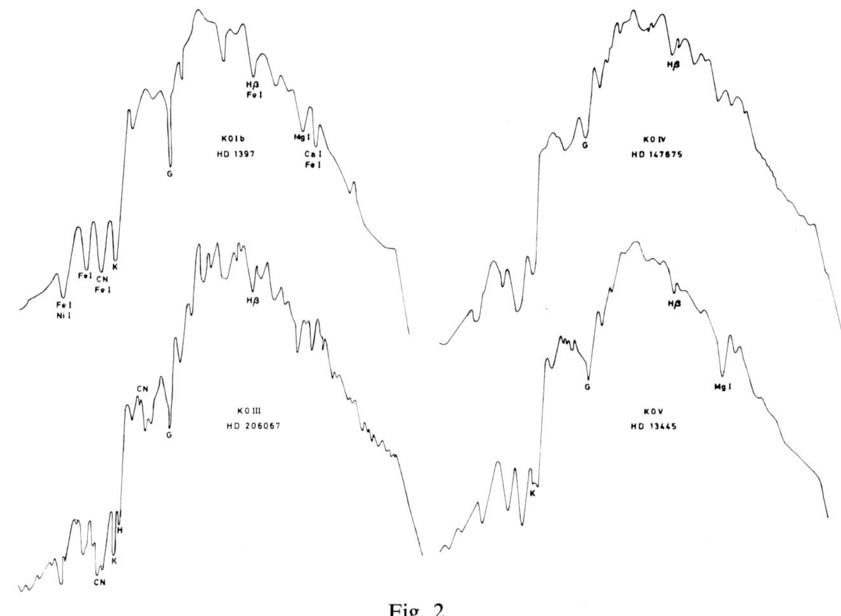

Fig. 2.

relative importance of the lines which contribute to the blend. It seems that the best results are obtained by using just the ratio of the intensities or the magnitude difference between two fixed points of the feature.

As an example, Figure 5 shows a preliminary classification scheme for M stars. The TiO band at 4955 Å has been used as a spectral type indicator, and the Ca I line at 4226 Å as a luminosity indicator. The indices used are:

$$T = I_{4945}/I_{5020} \tag{1}$$
$$C = I_{4190}/I_{4226}. \tag{2}$$

The M8V star has been classified by McCuskey (1949) using objective prism spectra. The spectrum scans suggest an earlier type.

Future work considers observations of as many stars as possible with a uniform MK classification, to fill in the gaps in Table I and to obtain homogeneous classification schemes. A computer program is being prepared for the accurate determination of intensity distributions. It includes the determination of atmospheric extinction using two approximations and considering its variation in the ultraviolet (Mendoza *et al.*, 1968). For the program stars, the computer program includes the determination of the outside-of-the-atmosphere intensity distributions, the reduction to a uniform system and the computation of the parameters selected for classification.

The possibility of studying peculiarities, such as differences of metal contents is being considered.

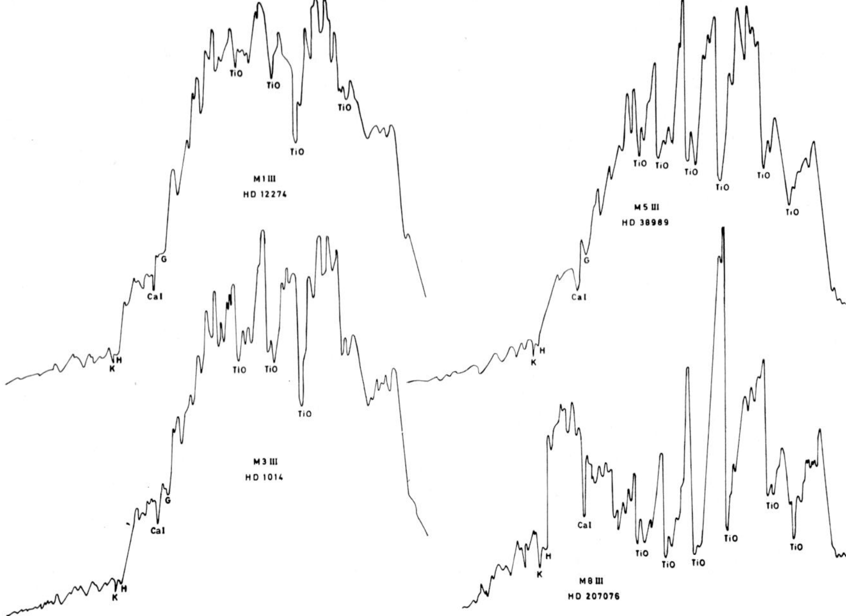

Fig. 3.

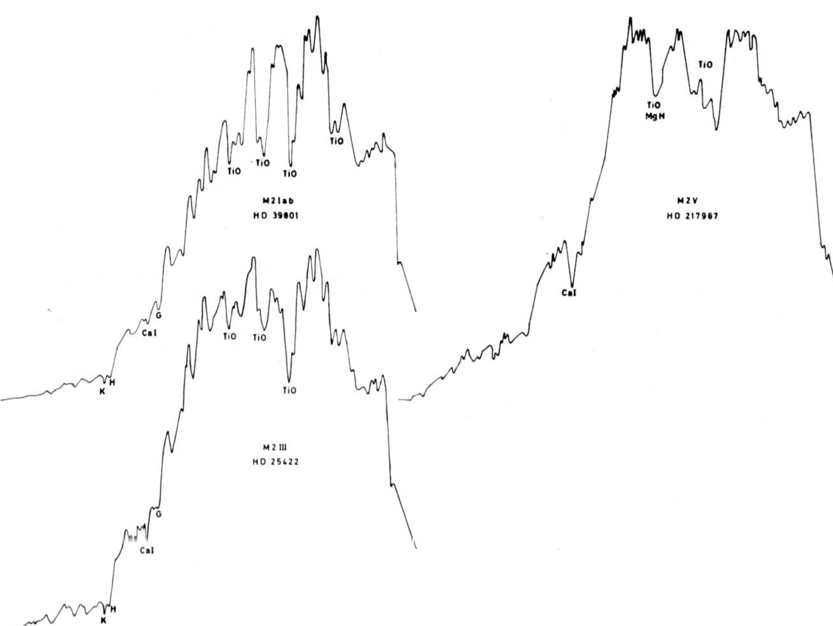

Fig. 4.

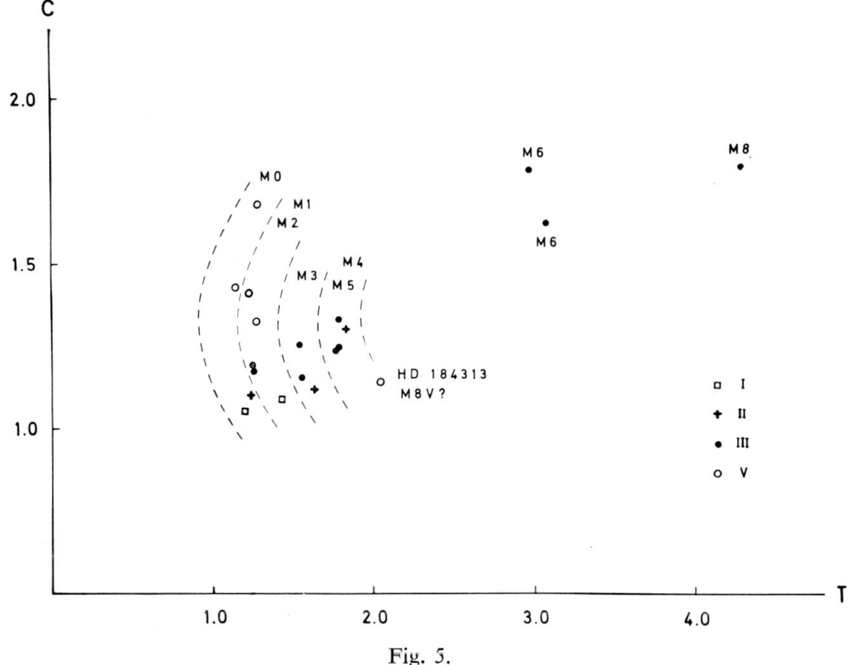

Fig. 5.

References

Elvius, T.: 1956, *Stockholm Obs. Ann.* **19**, No. 3.
Gutiérrez-Moreno, A., Moreno, H., and Stock, J.: 1966, *Publ. Dept. Astron. Univ. Chile* **1**, No. 2, 22.
Jaschek, C., Conde, H., and de Sierra, A. C.: 1964, *Obs. Astron. Univ. La Plata, Serie Astronomica* **28**, No. 2.
Lindblad, B. and Stenquist, E.: 1936, *Stockholm Obs. Ann.* **12**, No. 5.
McCuskey, S. W.: 1949, *Astrophys. J.* **109**, 426.
Mendoza, E., Moreno, H., and Stock, J.: 1968, *Publ. Dept. Astron. Univ. Chile* **1**, No. 5, 67.
Strömgren, B. and Gyldenkerne, Y.: 1955, *Astrophys. J.* **121**, 43.
Westerlund, B.: 1951, *Uppsala Astron. Obs. Ann.* **3**, No. 6; 1952, No. 8; 1953, No. 10.

one is likely to arrive at a figure of about $10\text{–}11 \times 10^3$ stars classified per year.

For MK classifications the figures are somewhat more difficult to obtain, because lists of stars published in this system contain very often only small numbers of objects. An earlier estimate (Jaschek, 1968) was that information doubles every 8 yr, which implies an annual increase of about 9%. Since the publication of the only existing catalogue of MK spectral types, which contained 15×10^3 stars, an increase of about 7000 stars is therefore expected. This makes it urgent to publish a second edition of the MK type catalogue, a task which we expect to start next year at La Plata. The importance of collecting the MK type information is underlined by the fact that this information refers mainly to bright stars, which are fundamental for photometry.

Let us turn next to color indices. It should be mentioned right at the start that the information in this field is rather poorly organized, despite the fact that color indices are being observed since many years ago.

The color indices used today are either photographic or photoelectric. Since in the past photographic color indices had a much lower accuracy than photoelectric ones, the modern tendency has been to disregard them completely, an opinion which I feel not qualified to discuss. In Table III are listed the most important lists of photographic color indices, omitting all lists containing less than thousand stars. Perhaps the most impressive fact is that except in the south (Jackson and Stoy, 1954–1962) where we have practically complete data between -30° and -80°, no comprehensive schemes equivalent to the HD in spectral classification were attempted. This is rather surprising, since photoelectric photometry cannot be expected to rival with photo-

TABLE III

Large projects of photographic color indices

Author	$N(10^3)$
Malmquist (1927)	2
Malmquist (1936)	4
Bertaud (1939)	3
Seares, *et al.* (1941)	2
McCuskey and Seyfert (1950)	2
Kharadse (1952)	14
Jackson and Stoy (1954–1962)	68
Brodskaia (1955)	6
McCuskey (1955)	2
Eklöf (1958)	2
Brodskaia and Shajn (1958)	3
Numerova (1958)	5
Pronik (1958)	4
Kotshlashvili (1958)	2
McCuskey (1959)	3
Brodskaia (1960)	3
Metik (1960)	3
Bartkus (1964)	2
Masnauskas (1964)	4
Total	136

graphic photometry in either speed or number of stars. This can be seen very clearly through the following figures. The catalogue of Blanco *et al.* (1968) contains 20×10^3 stars measured in the UBV system, a system which came into use in 1953 and which Blanco follows up to 1967. Our own La Plata reference catalogue which lists *all* photoelectric measurements in any color system and to which I shall refer later, includes some 25×10^3 stars, observed between 1913 and 1968. This implies that the duplication has been very large, which is, at least partially, due to the lack of a catalogue telling what was already observed. It should however be pointed out that neither one of the catalogues includes stars in clusters, in external galaxies or stars for which no positions were published. An estimate of this number gives about 10×10^3 additional stars.

We can therefore summarize the situation in the following way:

(a) no effort has been made to derive photographic color indices in a systematic sky survey;

(b) about 2×10^5 photographic color indices exist;

(c) photoelectric photometry exists for about 35×10^3 of the brighter stars, many of them having been measured several times in different color systems.

The next question concerns the growth of the information. For photographic colors we have estimated the growth according to the same lines as for spectral types, and the result comes close to about 10^4 stars per year. This comes mainly from the fact that our Soviet Union colleagues determine usually both the spectral type and a photographic color index.

For photoelectric colors we will use an earlier estimate, which gave a figure of about 12% annual, doubling thus the information every six years.

The results for both spectral types and color indices are summarized in Table IV. If one focuses on photoelectric colors and MK types, it seems that as an order of magnitude the information increases by about 10% annually, implying that information duplicates in about seven years. This of course is very satisfactory, provided that one can use the new information-but this is, alas, not so.

A simple perusal of the literature shows that most papers, either of spectral classification or of photoelectric photometry, contain data relating to something between

TABLE IV

Information growth

	N(1970)	a.g.	g
Spectral classifications, type I and II	7×10^5	1.1×10^4	2
type III	3.5×10^4	3×10^3	9
Photographic photometry	2×10^5	1×10^4	5
Photoelectric photometry	3.5×10^4	4×10^3	12

Notes: N(1970) = number of stars observed up to 1970,
a.g. = annual growth,
g = annual growth in percentage.

ten and hundred objects. Let us assume at the best that the average number of stars per paper is hundred. Then for increases of about 4×10^3 stars per year, we have a minimum of about 40 new papers. The real figures are different, because already in 1960 there were for MK types 55 papers, and for photoelectric photometry 127 papers, 54 of them giving data for more than 10 stars. This means that in order to locate the parameter of a given object one has to go through at least 40 papers per year.... I am quite certain that most of us will decide either to observe the star anew or to forget about it, before going to the library.

The inescapable conclusion is that updating catalogues have to be published rather rapidly.

If one accepts that a new catalogue is in time when more than two hundred papers were published on the subject, one arrives at a minimum time interval of five years. From the abovementioned quotation of a duplication every seven years it is evident on the other hand that this time interval should be nearer to three years than to five.

Speaking mainly from the experience of the three catalogues undertaken at La Plata, it seems that this is so short an interval that one has to proceed on a continuous basis, because usually catalogues take much longer than this to assemble. The MK catalogue took four years, the Be star catalogue four years and the photometric catalogue is taking six years. Similarly the photometric catalogue of Blanco *et al.* took six years. I think therefore that my first point is proved, namely that in order *to cope efficiently with the information inflow one must envisage a continuous publication system.*

This really implies the creation of a Data Center, because the traditional way of an astronomer near retirement publishing a catalogue and not worrying about its continuation, is clearly not in step with the times.

It is however important to stress that the idea of data centers is really nothing new. The Index of Numerical Data Projects (1969) shows that in 1968 there existed 157 data centers in the world compiling numerical data of different kinds in sciences and technology. Even in astronomy a preliminary report of IAU Commission 1 on *Data Centers Reported* showed the existence of eighteen institutions doing data center work, on the following subjects:

Atomic Energy Levels
Cross Sections for Collisions of Electrons and Photons with Atoms, Ions and Small Molecules
Eclipsing Binaries
Ephemerides, Planetary Data, Star Catalogues
Extra-galactic Objects
Globular Clusters
Observatories, Instruments
Planetary Research Center
Star Catalogues
Radio Sources
Spectroscopic Binaries
Transition Probabilities

to which at least two very important groups should be added: the Variable Star Catalogue group at Moscow and the Czechoslovakian group (Alter, Ruprecht and Vanyseka) working on clusters and associations.

Let us now explore in more detail what the most pressing needs are in each field and what can be done. We will start again with the spectra and discuss successively literature completeness, selectivity and values to be published.

Completeness is an ideal which can never be reached, except asymptotically and at a very high cost. Usually it suffices to obtain a 'reasonable' completeness. Is it 'reasonable' to try to assemble complete spectral information of all stars? Most of the astronomical community will probably expect something of this kind from a data center, but a little reflection shows that this is rather difficult. Although HD and the 'Luminous stars' are rather easy to deal with, because they both provide positions, most of the HDE is only on charts and so are practically all classifications in special regions, like for instance those of McCuskey and Brodskaia. After thinking it over some time I feel that the best policy is to assemble only the data for stars having a published position. For work until the tenth magnitude, the position can be rounded of to 0^m1 in α and $1'$ in δ. If fainter stars are to be included, 0^s1 and $1''$ must be kept; a system which implies among other things an improvement of the positions of all HD stars. The application of this criterium means however that a large section of the spectral classification literature will be left out and eventually be forgotten. So my second proposal is to establish a masterlist of spectral classification programs describing:

(a) the limits of the region studied (for some standard equinox);

(b) the magnitude limit;

(c) the classification system used;

(d) the purpose of the study (search for emission line stars, galactic structure, etc.).

This list should be supplemented by charts, so that a simple glance at it can tell if there is some special field in the region one is particularly interested in.

The next question concerns selectivity. What is one to do if a star has been classified in the HD, in the Bergedorfer Spektral-Durchmusterung and in the MK system? This problem is really part of a larger one, namely to decide between critical and bibliographic compilations. In order to answer this question I have made an inquiry among my colleagues and the general opinion is that it is preferable to set up a selection scheme, which starts at the top with MK, and passes next to Mt. Wilson, HD and others. If several classifications exist for a given object, only the topmost classification is given, whereas the others are kept in files for reference.

A complication arises when there are systematic differences between the series from which the particular classifications come. On one side it is better to have a single system, while on the other hand original values are always more valuable than reduced values. Up to now it is not entirely clear (at least to me) what is the best procedure.

The third question is related to the preceding ones, namely what to publish. What does one do for instance with MK types if several differing ones are available? In the La Plata catalogue we listed everything, feeling that this is the only fair procedure to all colleagues. In the introduction to the catalogue some general rules were laid down, of what to do when several discrepant classifications were available. Because of this 'bibliographic criterium' we have been under criticism from several sides, the main objection being that only a single classification – 'the best one' – should be provided. In order to see what could be done, I consulted again several colleagues and the opinion was that although difficult, a weighting could possibly be performed, in the sense that everything should be listed, but some type be underlined, stressing that *in the opinion of the compiler* this is the best type. This combines probably the best of two conflicting tendencies (and leaves one with the smallest number of enemies). As a practical consequence we expect to start in 1972 the continuation of our MK type catalogue from 1962 on, along these lines.

As for the other things mentioned, we will also attempt to publish soon the masterlist of regions where spectral classifications were done, but we will leave for the very last the attack upon the 'complete' bibliography.

Assuming that all this will be done in a reasonable time, one could think perhaps that the information problem is solved. But there is still one important aspect, namely the one of the non-normal spectra. In this case, bibliographic completeness should be attempted, because here even small bits of information can be important. A list of the partial catalogues made in this field is given in Table V.

TABLE V

Bibliographic catalogues of special groups of stars

Be stars	Merrill and Burwell (1933, 1943, 1949, 1950)
	Jaschek *et al.* (1971)
Emission line stars	Wackerling (1970)
	Bertiau and McCarthy (1970)
Late type emission line stars	Bidelman (1954)
Combination spectra	Hynek (1938)
Ap and Am stars	Bertaud (1959, 1960, 1965)

A glance at this table shows that several important groups were never covered like WR and Of stars, F peculiar star, late type giants, etc. This is an important omission and eventually it should be closed.

We will consider next color indices. Although I am not a photometrist, my justification for discussing this problem is that we are in the process of publishing the La Plata photometric catalogue, a project which has put me in touch with some of the problems one is likely to find in this area.

Let us start with the completeness problem. It has been doubted by many, if it is really a good idea to collect all the photographic color indices, mainly because of the large errors they have. On the other hand it would be only fair to say that the publication of a masterlist of all the regions in which photography photometry was once done, together with an assessment of the errors, would be a very valuable undertaking. To say the least, they could constitute a convenient starting point for photo-electric photometry. Since the effort to put up such a list is several orders of magnitude less than colllecting all measurements, it is probably a step in the right direction. Reduction to a common system, like for instance reduction to something like the $B-V$ index, would enhance certainly the value of published photographic color lists, but again here it would be necessary to examine if the effort is worthwhile, and this problem I leave to my photometric colleagues.

  If one turns now to photoelectric photometry, the situation is slightly more en-couraging. Of the 1900 papers listed up to 1968 in our bibliography, 870 deal with UBV photometry, 72 with multicolor photometries, 69 infrared measurements, 760 with unspecified systems and 84 with narrow band photometry. The majority of the stars which have been measured photoelectrically do have UBV, this being shown by Blanco's catalogue in which he lists 20000 stars measured in the UBV (and U_cBV) while our catalogue listing everything at a somewhat later date, contains only about 25000 stars. For some time in the future, UBV will be thus the most important photo-electric system; Strömgren's narrow band photometry following in second place and Golay's system third, sharing its place almost with Johnson's $UBVRI...$ and Stebbins, Whitford and Kron's six color photometry. I have quoted these details because of the need to underline that there are few photoelectric systems containing more than say three thousand measured stars.

  The implications for data centers are obvious, namely that one can concentrate on about half a dozen photometric systems and probably disregard the rest, except for very special purposes. This is essentially the idea of Dr B. Hauck and his group at Geneva. Since Dr Hauck will give later a report of his work, I will not go into more details.

  Until now we have not discussed the more technical aspects, namely the way the data are stored and the way they are distributed. We will now mention them briefly.

  With regard to data storage it seems that most people are inclined to punch cards, because this way the information can be subsequently handled easily, and even pub-lished without further complication. At La Plata we have always worked with hand-written cards, because it seemed simpler and because we had no punching unit the time we started. Since I have no experience with handling punched cards, I can only make the comment that whatever one does, cataloguing involves a great deal of cross checking, correcting (or trying to correct) misprints and misidentification, and it seems to me that this can be done with ease on handwritten cards. Of course the matter is quite different when it comes to printing, because then one has essentially to duplicate the work already done when writing the card. At this stage at least the punched card file is definitely superior.

  With regard to publication policy, the inquiry I mentioned before has not revealed a definite tendency among prospective users. In principle one can envisage a passive data center or an active center. The passive data center stores information and releases it only on request. This system is good only if the information is stored on punched card or tape. It is a cheap system, in the sense that the customer pays what he buys, but on the other hand the usefulness of the center is restricted to those who ask and can pay for the questions. For instance for countries with exchange restrictions this difficulty could spoil the usefulness of the center. Also it would make it more difficult for non-specialists to obtain the occasional data they need, because by experience one knows that if one has to write a letter for each object which comes up, generally one prefers to drop the subject. In active centers instead all information is published at intervals and distributed to all colleagues. This policy seems to be more in accord with the general psychology. Within this general policy, one can envisage a regular catalogue every X years and supplements at shorter intervals. The financiation could perhaps be covered by the sale of the catalogues.

Let me now come to the conclusion of my paper and state briefly what is being done in the field of our Commission on the data center problem.

European astronomers have decided to create, a few years ago, the 'Centre de Données Stellaires' (Stellar Data Center) at Strasbourg, through the collaboration of the Observatories of Paris, Heidelberg, Marseille, Lausanne, Strasbourg and La Plata. Each of these observatories will take one chapter of stellar data and act as a data center for this field. B. Hauck, Lausanne, will be in charge of the photometry and we, at La Plata, will do the same for spectral classification. F. Spite, Paris, will be in charge of the 'bibliographic file' which consists of an inventory of all bibliographic mentions of a given star in the literature. This very ambitious project will start with the publication of the mentions contained in a dozen leading astronomical journals in the last decade; later on it will be extended to more journals and back in time. T. Lederle, Heidelberg, will be in charge of the positions and proper motions and Mme. M. Barbier, Marseille, of the radial velocities. J. Jung, Strasbourg, finally will be in charge of the Center itself and of the combined files of all sub-centers.

Attention is called to the Information Bulletin of the Stellar Data Center, which provides the latest news about the project.

We all hope that this data center constitutes a first step toward our goal, which is to make available all stellar data to all colleagues.

References

Bartkus, R.: 1964, *Bull. Vilnius* **13**.
Bertaud, Ch.: 1939, *Ann. Obs. Paris, Sect. Meudon* **8**, f.3.
Bertaud, C.: 1959, *J. Obs.* **42**, 45.
Bertaud, C.: 1960, *J. Obs.* **43**, 129.
Bertaud, C.: 1965, *J. Obs.* **48**, 211.
Bertiau, F. C. and McCarthy, M. F.: 1969, *Specola Vatic. Ric Astron.* 7, 525.
Bidelman, W. P.: 1954, *Astrophys. J. Suppl.* **1**, 175.
Blanco, V. M., Demers, S., Douglass, G. G., and Fitzgerald, M. P.: 1968, *Publ. U.S. Naval Obs.,* 2nd Ser. **21**, 3.
Brodskaia, E. S.: 1955, *Comm. Crimea* **14**, 139.
Brodskaia, E. S. and Shajn, P. F.: 1958, *Comm. Crimea* **20**, 299.
Brodskaia, E. S.: 1960, *Comm. Crimea* **24**, 160.
Eklöf, O.: 1958, *Medd. Uppsala Astron. Obs.*, No. 120.

Haffner, H.: 1960, *Astron. Nachr.* **286**, 28.
Hynek, J. A.: 1938, *Contr. Perkins Obs.* **1**, No. 10, 185.
International Compendium of Numerical Data Projects: 1969, ed. Comm. data for Sci. and Tech.,
    Springer Verlag.
Jackson, J. and Stoy, R. H.: 1954–62, *Ann. Cape Obs.* **17**.
Jaschek, C.: 1968, *Publ. Astron. Soc. Pacific* **80**, 654.
Jaschek, C., Ferrer, L., and Jaschek, M.: 1971, *La Plata Publ. Ser. Astron.* **37**.
Kharadse, E. K.: 1952, *Bull. Abastumani* **12**.
Kotshlashvili, T. A.: 1958, *Bull. Abastumani* **22**, 67.
Malmquist, K. G.: 1927, *Lund. Medd, Ser. II* **37**.
Malmquist, K. G.: 1936, *Stockholm Obs. Ann.* **12**, No. 7.
Masnauskas, J.: 1964, *Bull. Vilnius* **12**.
McCuskey, S. W. and Seyfert, C. K.: 1950, *Astrophys. J.* **112**, 90.
McCuskey, S. W.: 1955, *Astrophys. J.* **122**, 359.
McCuskey, S. W.: 1959, *Astrophys. J. Suppl.* **4**, 1.
Merrill, P. W. and Burwell, G. G.: 1933, *Astrophys. J.* **78**, 87.
Merrill, P. W. and Burwell, G. G.: 1943, *Astrophys. J.* **98**, 153.
Merrill, P. W. and Burwell, G. G.: 1949, *Astrophys. J.* **110**, 387.
Merrill, P. W. and Burwell, G. G.: 1950, *Astrophys. J.* **112**, 72.
Metik, L. P.: 1960, *Comm. Crimea*, **23**, 60.
Numerova, A. B.: 1958, *Comm. Crimea* **19**, 230.
Pronik, I. I.: 1958, *Comm. Crimea* **20**, 208.
Seares, F. H., Ross, F. E., and Joyner, M. C.: 1941, Carnegie Inst. Washington, Publ. 532.
Wackerling, L. R.: 1970, *Mem. Roy. Astron. Soc.* **73**, 153.

PHOTOMETRIC DATA IN MACHINE-READABLE FORM

B. HAUCK

Institut d'Astronomie de l'Université de Lausanne et Observatoire de Genève

Abstract. For two years we have undertaken the collection of photoelectric data in view of their utilization by computer techniques. At present we are preparing a fine for each photometric system with the measurements published. As a second step, we shall form another file, alike for each system, with homogeneous measurements. This file will be obtained with weighed means of the published measurements. For that we need to compare each published list with a standard list before attributing the weights.

After that, we shall form a file indicating in which system a star is measured.

This work is undertaken in connection with the Data Center of Strasbourg and the group on spectroscopic and photometric data of the Commission 45.

1. Introduction

For the last two years we have been collecting photometric photoelectric measurements. It is very important to render this information available to processing by computer techniques. At the present we are setting up files for each photometric system. The second step of this work will be the establishing of a file with homogeneous data for each system. Finally, a last file will permit us to know in which systems a star has been measured.

This work is included in the study undertaken by the Stellar Data Center of Strasbourg (Jung, 1971).

On the other hand the constitution of the group on spectroscopic and photometric data of Commission 45 should allow us to bring the files up to date.

2. Homogenization of Data

The homogenization of data is the most delicate part of this work. It would certainly be an advantage to find homogeneous data in the literature! But this is not the case, and users of photometric data who are not specialists in this field can have difficulty in choosing the best value, somewhat like the photometrist who may find it difficult to choose the best spectral type.

The sources of inhomogeneity are numerous. However, the principal ones are:

(1) the photomultiplier;
(2) the filters;
(3) the methods of reduction outside the atmosphere;
(4) the lack, or the bad use of standard stars.

It would be dangerous to average all measurements concerning one star without examining the quality of each published list. To make this critical study, we establish, for each system, a publication or a set of publications, as reference. Then we go through the stars of the list to be studied and include those which are common to the reference.

This comparison is made color by color, or parameter by parameter. A color 'reference' vs color 'list' diagram is constructed by the computer and the regression line is calculated by the method of least squares.

If the connection with the reference is correct, then the slope of this line is 1 and the y intercept is zero. The standard error obtained is a quality index of the list.

In order to detect systematical errors, we then construct the diagram of the differences between the reference and the list for each color (or parameter) as function of the temperature parameter defined in the studied system.

With these comparisons it is possible to give a weight to the list.

Sometimes, the discrepancy will be too great and the list will not be included in the file. When this is the case the Data Center has to draw the user's attention to these discrepancies.

3. Common File

This file must enable one to know quickly in which systems a star is measured and to select thus the files of these systems (file with published data or weighted data).

With these three files we have the advantage of obtaining data without delay and loss of information since the published measurements are preserved. A second advantage is the existence of the mean values obtained after critical examination.

4. Identification of Stars

The identification of stars is a complex problem since a star can be identified in many ways. It is necessary to choose a means of identification allowing computer treatment and valid for all stars. For this point we shall adopt the same process as the Stellar Data Center of Strasbourg; namely the use of equatorial coordinates α, δ. It has been decided to use a code made with these two coordinates for the 1950, 0 epoch.

5. Comparison of Photometric Systems

When all the information concerning a system has been collected, it will be possible to compare it with others. The theoretical study of the comparison of two systems has been made by Golay (1959). For the comparison of the astrophysical information contained in two different systems, it is necessary to take some precautions. In particular, one must define for what types of stars a relation between two indices (or parameters) is valid. Then one must examine the validity of this relation for different samples, for example Am, subdwarfs, giants, supergiants, stars with interstellar reddening, etc. (Hauck, 1971). This work has been rarely done in the past.

6. State of Work at Present

We have established files 1 and 2 for the following systems: (1) six-colors of Stebbins and Whitford; (2) *uvby* β ($\sim$ 5000 stars).

Some systems can be considered as homogeneous, because all data come from the same observation group. They are the rather well-known systems by: (1) Borgman; (2) Kruszewski; (3) Argue; (4) Geneva Observatory; (5) Eggen (10200 Å, 6500 Å, 6200 Å); (6) Vilnius Observatory; (7) Tifft; (8) Bahng.

For the *UBV* system we have continued the work of Blanco *et al.* (1970), especially for the galactic star clusters (100 lists and 13000 measurements).

References

Blanco, V. M., Demers, S., Douglass, G. G., and Fitzgerald, P. M.: 1970, *U.S. Naval Obs. Publ., 2nd Ser.* **21**.

Golay, M.: 1959, *Publ. Obs. Genève Série A*, No. 60.

Hauck, B.: 1971, *Conf. Oss. Astron. Milano-Merate Serie I*, No. 12.

Jung, J.: 1971, *Inform. Bull. CDS Strasbourg*, Nos. 1 and 2.

REMARK IN REGARD TO BIDELMAN'S DATA-BANK PROJECT

W. P. BIDELMAN

Warner and Swasey Observatory, Case Western Reserve University, Cleveland, Ohio, U.S.A.

(Read by C. B. Stephenson)

The concept of a *general* stellar data bank is not a new idea, but the desirability of this project at the present time has been emphasized by a number of astronomers (Bidelman, Jaschek, Cayrel, Hauck and others) and work is proceeding both in the United States and elsewhere. The status of my own activities in this direction is as follows:

The general stellar data file now at the Warner and Swasey Observatory is continually being updated and much more of the astronomical literature is being systematically searched and the resulting data and references entered into the file. However this project is still largely unfunded and progress is far slower than desirable. In particular the file is still not available in machine-readable form, though several of the newer large data catalogues, such as the Warner and Swasey southern OB-star survey of Stephenson and Sanduleak, will be entered into the file in IBM card-form rather the entire file will be converted to punched cards.

Efforts during the past year to obtain increased funds for this project from the
Efforts during the past year to obtain increased funols for this project from the astronomy program of the NSF have proved of little avail, but it is intended to submit a proposal to the Scientific Information Office of the NSF, which is more particularly concerned with such a project, to support hiring one or more persons whose entire activity would be concerned with the data file. This should make it possible to bring the file essentially up to date within a year or two – at least for references of astrophysical interest – and to make a good start on converting the present file into a more useful one.

If financial support for this project is to be obtained from Federal sources it is clear that the work must have the enthusiastic support of the entire astronomical community; further, the agencies involved would like to be assured that the work will not be needlessly duplicated elsewhere and that the data will indeed be used. Expressions of support for our work will be much appreciated.

SPECTRAL AND *UBV* DATA FOR 583 STARS
(SUPERGIANTS AND FOREGROUND STARS)
IN THE DIRECTION OF THE LARGE MAGELLANIC CLOUD

A. ARDEBERG, J. P. BRUNET, E. MAURICE, and L. PRÉVOT

European Southern Observatory, Santiago de Chile, and Observatoire de Marseille, France

Abstract. The stars observed are mainly from the Catalogue by Fehrenbach and Duflot. Some other stars – overexposed on the objective-prism plates, faint blue stars in H II regions, stars with overlapping objective-prism spectra – have been added.

The photoelectric *UBV* photometry was done with the 1-m telescope at La Silla. The *UBV* filters have always been those recommended by Johnson. Normally 3 measurements or more on different nights have been made for each star. For a single observation of a program star the internal standard deviations are

$$\sigma_V = 0.017, \qquad \sigma_{(B-V)} = 0.013, \qquad \sigma_{(U-B)} = 0.017.$$

The spectra have been taken with a Cassegrain spectrograph mounted on the 1.52-m telescope at La Silla. The dispersion is 73 Å mm^{-1}, the spectral range 3250–5000 Å, and the limiting resolution 1.3 Å.

As a result of our observations, we present 409 stars as members of the LMC, 42 as possible members and 132 as galactic foreground stars (19 of these having high radial velocity). Remarkable features regarding velocities, spectra, magnitudes and colors are noted and commented.

All stars down to $B = 11.5$ have been observed spectrographically as well as photo-metrically. This is also the case for the majority of the high-velocity stars down to $B = 12.0$. For fainter objects the photometry is quite representative, whereas the spectroscopy is more scarce. The limiting magnitude is $B = 14.0$.

As radial velocities were used as membership criterion, our data do not suffer from any spectral type selection effects.

It was easy to apply the MK criteria to our dispersion using the Yerkes atlas and the Bidelman list of supergiants. A continuous series of supergiant standards in the LMC has been established from O6 to G2. Comparison with Radcliffe data and with recent results of Divan gives a dispersion in spectral type of less than one tenth of a spectral class. Also the luminosity classification shows good agreement.

In the classification several difficulties were met:

(1) The LMC stars are very luminous objects. In the Galaxy we do not have many corresponding stars. This is especially so for later spectral types.

(2) With the dispersion used, the differences between IaO and Ia are not very pronounced, especially not for F–G type stars.

(3) Many of the LMC supergiants are P Cygni stars. They show very sharp, often

asymmetric line profiles with a sharp rise to the long wavelength side. This might cause erroneous classification, and also affect radial velocities.

(4) The presence of an extended atmosphere, causing dilution effects, can give spurious spectral classification.

We have found 8 A type high-velocity stars. The *UBV* results confirm their classification. 11 G-K type stars have been found to be high-velocity stars, among which are two subdwarfs earlier discussed by Fehrenbach and Thackeray. The number of late type high-velocity stars is quite high.

Full accounts have been given in *Astron. Astrophys. Suppl.* **6** (1972), 249.

A STUDY OF INTERSTELLAR REDDENING AT HIGH GALACTIC LATITUDES

A. G. DAVIS PHILIP*

Dudley Observatory and The State University of New York at Albany

Abstract. A series of finding lists of A stars in regions at high galactic latitudes has been published in the last few years. Strömgren four-color and Hβ measures have been made of stars in five of the regions, allowing the interstellar absorption in each to be calculated. The two most important regions are the North Galactic Pole and the South Galactic Pole where color excesses of $E_{B-V}=0.00$ and $E_{B-V}=0.02$ were found respectively. An area in front of the Large Magellanic Cloud has a color excess of $E_{B-V}=0.04$.

1. Introduction

During the past few years a spectral survey has been made of areas at high galactic latitudes, spaced along four principal galactic longitudes ($l=0°, 76°, 180°,$ and $290°$)** and every 15° in galactic latitude. When a majority of the regions in the program have been analyzed, it will be possible to study the distribution of early type stars perpendicular to the galactic plane and to determine the tilts of lines of equal star density within a few kiloparsecs of the Sun.

2. The Observations

The distribution of areas is shown in Figure 1, where each of the areas under study is

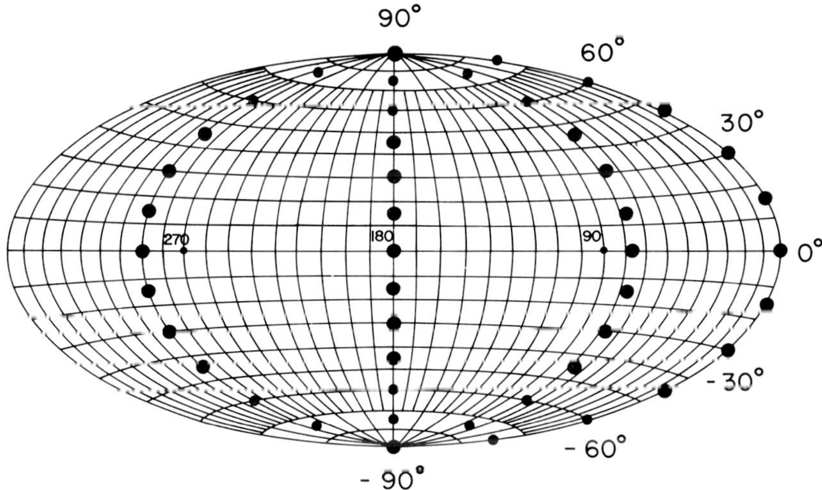

Fig. 1. The areas under investigation in a study of galactic structure perpendicular to the galactic plane plotted by their galactic longitudes and latitudes.

* Visiting astronomer, Kitt Peak National Observatory and Cerro Tololo Inter-American Observatory which are operated by the Association of Universities for Research in Astronomy, Inc.
** New galactic coordinates are used throughout this paper.

Ch. Fehrenbach and B. E. Westerlund (eds.), Spectral Classification and Multicolour Photometry, 295–301. All Rights Reserved.
Copyright © 1973 by the IAU.

indicated by a solid circle. A series of finding lists of early-type stars, which together constitute a catalogue of stars important in the study of galactic structure, has been published in the *Bol. Obs. Tonantzintla y Tacubaya.*

Schmidt spectral plates and a set of direct plates covering an area of 25 square degrees, are taken in each area. One hour spectral plates, taken at a dispersion of 280 Å mm^{-1}, on IIaO plates reach approximately $V = 13$–14th mag., depending on the Schmidt telescope used. Plates have been obtained with Schmidt telescopes at the Warner and Swasey Observatory, at the Tonantzintla Observatory, and the Cerro Tololo Inter-American Observatory (Michigan Curtis Schmidt telescope). All the stars with classifiable spectra are recorded and then measured for B or V magnitudes on the direct plates. In order to determine magnitudes from the images on the direct plates, sequences of B and A type stars are set up in each region. The spectral and magnitude data are then combined to yield the stellar density function in each of the directions indicated in Figure 1.

The interstellar reddening must be determined in each region before the stellar density analysis is made. The early-type stars in each area (spectral type A7 or earlier) are measured in the four-color and Hβ systems. The y magnitude transforms very well to the V magnitude of the UBV system and is used in the determination of the photographic magnitudes. The reddening in each region as a function of distance can be determined by means of a formula derived by Crawford (1970):

$$(b - y)_0 = 2.943 - \beta - 0.1\,(\delta m_1 + \delta c_1),$$

where $\delta m_1 = (m_1)_{\text{Hyades}} - (m_1)_{\text{observed}}$, and $\delta c_1 = (c_1)_{\text{observed}} - (c_1)_{\text{zams}}$. This formula is valid for A stars with $2.70 < \beta < 2.88$. The color excess, $E_{b-y} = (b-y)_{\text{observed}} - (b-y)_0$, can then be plotted as a function of the distance modulus, $m - M$, to determine the run of absorption with distance. Observations of this type have been made in a number of regions in the general program; this paper will summarize the reddenings derived in nine of these.

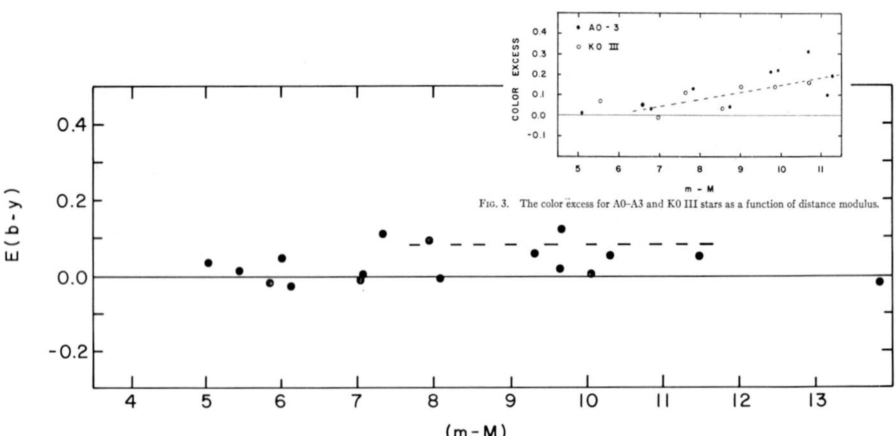

Fig. 3. The color excess for A0–A3 and K0 III stars as a function of distance modulus.

Fig. 2. The color excess in 1HLF2 plotted against the distance modulus. The insert in the upper right is an estimate (Philip, 1966) derived from Schmidt spectral types and UBV photometry.

3. The Reddening

The first area to be studied was 1HLF2 ($l=76°$, $b=-30°$) (Philip, 1966). In this study the reddening was derived by using the Schmidt spectral types to yield an intrinsic $B-V$ color, which was then subtracted from the observed $B-V$ color. At a distance modulus of 11, a color excess of 0.2 mag. was derived. More recently these same stars have been measured in the four-color and Hβ systems and a color excess of $E_{B-V}=0.07$ derived. (The A and B type stars in 1HLF2 are still being measured and the derived reddening estimate may change slightly in the final analysis.) The two estimates of the reddening are shown in Figure 2 in which the color excess, E_{b-y}, is plotted against the distance modulus. The small insert in the upper right is a reproduction of Figure 3 from Philip (1966) in which E_{B-V} is plotted vs $m-M$. E_{b-y} is plotted vs $m-M$ in the large figure. ($E_{b-y}=0.7E_{B-V}$). The observations reported here were made in the four-color system; the color excesses have been converted to E_{B-V} since most of the reddenings reported in the literature are in the UBV system.

The reddening in 4HLF6 ($l=180°$, $b=-60°$) is presented in Figure 3. In this region $E_{B-V}=0.013$. The reddening at the SGP is presented in Figure 4 ($E_{B-V}=0.002$). The

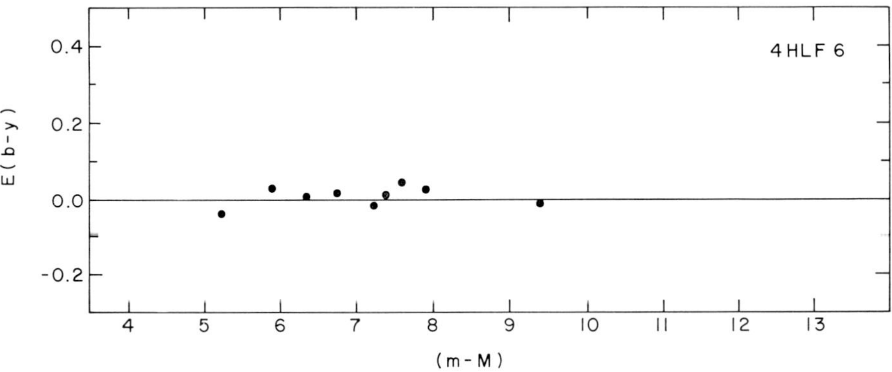

Fig. 3.   The color excess in 4HLF6.

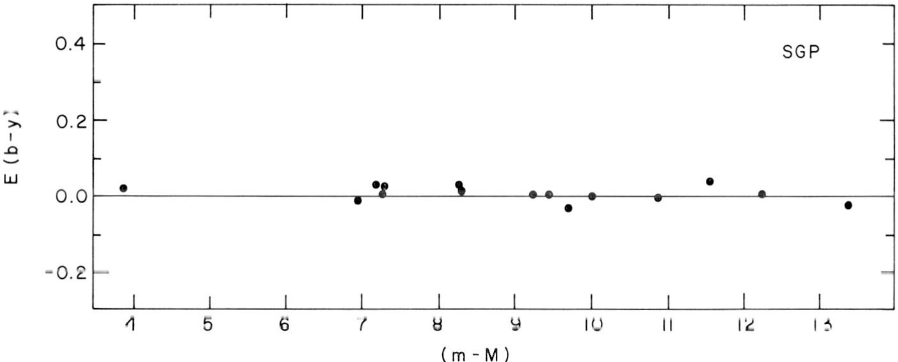

Fig. 4.   The color excess at the SGP.

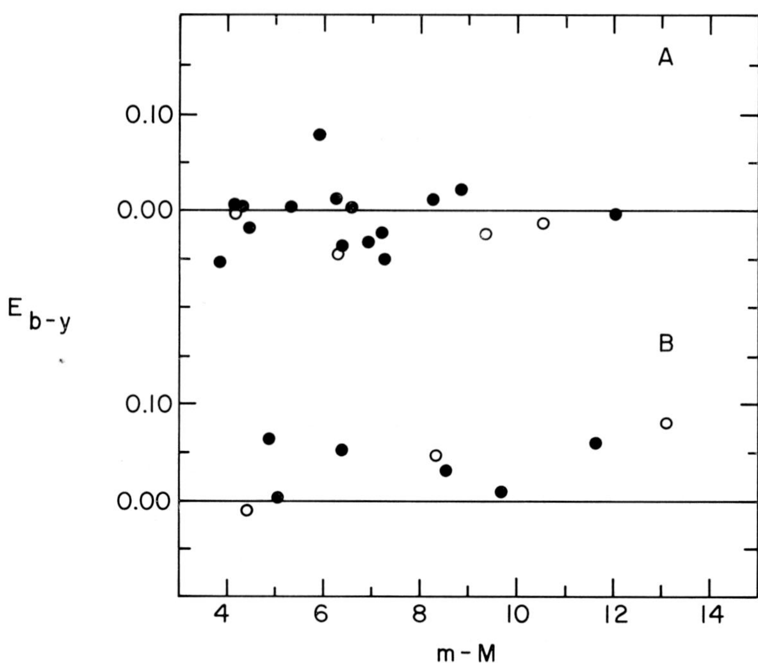

Fig. 5.   The color excess at the NGP. The regions A and B are defined in Figure 6.

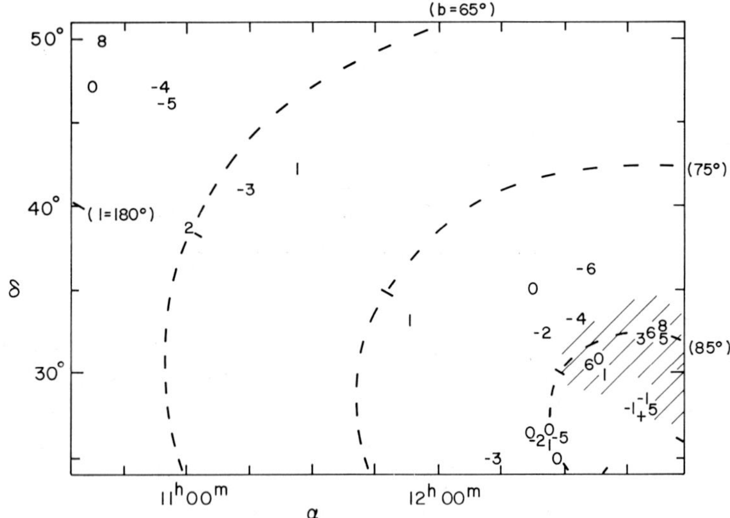

Fig. 6.   The distribution of color excesses at the NGP. Area B is a small area just North of the NGP.

reddening at the NGP is presented in Figure 5 and 6, (Figures 1 and 2 of Philip and Tifft, 1971). The color excesses in two regions A and B are shown as a function of distance modulus in Figure 5. In A, $E_{B-V}=0.05$, in B, $E_{B-V}=0.00$. (Region B is indi-

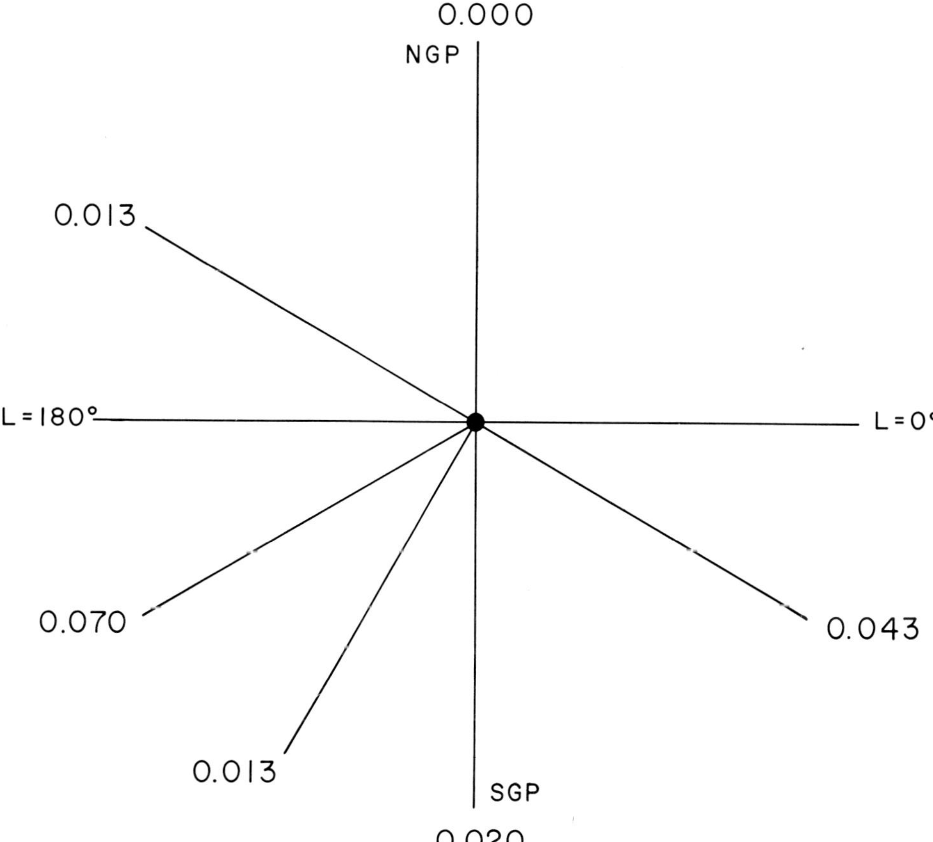

Fig. 7. E_{B-V} for certain galactic latitudes at galactic longitudes 0° and 180°.

cated in Figure 6 as the diagonally shaded area.) The hollow circles in Figure 5 indicate stars with $2.88 < \beta < 2.91$ which are slightly outside the limits set by Crawford (1971). In Figure 6, the NGP is marked by a small cross in the lower right. The numbers indicate the values of $b-y$ to the nearest hundredth.

The marked increase in accuracy of estimates of the color excess, made by means of the $b-y$, β relation as a function of distance modulus relative to estimates of the color excess made from knowledge of the Schmidt spectral type and $B-V$ color, can be seen by comparing the scatter of the points in Figures 2–5 with the scatter in the small insert in Figure 2. The major cause of the increased scatter is due to the fact that the grid of possible spectral types is not fine enough at the low dispersion of Schmidt spectral types. The difference also may be due partly to small systematic errors in the Schmidt spectral types, both as a function of magnitude and temperature class (in the range A0–A7) which are now under study.

The estimates of E_{B-V} in nine regions are summarized in Figures 7 and 8. Areas at $l=0°$ and $b=-30°$, $\pm 90°$, $l=180°$ and $b=\pm 30°$ and $-60°$, are summarized in Figure

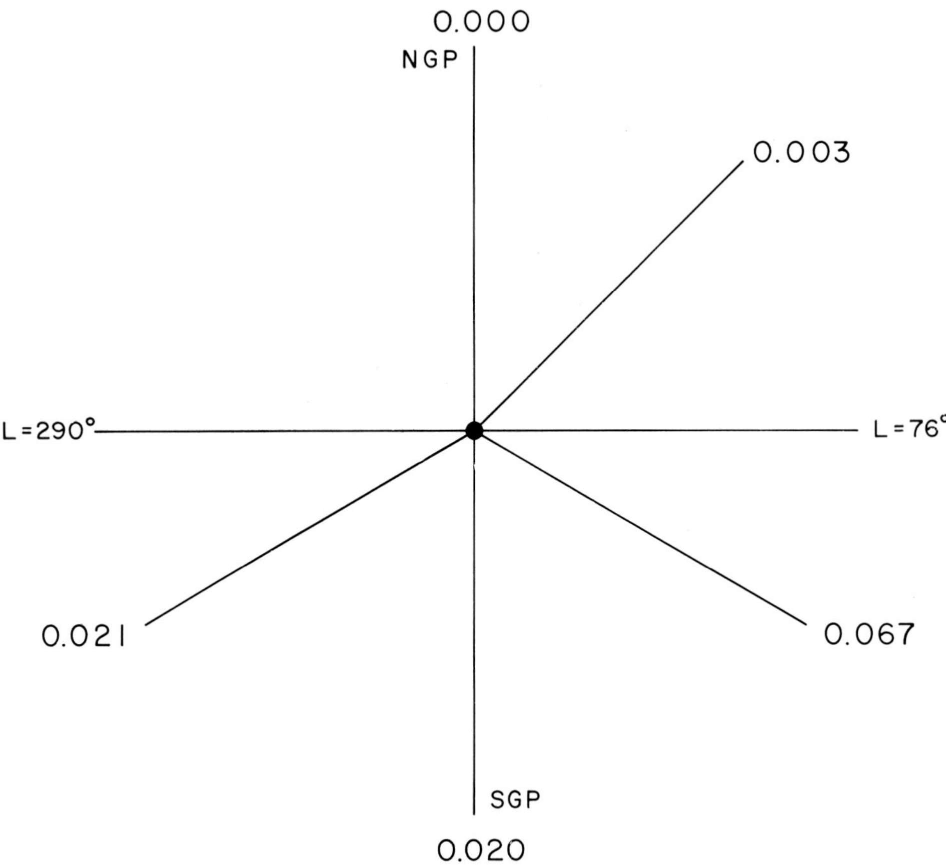

Fig. 8. E_{B-V} for galactic latitudes at galactic longitudes 76° and 290°.

7. The number at each position is the estimate of E_{B-V} in that region. Areas at $l=76°$ and $b=+45°$, and $-30°$, and $l=290°$, $b=-30°$ are shown in Figure 8.

There is a smooth progression from very low reddenings at the galactic poles to estimates of a few hundredths at latitudes of $\pm30°$. As more finding lists of early-type stars are completed and the stars in them measured in the four-color and Hβ systems, the reddening at high galactic latitudes can be mapped more exactly. At the present time, it can be seen that the reddening at the galactic poles is very small which means that measurements of colors of stars in these areas are very nearly the intrinsic colors. In agreement with current ideas that the Sun is a few tens of parsecs north of the galactic plane, the data suggest that absorption north of the plane is less than that to the south.

Acknowledgements

The directors of Kitt Peak National Observatory, Cerro Tololo Inter-American Observatory, and the Tonantzintla Observatory are thanked for their generous allotment of telescope time for this project. The project was partially supported by the National Science Foundation and the SUNYA Research Foundation.

Appendix

A summary of the finding lists of early-type stars published and in preparation will be found below

Paper	Reference	Area	l	b	No. of stars	Area of survey sq deg
I	Philip (1967)	1HLF4	76°	−45°	33	25
II	Philip and Sanduleak (1968)	SGP	−90°		180	230
III	Philip and Drilling (1970)	4HLF4	180°	−45°	88	45
IV	Drilling and Philip (1970)	3HLF4	0°	−45°	146	45
V	Philip and Relyea (1971)	1HLF3	76°	+45°	16	19
VI	Philip and Stock (1972)	Extended SGP	22° to 225°	−45° −50°	539	434
					1002	798

References

Crawford, D. L.: 1970, in A. Slettebak (ed.), *Stellar Rotation*, D. Reidel, Dordrecht, Holland, p. 144.
Drilling, J. S. and Philip, A. G. D.: 1970, *Bol. Obs. Tonantzintla Tacubaya* **4**, 307.
Philip, A. G. D.: 1966, *Astrophys. J. Suppl.* **12**, 391.
Philip, A. G. D.: 1967, *Bol. Obs. Tonantzintla y Tacubayay* **4**, 215.
Philip, A. G. D. and Sanduleak, N.: 1968, *Bol. Obs. Tonantzintla y Tacubayay* **4**, 253.
Philip, A. G. D. and Drilling, J. A.: 1970, *Bol. Obs. Tonantzintlay y Tacubayay* **4**, 297.
Philip, A. G. D. and Relyea, L. J.: 1971, *Bol. Obs. Tonantzintla y Tacubayay* **5**, 69.
Philip, A. G. D. and Tifft, L. F.: 1971, *Astron. J.* **76**, 567.
Philip, A. G. D. and Stock, J.: 1972, *Bol. Obs. Tonantzintla y Tacubaya* **6**, 201.

THE CONSTRUCTION AND DOCUMENTATION
OF THE CELESCOPE CATALOG

K. HARAMUNDANIS

Smithsonian Institution, Astrophysical Observatory, Cambridge, Mass. 02138, U.S.A.

Abstract. In this paper, I describe briefly the instrument and the reduction system we have used for obtaining ultraviolet magnitudes and present some information concerning the sky areas and types of stars we have observed. I shall also comment on our use of catalogs in machine-accessible form and make suggestions concerning the most useful identifications to be assigned to observed objects, particularly those in the southern hemisphere.

1. The Instrument

In 1968, the Orbiting Astronomical Observatory OAO 2 was launched. Figure 1 is a picture of the satellite taken shortly before launch to give an idea of sits ize. The four Celescope photometers of the Smithsonian Astrophysical Observatory (SAO) were placed in one end of the satellite; the University of Wisconsin's experiment was in the other. In the photograph, the large solar paddles are fully extended and the upper sunshade is open.

Fig. 1.   OAO 2 during checkout procedures shortly before launch (December 1968) (courtesy of NASA).

Ch. Fehrenbach and B. E. Westerlund (eds.), Spectral Classification and Multicolour Photometry, 302–314. All Rights Reserved.
Copyright © 1973 by the IAU.

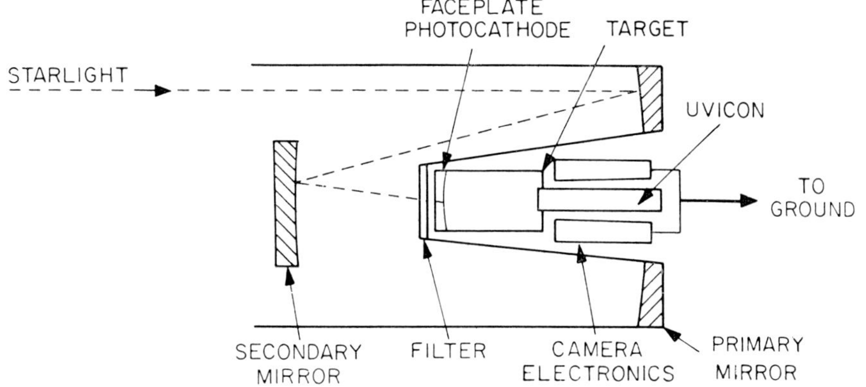

Fig. 2. Schematic diagram of Celescope photometer and telescope (from Nozawa, 1969).

A schematic diagram of one of the photometers is shown in Figure 2. The primary mirror is 12 in., and the filter is split into two sections in each of the four photometers so as to continue to provide data in more than a single broad passband in case of failure of any one of the cameras. The heart of the system is the Uvicon (photocathode,

TABLE I

Effective wavelength and half-width of Celescope filters

Filter	λ_{eff}	$\Delta\lambda$
U1	2582	550
U2	2308	850
U3	1621	325
U4	1537	450

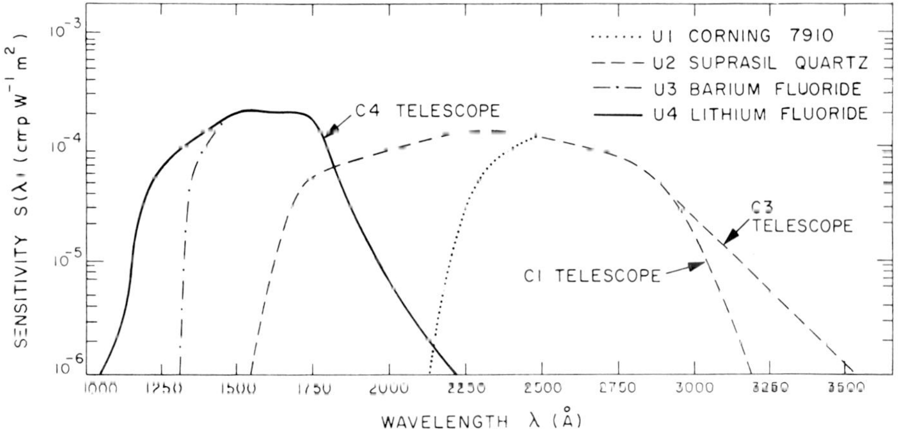

Fig. 3. Spectral response curves of Celescope (from Davis, 1968).

 K.HARAMUNDANIS

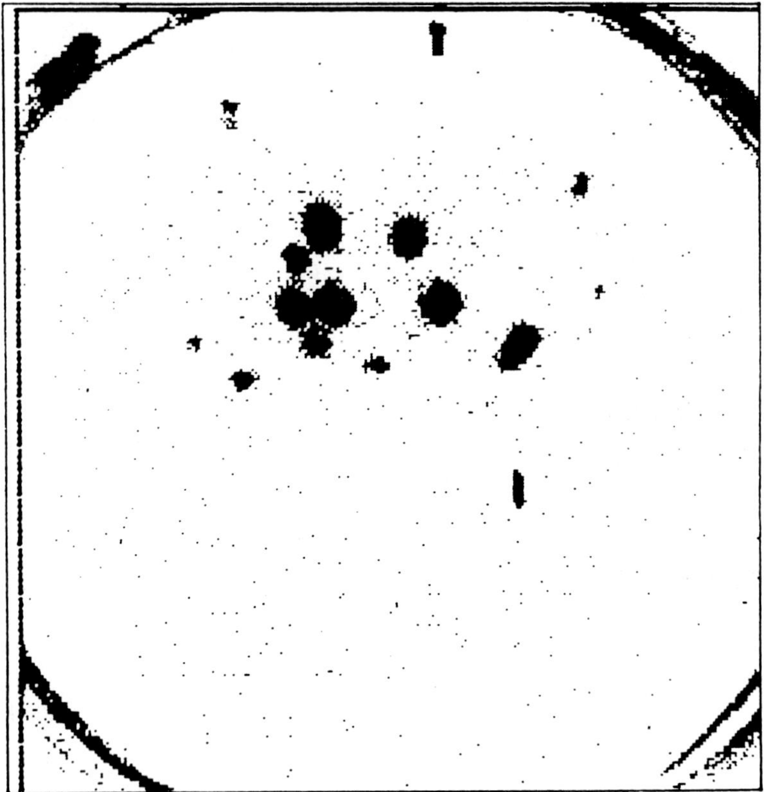

Fig. 4a. The Pleiades taken by Celescope.

target, and high-voltage power supply) described in numerous papers (cf. Nozawa, 1969; Celescope Staff, 1971).

With the four photometers, we observed stars over four passbands, two in each photometer for redundancy. The effective wavelength and half-width of each filter are given in Table I. The sensitivity of camera and filter combined is indicated in Figure 3. Note the difference between the U1 filters in cameras 1 and 3 (labeled C1 and C3). Figure 4 gives a single example of a Celescope picture. I have selected this picture of the Pleiades because it was taken midway in the lifetime of the experiment and was an exposure of only 5 s. For comparison, a ground-based photograph of the cluster is also shown. All the stars visible in the Celescope picture are in the U4 filter, and all are of B spectral class. The faintest object in the picture is a 6.16 mag. B9V star just above the pair Atlas and Pleione. Recall that this is a 5 s exposure; our usual exposure time is 60 s, which in this wavelength unfortunately is swamped by radiation from the geo-corona. This swamping initially proved a major obstacle in our processing of the data but was effectively overcome by some program refinements; we have as yet, however, been unable to reduce adequately the magnitudes of stars observed in such a swamped picture.

Fig. 4b.   The Pleiades taken by the Crossley reflector (Lick Observatory).

2. The Reductions

Figure 5 is a diagram of the system we have developed for reducing the observations
to ultraviolet magnitudes and for obtaining Durchmusterung identifications. The data,
originally transmitted from the satellite to various ground stations are sent to Goddard
Space Flight Center, where they are digitized. From there they are brought to SAO for
our processing. It may be of interest that our processing at SAO is done on a CDC
6400 computer (maximum user capacity 140 K). Our computer programs are largely
written in FORTRAN, with some special routines in the CDC machine language
COMPASS.

After verification of auxiliary information relating to pointing, temperatures, exposure time, and cameras, the data are sent out for conversion to picture displays.
Some resolution is thereby lost, since the original data are quantized in 127 intensity
levels while the displays comprise only 16 steps.

The digitized data in our main reduction system are passed into our signal-proces-
sing program, which treats each picture separately to determine background intensity
and locate all non-noise objects. The significance level is allowed to differ for each
picture and for different locations in the picture. The digitized objects are displayed

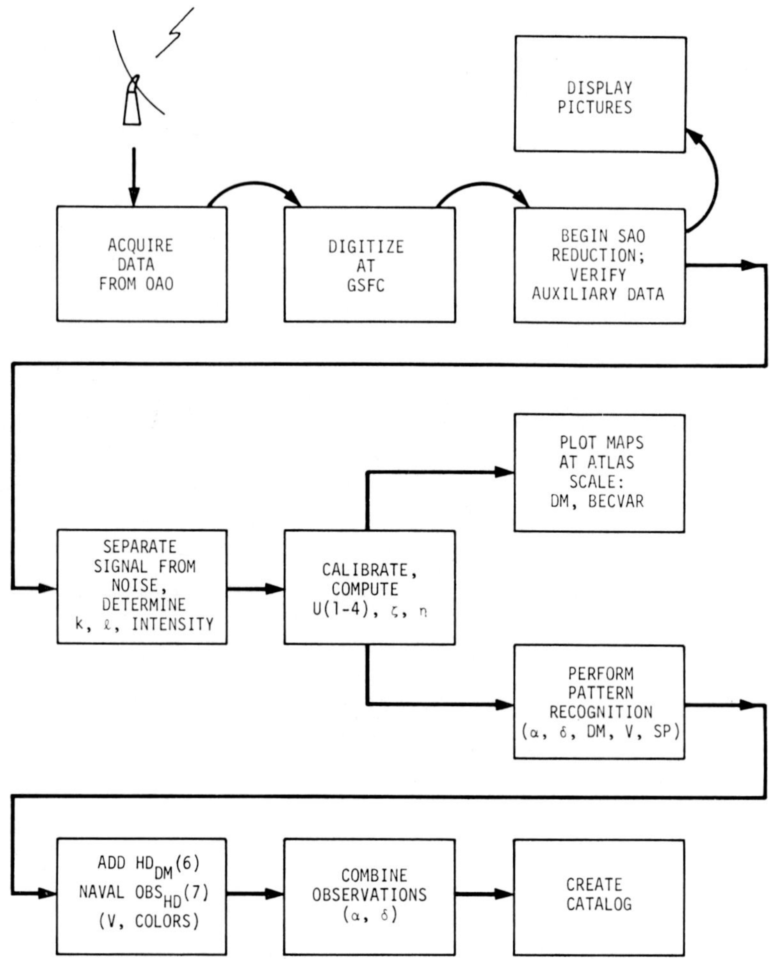

Fig. 5.   The Celescope data reduction system.

in several ways to facilitate the inspection necessary for a uniform quality of output from the system. Our signal processing program produces a record for every object, including all its known parameters, its coordinates in the picture matrix (k, l), and its observed intensity (I), which is defined as the sum of all non-noise contiguous elements making up the object, minus the background intensity for each such element.

After the computation of relative intensity, the data are reduced to ultraviolet magnitudes by using calibration data prepared before launch in laboratory tests and verified with orbital data taken throughout the satellite's lifetime. Principal features of the calibration system are an optimization procedure by the Fletcher-Powell method, used to refine the prelaunch calibration tables, and the utilization of repeated orbital observations of special sky areas to determine the time degradation of the system in orbit. The early calibration has been extensively discussed in Davis (1968); special techniques in current use are treated in Lundquist *et al.* (1971). The most recent determination of the standard error of unit weight, using only those observations accumulated before June of this year, gave the values reported in Table II.

Part of the prelaunch calibration was devoted to the conversion of picture matrix elements (k, l) to a pseudoequatorial system (ζ, η) that, with knowledge of the satellite pointing (α, δ) and of the moment of exposure, would be directly comparable to right ascension and declination. With the ζ, η coordinates, a small plot is made of each picture at the scale of the most useful atlases, the BD ,CD, and Bečvář. These plots are often useful in resolving difficult identifications. The ζ, η coordinates alone are capable of representing the α, δ of an object within $\pm 10'$.

TABLE II

RMS magnitude devia-
tion for each camera

Camera	RMS Δm
1	0.17
3	0.20
4	0.19

The ζ, η coordinates and the necessary information on pointing and moment of exposure are used in a pattern-recognition program that matches the observatione against a specially prepared list of early-type stars. Currently, by making composits 'pictures' from several actual pictures and matching stars in areas several degrees in diameter, we achieve satisfactory identifications for about 80% of our observations. The pattern-recognition program that we use was adapted from that employed for many years at SAO and based on the Christie-Dyson method of 'plate constants' (Arnold, 1970).

After satisfactory identifications are determined and checked for all our data, we add – for our own use and for the aid of others who will use the final catalog – magnitudes and colors in the UBV system, spectra, HD numbers, DM numbers, and magnitudes (m_v, m_{pg}) if no UBV data are available. The major sources we utilized to accumulate these ground-based data, all in machine-accessible form, are as follows:

(1) The Celescope identification list, a compilation of several hundred references specially selected as containing data on the objects most likely to be seen by the instrument.

(2) The Henry Draper Catalogue (HD), obtained on 10 tapes from Yale University Observatory and reduced, for ease in handling, to a single tape and sorted by DM number.

(3) The Naval Observatory Catalogue of Blanco *et al.* (1968), which originally contained several observations for each entry and which we have reduced to a single entry per star.

(4) The SAO Catalog, which we have also reduced to a single tape and sorted by DM number.

An important feature of our system is the ease with which we can rearrange all observations and catalog data so as to make the programs that operate on them efficient. For example, our initial processing through the calibration phase progresses one *picture* at a time; pattern recognition is performed with the data sorted into convenient blocks by *pointing*; ground-based data are added in order either by *DM number* or by *HD number*. Final combination of observations will be carried out with the data in order by *right ascension* and *declination*. The ease with which we can perform these rearrangements depends largely on our adherence to a uniform format for the magnetic tapes that are internal to our system.

3. The Observations

With these photometers, we have taken close to 9000 pictures, from about 50% of which we have currently reduced about 22000 observations of 5000 stars. The sky area covered during the exposures taken during our 16 months of operation is shown in Figure 6. Aside from a few pictures taken at high galactic latitude, we have concentrated principally on the galactic plane. As an illustration of the characteristics of our survey, I have compared our data on early-type stars with data from the HD (Cannon and Pickering, 1918–1924). Figure 7 compares the number of stars observed by Celescope and listed in the HD for spectral classes B and A. For these classes, it is apparent that the HD, with its 60 min exposures, is virtually complete to $m_v = 8.3$ and that, for the areas our experiment observed with its 60 s exposures, Celescope has a faint limit around $m_v = 8.1$. Only stars with known spectral class have been included in this diagram; a considerable number of faint stars in our data are of unknown spectral classes.

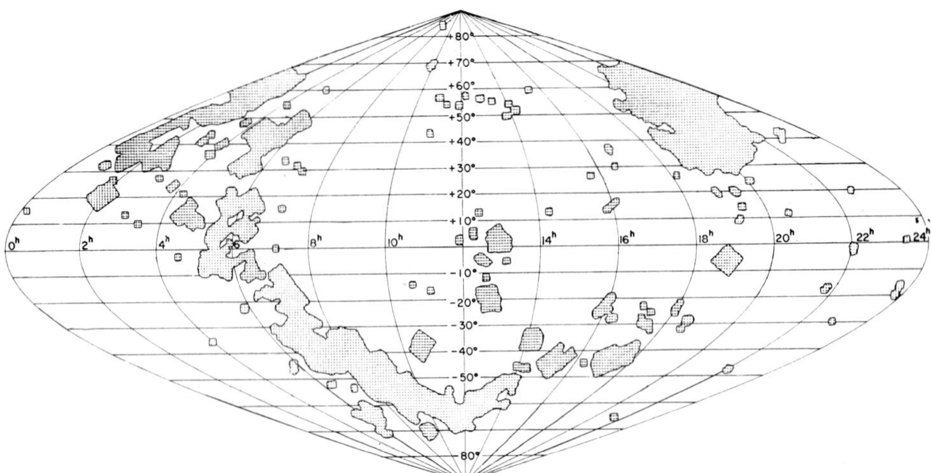

Fig. 6.   Sky areas observed by Celescope, by (α, δ) (1950).

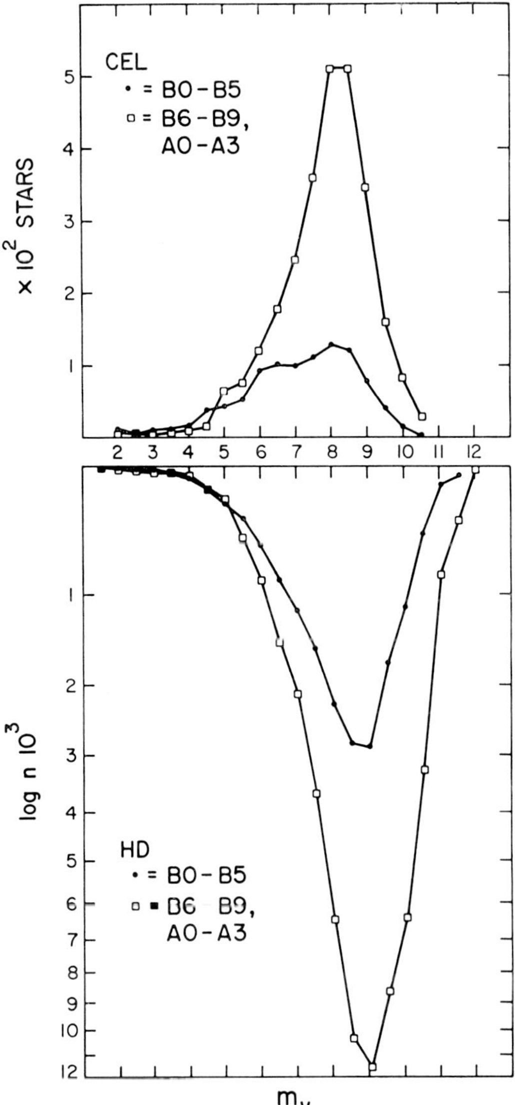

Fig. 7. A comparison of the number of B and A stars observed by Celescope and listed in the HD:
● = B0–B5; □ = B6–B9, A0–A3.

A comparison of the O and Wolf-Rayct stars seen by Celescope and recorded in the HD is shown in the pair of diagrams in Figure 8. The two diagrams are drawn in a three-dimensional Cartesian system, with axes corresponding to $(\alpha, \delta)_{1950}$ and with the number of stars in each sky area bounded by the limits represented. The Great Rift and the Magellanic Clouds are prominent. Dashed lines on the HD part of the diagram indicate an area not observed by Celescope. It may be of interest to note that

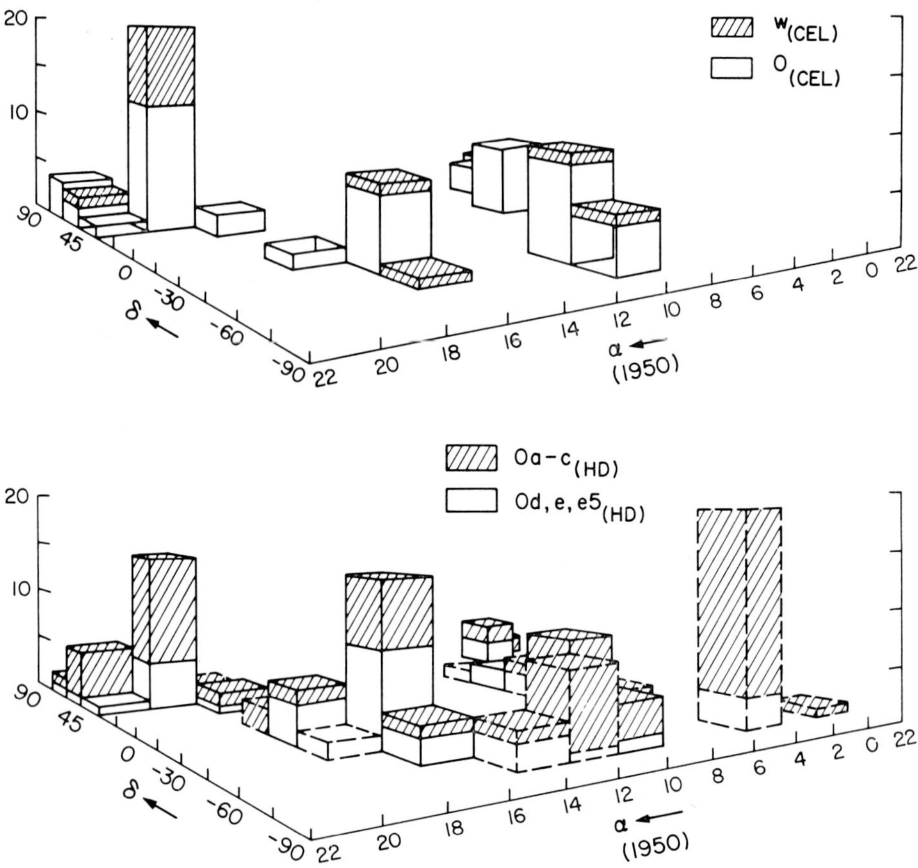

Fig. 8. The O and Wolf-Rayet stars observed by Celescope (above) and listed in the HD (below). Right ascension, declination, and number of stars are shown.

Shapley and Cannon (1924) in their discussion of the data from the HD did not give information concerning the O stars, which I furnish here, and that their numerical results for B and A stars differ from mine by several percent (see Table III). Both tallies were, of course, performed on the same catalog, theirs for 220 570 stars while the catalog was in manuscript and mine by computer on 225 300 stars in the published HD.

TABLE III

Early-type HD stars tabulated by Shapley and Cannon (1924) and
in the present paper

Spectral class	O	B0–B5	B8, B9, A0–A3
Shapley and Cannon	–	3567	64259
This paper	168	3230	62477
$\Delta\%$	–	9	3

4. Documentation of Data and Methods

My attention has been brought many times to the difficult task of adequate documentation of the data and results of the experiment. Figure 9 illustrates the amounts of material we have accumulated, first in the laboratory and later in orbit. The largest are the data themselves; the supporting documents are more manageable, perhaps 10% of the original; and the summary documents contain only the vital information and reduce the sheer weight of paper to a fraction of 1% of the original. It can be extremely difficult to document adequately an enterprise of this size, but we have found that certain principles can be profitably followed:

(1) Any document or related group of documents of more than 200 to 300 pages

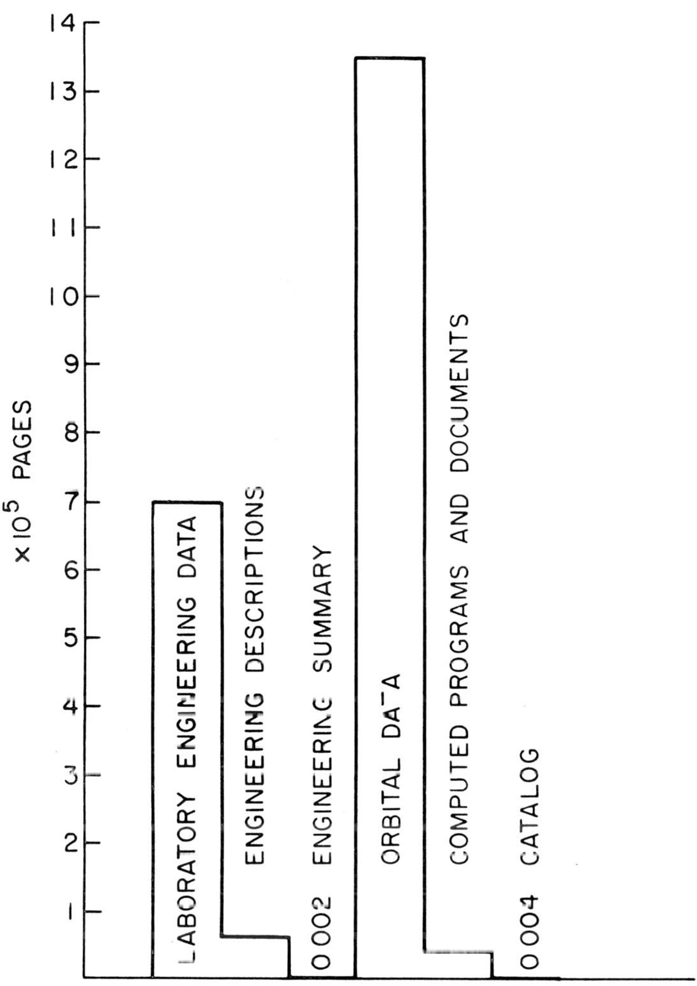

Fig. 9.   Celescope documentation.

needs a *summary* not longer than a page or so. Failure to provide such a summary may result in misplacement or loss of important information.

(2) Documents of the same kind should be handled together in the documentation; engineering documents requiring special treatment are not always needed for data analysis, and vice versa. It can be useful to have an individual knowledgeable in a special field in charge of the documentation for his type of information. His documentation should conform to certain standards and dovetail with the complementary documentation from the other parts of the project.

(3) Data (in the digitized form in which we use it) and computer programs, if supplied with consistent and appropriate headings and comments, are often their own best documentation. However, a computer program without the name of its author, dates of creation and revisions, and a multitude of explanatory comments placed within the listing is virtually useless to anyone else who may need to use or revise it.

(4) In the description of the essentials of a catalog or compendium of data such as ours of ultraviolet observations, it is important to the user that the compiler specify as completely as possible what was gone into the catalog:

(a) First, a well-organized description of the *purpose*, *design*, and *preparation* of the list should be given.

(b) Second, the compiler should try to remember that *what is familiar to him* is not necessarily familiar to those who will use his work.

(c) Third, it would be extraordinarily useful if *original data* presented for the first time in a list could be *distinguished* from the data merely copied from somewhere else. In this age of lists often taken directly from computer printout or CRT, no easy substitute for italics may be available, but column headings or perhaps single and double vertical lines scribed on the page might be effectively utilized.

(d) Finally, if the introduction to the list is necessarily lengthy and complex, an *index* would be a useful addition.

Two excellent examples of list specification come to mind: the Henry Draper Catalogue, where all the spectral classes are explicitly defined and, where room did not allow greater discussion of special methods, a specific reference is given; and the N30 (Morgan, 1950), where Morgan discussed not only the entire reasoning behind his catalog but also his whole system of reduction, in a completely clear and understandable fashion. Some catalogs do not come up to these standards because they give either so little information that their scheme is insufficiently explained or so much that it actually impedes use of the catalog. The user must in effect analyze the catalog before he can use it!

In this light, I have one last figure that illustrates a problem well known to those working with stars and their identifications. In Figure 10 I have plotted, for every $1°$ of declination, the number of stars in each of the four Durchmusterungen. When the HD was constructed, Pickering and Cannon decided that south of the last complete zone of the SBD they would use the CD, and from $-52°$ to the south pole, the CPD for their identifications. If they based their choice on the number of stars available from the two southern-hemisphere catalogs, then their choice was very wise.

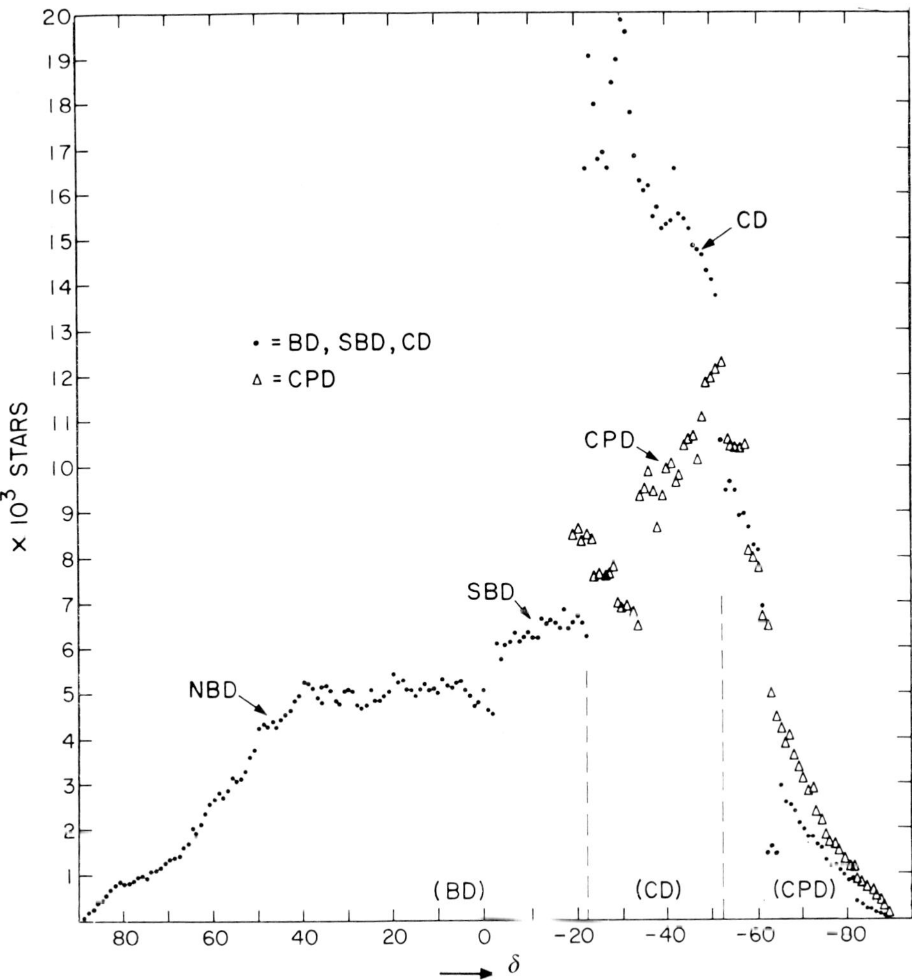

Fig. 10. The number of stars in each Durchmusterung, by declination; HD convention indicated in parentheses.

They did not, however, use the CD in the zones where the BD overlapped with it. The existence of the two DM in the south, however, has caused some difficulty, and I should like to hope that future list and catalog compilers adhere to a single convention in this regard. Three choices suggest themselves:

(1) Always use the CD.

(2) Always give both CD and CPD numbers.

(3) Adopt the Henry Draper convention, using CD numbers north of $-52°$ and CPD numbers south of it.

My personal preference is toward the HD convention (even to the exclusion of the somewhat more numerous CD stars between $-19°$ and $-22°$), but I vigorously recommend that, at the least, in any list a DM number, where it exists, should always accompany data reported for an object.

Acknowledgements

I am indebted to Peter Collins for his assistance in preparing some of the data for this paper, and to C. Payne-Gaposchkin for helpful discussions. This work was supported in part by contract NASA–1535 from the National Aeronautics and Space Administration.

References

Arnold, D.: 1970, Photoreduction Manual, internal SAO document.

Blanco, V. M., Demers, S., Douglass, G. G., and Fitzgerald, M. P.: 1968, 'Photoelectric Catalogue: Magnitudes and Colors of Stars in the *U*, *B*, *V*, and *U*$_c$, *B*, *V* Systems', *Publ. Naval Obs.*, *2nd Ser.* **XXI**.

Cannon, A. J. and Pickering, E. C.: 1918–1924, 'The Henry Draper Catalogue', *Ann. Astron. Obs. Harvard Coll.* **91–99**.

Celescope Staff 1971, 'Performance Evaluation of the Celescope Experiment', *Rpt. to NASA*, July.

Davis, R. J.: 1968, *Smithsonian Astrophys. Obs. Spec. Rpt.*, No. 282.

Lundquist, C. A., Deutschman, W. A., and Eng Young, R.: 1971, 'Digital Processing Techniques Used in Determining Stellar Ultraviolet Magnitudes from OAO-Celescope Image Data', presented at the *Symposium on the Processing of Telescopic Images*.

Morgan, H. R.: 1950, 'Catalog of 5,268 Standard Stars, 1950.0. Based on the Normal System N30', *Astronomical Papers of the American Ephemeris and Nautical Almanac* **XIII**.

Nozawa, Y.: 1969, in T. D. McGee, D. McMullan, and E. Kahan (eds.), *Advances in Electronics and Electron Physics* **28B**, Academic Press, New York, p. 891.

Shapley, H and Cannon, A. J.: 1924, Harvard Reprint No. 6.